Editorial Policy

§ 1. Lecture Notes aim to report new developments - quickly, informally, and at a high level. The texts should be reasonably self-contained and rounded off. Thus they may, and often will, present not only results of the author but also related work by other people. Furthermore, the manuscripts should provide sufficient motivation, examples and applications. This clearly distinguishes Lecture Notes manuscripts from journal articles which normally are very concise. Articles intended for a journal but too long to be accepted by most journals, usually do not have this "lecture notes" character. For similar reasons it is unusual for Ph. D. theses to be accepted for the Lecture Notes series.

§ 2. Manuscripts or plans for Lecture Notes volumes should be submitted (preferably in duplicate) either to one of the series editors or to Springer- Verlag, Heidelberg . These proposals are then refereed. A final decision concerning publication can only be made on the basis of the complete manuscript, but a preliminary decision can often be based on partial information: a fairly detailed outline describing the planned contents of each chapter, and an indication of the estimated length, a bibliography, and one or two sample chapters - or a first draft of the manuscript. The editors will try to make the preliminary decision as definite as they can on the basis of the available information.

§ 3. Final manuscripts should preferably be in English. They should contain at least 100 pages of scientific text and should include
- a table of contents;
- an informative introduction, perhaps with some historical remarks: it should be accessible to a reader not particularly familiar with the topic treated;
- a subject index: as a rule this is genuinely helpful for the reader.

Further remarks and relevant addresses at the back of this book.

Lecture Notes in Mathematics　　　1701

Editors:
A. Dold, Heidelberg
F. Takens, Groningen
B. Teissier, Paris

Springer

Berlin
Heidelberg
New York
Barcelona
Hong Kong
London
Milan
Paris
Singapore
Tokyo

Ti-Jun Xiao
Jin Liang

The Cauchy Problem
for Higher-Order Abstract
Differential Equations

Springer

Authors

Ti-Jun Xiao
Jin Liang
Department of Mathematics
University of Science and Technology of China
Hefei 230026, Anhui
People's Republic of China
e-mail: xiaotj@ustc.edu.cn
        jliang@ustc.edu.cn

This work was supported by the National Natural Science Foundation of China

Cataloging-in-Publication Data applied for

Die Deutsche Bibliothek - CIP-Einheitsaufnahme

Xiao, Ti-Jun:
The Cauchy problem for higher order abstract differential equations /
Ti-Jun Xiao and Jin Liang. - Berlin ; Heidelberg ; New York ;
Barcelona ; Budapest ; Hong Kong ; London ; Milan ; Paris ; Santa
Clara ; Singapore ; Tokyo : Springer, 1998
   (Lecture notes in mathematics ; 1701)
   ISBN 3-540-65238-8

Mathematics Subject Classification (1991): Primary: 34G10, 47D06;
Secondary: 47N20, 35G10, 47D09, 93C25, 47F05

ISSN 0075-8434
ISBN 3-540-65238-8 Springer-Verlag Berlin Heidelberg New York

© Springer-Verlag Berlin Heidelberg 1998
Printed in Germany

Typesetting: Camera-ready T$_E$X output by the author
SPIN: 10650166   41/3143-543210 - Printed on acid-free paper

To Our Motherland

To our parents and teachers

To Xiao Liang

Preface

The main purpose of this book is to present the basic theory and some recent developments concerning the Cauchy problem for higher order abstract differential equations

$$\begin{cases} u^{(n)}(t) + \sum_{i=0}^{n-1} A_i u^{(i)}(t) = 0, \quad t \geq 0, \\ u^{(k)}(0) = u_k, \quad 0 \leq k \leq n-1. \end{cases} \qquad (ACP_n)$$

where A_0, A_1, $\cdots$, A_{n-1} are linear operators in a topological vector space E.

Many problems in nature can be modeled as (ACP_n). For example, many initial value or initial-boundary value problems for partial differential equations, stemmed from mechanics, physics, engineering, control theory, etc., can be translated into this form by regarding the partial differential operators in the space variables as operators A_i $(0 \leq i \leq n-1)$ in some function space E and letting the boundary conditions (if any) be absorbed into the definition of the space E or of the domain of A_i (this idea of treating initial value or initial-boundary value problems was discovered independently by E. Hille and K. Yosida in the forties). The theory of (ACP_n) is closely connected with many other branches of mathematics. Therefore, the study of (ACP_n) is important for both theoretical investigations and practical applications.

Over the past half a century, (ACP_n) has been studied extensively. Especially for (ACP_1), the theory (or closely related operator semigroup theory) has evolved comparatively perfect since the well-known Hille-Yosida theorem came out in 1948, and is well documented in the monographs of, e.g., Brézis [1], Davies [1], deLaubenfels [9], Fattorini [6], Goldstein [7], Hille-Phillips [1], Nagel [2], Pazy [2], van Casteren [1], van Neerven [1], Yosida [4]. On the other hand, since the work of Lions [1] in 1957, the study of higher order (ACP_n) $(n \geq 2)$ has also received much attention (cf., e.g., Fattorini [7], Goldstein [7], Krein [1], Xiao [2] and references therein). So far, a rich theory of (ACP_n), including the Hille-Yosida type characterization for wellposed (ACP_n) of higher order, has unfolded before us.

A survey of the research history of (ACP_n) $(n \geq 2)$ shows that one popular approach is to reduce the higher order problem to a first order system in a suitable phase space and use operator semigroup theory. The disadvantage of this approach is that, finding an ideal phase space is generally difficult, and the structure of the phase space (if any) may be complicated so that inconvenient to application; also some inherent properties of higher order problems can not always be reflected precisely from the corresponding first order systems. The strong desire to establish concise, convenient and more inclusive theories about (ACP_n) therefore gives rise to another approach — direct treatment of (ACP_n) $(n \geq 2)$. This book depends heavily on this second idea, with the aid of the first one when needed.

The main material in this book is taken from the authors' work on this topic. We have tried to give a systematic exposition of the abstract theory of (ACP_n), but no attempt at completeness can be made at this time, either in the text or in the references. Actually, many results and papers have not been mentioned. Also we do not attempt to give detailed applications, although many results are illustrated with concrete examples.

As prerequisites for the reading of this book we assume the reader to have a sound knowledge of complex and functional analysis. Familiarity with the basic theory of operator semigroups is desirable but not necessary. Some basic facts for the fractional powers of closed operators, and for the classical strongly continuous operator semigroups as well as cosine operator functions, which are needed in this book, are gathered in Appendix. For other preliminaries, we have dispensed with a special chapter of them in favour of reminders in the body of the text and where necessary we refer to other books and papers for background material.

Chapters 1 and 2 are presented mainly in the setting of sequentially complete locally convex spaces, while other chapters in Banach spaces. Throughout, the method of Laplace transforms will be a fundamental tool. So we firstly in Chapter 1 discuss basic properties of Laplace transforms, especially the integrated version of the Widder's classical representation theorem for Laplace transforms (Theorem 2.1). In addition, we give a brief introduction of the basics of integrated and regularized semigroups or cosine functions, as well as their relationship to abstract Cauchy problems. Chapter 2 is devoted to the establishment of the Hille-Yosida type characterization of strongly wellposed (ACP_n), and others. Chapter 3 selects to deal with several types of (ACP_n) which are not wellposed in a standard sense. Chapters 4 – 7 investigate various properties of the propagators or solutions of (ACP_n), including analyticity, parabolicity, exponential growth bound, exponential stability, differentiability, norm continuity, and almost periodicity; corresponding characterizations are given.

Within each chapter definitions, lemmas, theorems, corollaries, etc. are numbered consecutively as 1.1, 1.2, $\cdots$, in section x.1 (x=1, 2, $\cdots$, 7), as 2.1, 2.2, $\cdots$, in section x.2 (x=1, 2, $\cdots$, 7) and so on. When making a reference to another chapter we always add the number of that chapter, e.g., 2.1.2.

Throughout this book, N, N_0, R, R^+, C denote the positive integers, the nonnegative integers, the real numbers, the nonnegative real numbers, the complex plane, respectively. For $b \in R$, $[b]$ will be the least integer$> b-1$. Let E and X be topological vector spaces. $L(E, X)$ denotes the space of all continuous linear operators from E to X, and $L(E, E)$ will be abbreviated to $L(E)$. If E is a locally convex space topologized by the family Γ of seminorms, we denote by $\mathcal{B}_{\Gamma}(E)$ the subspace of $\mathbf{L}(E)$ whose elements B satisfy $\|B\|_{\mathbf{\Gamma}} := \sup\{p(Bx); \; p \in \mathbf{\Gamma}, \; x \in E$ with $p(x) \leq 1\} < \infty$; $\mathcal{B}_{\mathbf{\Gamma}}(E)$ with norm $\|\cdot\|_{\mathbf{\Gamma}}$ is a normed algebra. For $k \in N_0$, $C^k(R^+, \; E)$ is the set of all k-times continuously differentiable E-valued functions in R^+; $C(R^+, \; E) := C^0(R^+, \; E)$; $C^\infty(R^+, E) := \bigcap_{k=0}^\infty C^k(R^+, E)$. For a linear operator A, we will write $\mathcal{D}(A), \mathcal{R}(A), \mathcal{N}(A), \sigma(A), \sigma_p(A), \rho(A), R(\lambda; A),$ A^*, respectively, for the domain, the image, the kernel, the spectrum, the point spectrum, the resolvent set, the resolvent, the adjoint operator. Finally, the characteristic polynomial of the equation in (ACP_n) is denoted by

$$P_\lambda := \lambda^n + \sum_{i=0}^{n-1} \lambda^i A_i, \quad \lambda \in \mathbf{C},$$

and

$$R_\lambda := P_\lambda^{-1},$$

if the inverse exists. For each $0 \leq k \leq n-1$, A_k will denote the restriction of A_k on $\bigcap_{i=0}^k \mathcal{D}(A_i)$. Sometimes we write $A_n := I$ (the identity operator).

It is a great pleasure to acknowledge the help, advice or encouragement we have received from many persons, and in particular from Kongcing Chang, Edward Brian Davies, Falun Huang, Bingren Li, Cunjun Li, Beyu Liao, Yingming Liu, Zhiming Ma, Kezhi Wang, Chenlun Yang, Heliang Yang and Le Yang. We would like to thank Sichuan University (the Sichuan Union University now), King's College London, Institute of Mathematics of the Chinese Academy of Sciences, Beijing University, and Institute of Applied Mathematics of the Chinese Academy of Sciences for their hospitality during our stay there. Finally, we are indebted to the National Nature Science Foundation of China, Applied and Basic Science Foundation of Yunnan Province, and Science Foundation of China Nonferrous Metals Industry Corporation for their support all these years.

Ti-Jun Xiao
Jin Liang

Contents

Chapter 1

Laplace transforms and operator families in locally convex spaces

Summary

Section 1.1 collects the basic facts which one needs to know about Laplace transforms. Among others are three inversion formulas and two representation theorems (Theorems 1.5, 1.7 and 1.8, Theorems 1.11 and 1.12); Theorem 1.11 will be widely used later in the treatment of perturbation problems.

Section 1.2 is devoted to the statement of an integrated version of the classical Widder's representation theorem of Laplace transforms and its proof in a sequentially complete locally convex space (in short SCLCS). This theorem (Theorem 2.1) is useful in the treatment of operator families and operator differential equations. In particular, it will be used in the proof of Theorem 2.2.2.

In Section 1.3, we introduce (exponentially equicontinuous) r-times integrated, C-regularized semigroups in SCLCS for any $r \geq 0$, and characterize their generators in terms of the estimates of the resolvents using Theorem 2.1. The resulting theorems are generalizations of the corresponding results for strongly continuous semigroups in SCLCS. Also, some elementary properties about these semigroups are given.

Section 1.4 is an analogue of Section 1.3 for r-times integrated, C-regularized cosine functions in SCLCS for any $r \geq 0$.

We consider in Section 1.5 a large class of differential operators on certain function spaces $L_l^p(R^n)$ ($1 < p < \infty$, $l = 0, \cdots, n$; $L_0^p(R^n)$ is just the usual Banach space $L^p(R^n)$), and show that they generate integrated or regularized semigroups. This is meaningful in considering that very few of these operators generate the classical strongly continuous semigroups. Actually, as made clear by Hörmander [1] in 1960, the Schrödinger operator $i\Delta$ generates a strongly continuous semigroup on $L^p(R^n)$ $(1 \leq p \leq \infty)$ only if $p = 2$.

The study of integrated, regularized semigroups and other operator families provides us with unified techniques for dealing with both wellposed and illposed Cauchy problems. In Section 1.6 finally, we exhibit the connection between integrated, regularized semigroups (resp. cosine functions) and the abstract Cauchy problems.

1.1   Laplace transforms

Throughout this chapter, E will be a sequentially complete locally convex space (in short SCLCS) topologized by the family Γ of seminorms. If an E-valued function $g(s)$ is continuous on a finite interval $[a, b]$, the integral $\int_a^b g(s)ds$ is defined by means of Riemann sums in the same way as for numerical functions. Improper integrals like $\int_a^\infty g(s)ds$ is defined as $\lim\limits_{b \to \infty} \int_a^b g(s)ds$ if the limit exists. It is clear that if $g(\cdot)$ is a continuous E-valued function on $[a, \infty)$ such that $p(g(\cdot))$ is Lebesgue integrable on $[a, \infty)$ for each $p \in \Gamma$, then $\int_a^\infty g(s)ds$ exists by noting the sequential completeness of E.

Definition 1.1. A family $\mathcal{F}$ of linear operators on E is equicontinuous if, for each $p \in \Gamma$, there is a continuous seminorm q on E such that

$$p(Au) \leq q(u), \quad A \in \mathcal{F}, \ u \in E.$$

Definition 1.2. A family $\mathcal{F}$ of linear operators on E is called $\Gamma(M_p)$-equicontinuous if, for each $p \in \Gamma$, there is a constant M_p such that

$$p(Au) \leq M_p p(u), \quad A \in \mathcal{F}, \ u \in E;$$

$\mathcal{F}$ is called $\Gamma(M)$-equicontinuous if, there is a constant M such that

$$p(Au) \leq Mp(u), \quad p \in \Gamma, \ A \in \mathcal{F}, \ u \in E.$$

Definition 1.3. Let $a \in R$.

(i) A function $f : (a, \infty) \to E$ is in the class $LT - E$, if there exists a function $h(\cdot) \in C(R^+, E)$ such that, to each $p \in \Gamma$ corresponds a constant M_p satisfying $p(h(t)) \leq M_p e^{at}$ $(t \geq 0)$, and

$$f(\lambda) = \int_0^\infty e^{-\lambda t}h(t)dt \quad (\lambda > a).$$

$f(\lambda)$ is called the Laplace transform of $h(t)$ and $h(t)$ is called the determining function of $f(\lambda)$. Sometimes, we denote by $\mathcal{L}$ the Laplace transform

$$\mathcal{L}[h(t)](\lambda) = \int_0^\infty e^{-\lambda t}h(t)dt \quad (\lambda > a).$$

(ii) A function $F : (a, \infty) \to \mathbf{L}(E)$ is in the class $LT - \mathbf{L}(E)$, if there exists a function $H(\cdot) : R^+ \to \mathbf{L}(E)$ such that $H(\cdot)$ is strongly continuous (i.e., $H(\cdot)u \in C(R^+, E)$ for any $u \in E$), and $\{e^{-at}H(t); \ t \geq 0\}$ is equicontinuous with

$$F(\lambda)u = \int_0^\infty e^{-\lambda t} H(t) u \, dt \quad (\lambda > a, \ u \in E).$$

Immediately, we have

Lemma 1.4. *If $M(\cdot) \in LT - E$ (resp. $LT - \mathbf{L}(E)$), then $\lambda^{-j} M(\lambda) \in LT - E$ (resp. $LT - \mathbf{L}(E)$), for each $j \in N$.*

Next, we present several important inversion formulas for Laplace transforms.

Theorem 1.5. *Let $h \in C(R^+, E)$ such that $\int_0^\infty e^{-ct} h(t) dt$ exists for some positive c. Then as $k \to \infty$,*

$$M_k(t) := \left(\frac{k}{t}\right)^k \frac{1}{(k-1)!} \int_0^\infty e^{-ks/t} s^{k-1} h(s) ds \longrightarrow h(t),$$

$$N_k(t) := \left(\frac{k}{t}\right)^{k+1} \frac{1}{k!} \int_0^\infty e^{-ks/t} s^k h(s) ds \longrightarrow h(t),$$

uniformly on compacts of $t > 0$.

The proof of the first expression is essentially the same as the one of Widder [1, p. 285, Theorem 5a] and we omit it here. The second follows immediately from $M_k(t) = N_{k-1}\left(\frac{(k-1)t}{k}\right)$.

As an immediate consequence, we get the following uniqueness theorem for Laplace transforms.

Theorem 1.6. *Let $h_1, h_2 \in C(R^+, E)$ such that for some $a \in R$,*

$$\mathcal{L}[h_1](\lambda) = \mathcal{L}[h_2](\lambda), \quad \lambda > a.$$

Then $h_1(t) \equiv h_2(t)$ on R^+.

Theorem 1.7. *Let $h \in C(R^+, E)$. If there is $a > 0$ such that*

$$\sup_{t \geq 0} p\left(e^{-at} h(t)\right) < \infty$$

for any $p \in \Gamma$, then for each $t \geq 0$,

$$\int_0^t h(s) ds = \lim_{\lambda \to \infty} \sum_{n=1}^\infty \frac{(-1)^{n-1}}{n!} e^{n\lambda t} \int_0^\infty e^{-n\lambda r} h(r) dr.$$

The proof is completely the same as the first part in the proof of Phragmén's representation theorem; see Neubrander [3]. We also omit it.

Theorem 1.8. *Let $h \in C^1(R^+, E)$. If there is $a > 0$ such that*

$$\sup_{t \geq 0} p\left(e^{-at}h(t)\right) < \infty$$

for any $p \in \Gamma$, then for $\bar{a} > a$,

$$h(t) = \frac{1}{2\pi i} \int_{\bar{a}-i\infty}^{\bar{a}+i\infty} e^{\lambda t} \left[\int_0^\infty e^{-\lambda s} h(s) ds\right] d\lambda, \quad t > 0.$$

Proof. We can show that the integral

$$\int_{\bar{a}-i\infty}^{\bar{a}+i\infty} e^{\lambda t} \left[\int_0^\infty e^{-\lambda s} h(s) ds\right] d\lambda$$

converges using arguments similar to those of Widder [1, p. 68, Theorem 7.5], noting that the Riemann-Lebesgue lemma is applicable to vector-valued functions (cf. Hewitt [1, Proof of (21.39)] and noting Hille-Phillips [1, Theorem 3.8.3]).

On the other hand, Theorem 6.3.1 of Hille-Phillips [1] yields that for each $u^* \in E^*$ (the dual space of E),

$$
\begin{aligned}
u^*(h(t)) &= \frac{1}{2\pi i} \int_{\bar{a}-i\infty}^{\bar{a}+i\infty} e^{\lambda t} \left[\int_0^\infty e^{-\lambda s} u^*(h(s)) ds\right] d\lambda \\
&= u^* \left\{ \frac{1}{2\pi i} \int_{\bar{a}-i\infty}^{\bar{a}+i\infty} e^{\lambda t} \left[\int_0^\infty e^{-\lambda s} h(s) ds\right] d\lambda \right\}, \quad t > 0.
\end{aligned}
$$

Then we obtain the desired formula by the arbitrariness of u^*.

Theorem 1.9. *Let h_1, $h_2 \in C(R^+, E)$ satisfying that for any $p \in \Gamma$,*

$$\sup_{t \geq 0} p\left(e^{-at}h_i(t)\right) < \infty \quad (i = 1, 2)$$

for some $a > 0$. Suppose that for $\lambda > a$,

$$\int_0^\infty e^{-\lambda t} h_2(t) dt = \lambda \int_0^\infty e^{-\lambda t} h_1(t) dt - v$$

for some $v \in E$. Then $h_1'(t) = h_2(t)$ $(t \geq 0)$, $h_1(0) = v$.

Proof. It follows easily from Theorem 1.6.

Theorem 1.10. *Let* h_1, h_2 *be as in Theorem 1.9. Suppose that A is a closed linear operator in E satisfying that for $\lambda > a$,*

$$\int_0^\infty e^{-\lambda t} h_1(t) dt \in \mathcal{D}(A)$$

such that

$$A \int_0^\infty e^{-\lambda t} h_1(t) dt = \int_0^\infty e^{-\lambda t} h_2(t) dt \quad (\lambda > a).$$

Then for each $t \geq 0$, $h_1(t) \in \mathcal{D}(A)$ and $Ah_1(t) = h_2(t)$.

Proof. In view of Theorem 1.5, we obtain that for $t > 0$,

$$\lim_{m \to \infty} \frac{(-1)^m}{m!} \left(\frac{m}{t}\right)^{m+1} f^{(m)} \left(\frac{m}{t}\right) = h_1(t),$$

$$\lim_{m \to \infty} \frac{(-1)^m}{m!} \left(\frac{m}{t}\right)^{m+1} A f^{(m)} \left(\frac{m}{t}\right) = h_2(t),$$

where

$$f(\lambda) := \int_0^\infty e^{-\lambda t} h_1(t) dt \quad (\lambda > a).$$

The two equalities together imply that for each $t > 0, h_1(t) \in \mathcal{D}(A)$ and $Ah_1(t) = h_2(t)$ due to the closedness of A. It follows (using the closedness of A again) that $h_1(0) \in \mathcal{D}(A)$ and $Ah_1(0) = h_2(0)$ since h_1, h_2 are continuous at the point 0.

Theorem 1.11. *Let E be a Banach space, and let $F_0(\cdot)$, $F_1(\cdot) \in LT - L(E)$, $f(\cdot) \in LT - E$. Then*

$$[I - F_0(\cdot)]^{-1} - I \in LT - L(E);$$

$$F_1(\cdot)[I - F_0(\cdot)]^{-1}, \quad [I - F_0(\cdot)]^{-1} F_1(\cdot) \in LT - L(E);$$

$$[I - F_0(\cdot)]^{-1} f(\cdot) \in LT - E.$$

Proof. By hypothesis, for $i = 0, 1$, there is a strongly continuous function $H_i(\cdot) : R^+ \to L(E)$ satisfying $\|H_i(t)\| \leq C_i e^{at}$ $(t \geq 0)$ for some C_i, $a > 0$ such that

$$F_i(\lambda)u = \int_0^\infty e^{-\lambda t} H_i(t) u dt \quad (\lambda > a, \ u \in E).$$

Therefore for $\lambda > C_0 + a$, $u \in E$,

$$
\begin{cases}
[I - F_0(\lambda)]^{-1}u - u = \sum_{m=1}^{\infty} [F_0(\lambda)]^m u \\
\\
\qquad\qquad\qquad = \sum_{m=1}^{\infty} \int_0^{\infty} e^{-\lambda t} [H_0(t)]^{*m} u \, dt, \\
\\
\|[H_0(t)]^{*m}\| \le C_0^m e^{at} \dfrac{t^{m-1}}{(m-1)!} \quad (t \ge 0,\ m \in N),
\end{cases}
$$

where $*m$ indicates the mth convolution power. This shows that

$$
H(t) := \sum_{m=1}^{\infty} [H_0(t)]^{*m} \quad (t \ge 0)
$$

defines a strongly continuous $\mathbf{L}(E)$−valued function satisfying

$$
\|H(t)\| \le \sum_{m=1}^{\infty} C_0^m e^{at} \frac{t^{m-1}}{(m-1)!} = C_0 e^{(C_0+a)t} \quad (t \ge 0).
$$

In conclusion, for $\lambda > C_0 + a$, $u \in E$,

$$
[I - F_0(\lambda)]^{-1}u - u = \int_0^{\infty} e^{-\lambda t} H(t) u \, dt, \tag{1.1}
$$

and therefore $[I - F_0(\cdot)]^{-1} - I \in LT - \mathbf{L}(E)$. Next, we have by (1.1) that for $\lambda > C_0 + a$, $u \in E$, $t \ge 0$,

$$
\begin{aligned}
[I - F_0(\lambda)]^{-1} F_1(\lambda) u &= F_1(\lambda) u + \{[I - F_0(\lambda)]^{-1} - I\} F_1(\lambda) u \\
\\
&= \int_0^{\infty} e^{-\lambda t} [H_1(t) + H(t) * H_1(t)] u \, dt,
\end{aligned}
$$

$$
\begin{aligned}
\|H_1(t) + H(t) * H_1(t)\| &\le C_1 e^{at} + C_1 e^{at} \left(e^{C_0 t} - 1\right) \\
\\
&= C_1 e^{(C_0+a)t}.
\end{aligned} \tag{1.2}
$$

Consequently,

$$
[I - F_0(\cdot)]^{-1} F_1(\cdot) \in LT - \mathbf{L}(E).
$$

Similar reasoning implies that

$$
F_1(\cdot)[I - F_0(\cdot)]^{-1} \in LT - \mathbf{L}(E),
$$

$$
[I - F_0(\cdot)]^{-1} f(\cdot) \in LT - E.
$$

Theorem 1.12. *Let $r \in R$, $a > 0$, $M_p > 0$ for each $p \in \Gamma$, and let*

$$f : \{\lambda \in C; \ \mathrm{Re}\lambda > a\} \longrightarrow E$$

be an analytic function satisfying

$$p(f(\lambda)) \leq M_p |\lambda|^r, \quad p \in \Gamma, \ \mathrm{Re}\lambda > a. \tag{1.3}$$

Then for each $\alpha > 1$ there exists a function $h_\alpha \in C(R^+, E)$ with $h_\alpha(0) = 0$ such that

$$p(h_\alpha(t)) \leq M_\alpha M_p e^{at}, \quad p \in \Gamma, \ t \geq 0, \tag{1.4}$$

$$f(\lambda) = \lambda^{r+\alpha} \int_0^\infty e^{-\lambda t} h_\alpha(t) dt, \quad \mathrm{Re}\lambda > a,$$

where M_α is a constant independent of p, f.

Proof. Let $\bar{a} > a$ and

$$h_\alpha(t) = \frac{1}{2\pi i} \int_{\bar{a}-i\infty}^{\bar{a}+i\infty} e^{\mu t} \mu^{-r-\alpha} f(\mu) d\mu, \quad t \geq 0.$$

Then by (1.3)

$$p(h_\alpha(t)) \leq \frac{M_p e^{\bar{a}t}}{2\pi} \int_{-\infty}^\infty (a^2 + |\mu|^2)^{-\frac{\alpha}{2}} d\mu, \quad t \geq 0, \ p \in \Gamma.$$

Accordingly, letting $\bar{a} \to a$ gives (1.4). Next, let $\mathrm{Re}\lambda > \bar{a}$. We have

$$\int_0^\infty e^{-\lambda t} h_\alpha(t) dt = \int_0^\infty e^{-\lambda t} \frac{1}{2\pi i} \int_{\bar{a}-i\infty}^{\bar{a}+i\infty} e^{\mu t} \mu^{-r-\alpha} f(\mu) d\mu dt$$

$$= \frac{1}{2\pi i} \int_{\bar{a}-i\infty}^{\bar{a}+i\infty} \int_0^\infty e^{-(\lambda-\mu)t} dt \left(\mu^{-r-\alpha} f(\mu) \right) d\mu$$

$$= \frac{1}{2\pi i} \int_{\bar{a}-i\infty}^{\bar{a}+i\infty} \frac{\mu^{-r-\alpha} f(\mu)}{\lambda - \mu} d\mu$$

$$= \lambda^{-r-\alpha} f(\lambda) + \lim_{\tau \to +\infty} \frac{1}{2\pi i} \int_{\gamma(\tau)} \frac{\mu^{-r-\alpha} f(\mu)}{\lambda - \mu} d\mu$$

$$= \lambda^{-r-\alpha} f(\lambda) \quad \text{(by Cauchy's formula)},$$

where

$$\gamma(\tau) := \left\{ \bar{a} + \tau e^{i\theta}; \ -\frac{\pi}{2} \leq \theta \leq \frac{\pi}{2} \right\}.$$

Finally, we obtain

$$h_\alpha(0) = \lim_{\tau \to +\infty} \frac{1}{2\pi i} \int_{\gamma(\tau)} \mu^{-r-\alpha} f(\mu) d\mu$$

$$= 0.$$

This completes the proof.

1.2   An integrated version of Widder's theorem in SCLCS

Theorem 2.1. *Let $r \in (0, 1]$, $a \geq 0$, $\omega \in (-\infty, a]$, $M_p > 0$ for each $p \in \Gamma$, and let $f : (a, \infty) \to E$ be a function. Then the following assertions are equivalent.*
(i) *f is infinitely differentiable and*

$$p\left(f^{(k)}(\lambda)\right) \leq M_p k!(\lambda - \omega)^{-k-1}, \quad p \in \Gamma, \ \lambda > a, \ k \in N_0.$$

(ii) *There exists a function $F_r : R^+ \to E$ satisfying $F_r(0) = 0$ and*

$$f(\lambda) = \lambda^r \int_0^\infty e^{-\lambda t} F_r(t) dt, \quad \lambda > a,$$

and for each $p \in \Gamma$, $t, \ h \geq 0$,

$$p\left(\widetilde{F}_r(t + h) - \widetilde{F}_r(t)\right) \leq M_p e^{\omega t} \max\left\{e^{\omega h}, \ 1\right\} h, \tag{2.1}$$

where

$$\widetilde{F}_r(t) := \begin{cases} \displaystyle\int_0^t \frac{(t - s)^{-r}}{\Gamma(1 - r)} F_r(s) ds, & \text{if } r < 1, \\[2mm] F_r(t), & \text{if } r = 1. \end{cases} \tag{2.2}$$

Moreover, in this case, F_r satisfies that for each $p \in \Gamma$, $t, \ h \geq 0$,

$$p(F_r(t + h) - F_r(t)) \leq \frac{2M_p}{r\Gamma(r)} \max\left\{e^{\omega(t+h)}, \ 1\right\} h^r. \tag{2.3}$$

Proof. (i)$\Rightarrow$ (ii). By hypothesis,

$$\Omega_0 := \left\{\frac{1}{k!}(\lambda - \omega)^{k+1} f^{(k)}(\lambda); \ \ \lambda > a, \ k \in N_0\right\}$$

is a bounded subset of E. Let E^* be the dual space of E, topologized by the class of seminorms $\Gamma^* := \{p_\Omega; \ \Omega \text{ is a bounded subset of } E\}$; p_Ω is defined by

$$p_\Omega(u^*) = \sup_{v \in \Omega} |\langle v, \ u^* \rangle|, \quad u^* \in E^*.$$

Let $u^* \in E^*$. Then

$$\sup\left\{\left|\left\langle \frac{1}{k!}(\lambda - \omega)^{k+1} f^{(k)}(\lambda), \ u^* \right\rangle\right|; \ \ \lambda > a, \ k \in N_0\right\} \leq p_{\Omega_0}(u^*).$$

Widder's classical theorem therefore assures the existence of a numerical function $g(\cdot, \ u^*)$ with

$$|g(t, \ u^*)| \leq p_{\Omega_0}(u^*) e^{\omega t}, \quad \text{for } t \geq 0, \tag{2.4}$$

such that

$$\langle f(\lambda),\ u^* \rangle = \int_0^\infty e^{-\lambda t} g(t,\ u^*) dt, \quad \lambda > a. \tag{2.5}$$

Set

$$F_r(t,\ u^*) = \int_0^t \frac{(t-s)^{r-1}}{\Gamma(r)} g(s,\ u^*) ds, \quad t \ge 0,\ u^* \in E^*.$$

Then we get from (2.4) that

$$|F_r(t,\ u^*)| \le \frac{t^r}{r\Gamma(r)} \max\{e^{\omega t},\ 1\} p_{\Omega_0}(u^*), \quad t \ge 0,\ u^* \in E^*. \tag{2.6}$$

Also, in view of the convolution theorem for Laplace transforms, we obtain by (2.5),

$$\langle f(\lambda),\ u^* \rangle = \lambda^r \int_0^\infty e^{-\lambda t} F_r(t,\ u^*) dt, \quad \lambda > a,\ u^* \in E^*. \tag{2.7}$$

Thus, an application of the uniqueness theorem for Laplace transforms (noting that $F_r(\cdot,\ u^*)$ for each $u^* \in E^*$ is continuous) shows that

$$\text{for each } t \ge 0,\ F_r(t,\ u^*) \text{ is linear in } u^* \in E^*. \tag{2.8}$$

We now denote by $\widetilde{E}$ the space of all linear functionals $\widetilde{u}$ on E^* satisfying

$$p_G(\widetilde{u}) := \sup_{u^* \in G} |\langle u^*,\ \widetilde{u} \rangle| < \infty$$

for each equicontinuous family $G \subset E^*$. The family $\widetilde{\Gamma}$ of seminorms p_G for all G as above induces a locally convex topology τ on $\widetilde{E}$. We claim that $\left(\widetilde{E},\ \tau \right)$ is sequentially complete. Indeed, if $\{\widetilde{u}_k\}$ is a Cauchy sequence in $\widetilde{E}$, then for each $\varepsilon > 0$, $p_G \in \widetilde{\Gamma}$, there exists $k_0 > 0$ such that for $k,\ j > k_0$, $p_G(\widetilde{u}_k - \widetilde{u}_j) < \varepsilon$, and therefore

$$|\langle u^*,\ \widetilde{u}_k \rangle - \langle u^*,\ \widetilde{u}_j \rangle| < \varepsilon \quad (u^* \in G). \tag{2.9}$$

This implies the existence of a linear functional $\widetilde{u}_0$ on E^* such that for each $u^* \in E^*$,

$$\lim_{j \to \infty} \langle u^*,\ \widetilde{u}_j \rangle = \langle u^*,\ \widetilde{u}_0 \rangle.$$

It follows from (2.9) that

$$p_G(\widetilde{u}_k - \widetilde{u}_0) = \sup_{u^* \in G} |\langle u^*,\ \widetilde{u}_k \rangle - \langle u^*,\ \widetilde{u}_0 \rangle| \le \varepsilon \text{ for all } k > k_0.$$

Consequently, $\widetilde{u}_0 \in \widetilde{E}$ and $\widetilde{u}_k \longrightarrow \widetilde{u}_0$ in $\widetilde{E}$, which verifies our claim.

Next, (2.6) and (2.8) together tell us that for each $t \ge 0$, there exists $F_r(t) \in \widetilde{E}$ such that

$$F_r(t,\ u^*) = \langle F_r(t),\ u^* \rangle, \quad u^* \in E^*.$$

From (2.4) we get that for each $p_G \in \widetilde{\Gamma}$, t, $h \geq 0$,

$$p_G(F_r(t+h) - F_r(t))$$

$$= \sup_{u^* \in G} |F_r(t+h, u^*) - F_r(t, u^*)|$$

$$= \sup_{u^* \in G} \left\{ \frac{1}{\Gamma(r)} \left(\left| \int_0^t [(t+h-s)^{r-1} - (t-s)^{r-1}] g(s, u^*)ds \right. \right. \right.$$

$$\left. \left. \left. + \int_t^{t+h} (t+h-s)^{r-1} g(s, u^*)ds \right| \right) \right\} \tag{2.10}$$

$$\leq \sup_{u^* \in G} p_{\Omega_0}(u^*) \begin{cases} \dfrac{2\max\left\{e^{\omega(t+h)}, 1\right\}}{r\Gamma(r)} h^r, & \text{if } r < 1 \\[2ex] e^{\omega t} \max\left\{e^{\omega h}, 1\right\} h, & \text{if } r = 1 \end{cases}$$

$$\longrightarrow 0, \quad \text{as } h \longrightarrow 0;$$

Accordingly,

$$F_r(\cdot) \in C\left(R^+, \widetilde{E}\right). \tag{2.11}$$

Also, we obtain by (2.6) again that for each $p_G \in \widetilde{\Gamma}$, $t \geq 0$,

$$p_G(F_r(t)) = \sup_{u^* \in G} |F_r(t, u^*)|$$

$$\leq \sup_{u^* \in G} p_{\Omega_0}(u^*) \frac{t^r}{r\Gamma(r)} \max\left\{e^{\omega t}, 1\right\}. \tag{2.12}$$

These facts in combination with (2.7) indicate that in $\widetilde{E}$,

$$f(\lambda) = \lambda^r \int_0^\infty e^{-\lambda t} F_r(t) dt \quad (\lambda > a), \tag{2.13}$$

recalling that $\widetilde{E}$ is sequentially complete.

We now identify E with a subspace of $\widetilde{E}$ via evaluation. For each $p \in \Gamma$, set

$$Q_p = \{u \in E; \ p(u) \leq 1\},$$

$$Q_p^0 = \left\{u^* \in E^*; \ \sup_{v \in Q_p} |\langle v, u^* \rangle| \leq 1\right\}.$$

Clearly, Q_p^0 is an equicontinuous subset of E^*. We observe that for each fixed $v \in E$, $\frac{v}{p(v)} \in Q_p$, and so $\sup_{u^* \in Q_p^0} |\langle v, u^* \rangle| \leq p(v)$; moreover, the Hahn-Banach theorem assures the existence of an $u_0^* \in Q_p^0$ such that $\langle v, u_0^* \rangle = p(v)$. Accordingly,

$$p(v) = \sup_{u^* \in Q_p^0} |\langle v, u^* \rangle| = p_{Q_p^0}(v), \quad \text{for } v \in E, \ p \in \Gamma. \tag{2.14}$$

On the other hand, to each $p_G \in \tilde{\Gamma}$ there corresponds to $p \in \Gamma$ and $b > 0$ such that $G \subset b Q_p^0$. From these observations, we can see that

the topology τ on $\tilde{E}$ induces the original topology on E. $\qquad$ (2.15)

Next, we set

$$\Phi_r(t; \ k) = (-1)^k \left(\frac{k}{t}\right)^{k+1} \frac{1}{k!} [\lambda^{-r} f(\lambda)]^{(k)} \bigg|_{\lambda = \frac{k}{t}}, \qquad k \in N, \ t > 0.$$

Then applying Theorem 1.5, we obtain by (2.11) and (2.13) that

$$F_r(t) = \lim_{k \to \infty} \Phi_r(t; \ k) \quad (t > 0).$$

It follows, from (2.15) and the fact that $\Phi_r(t; \ k) \in E$ and E is sequentially complete, that $F_r(t) \in E$ for each $t \geq 0$, noting $F_r(0) = 0$.

Observe by hypothesis that for each $v \in \Omega_0$, $M_p^{-1} v \in Q_p$, and therefore

$$M_p^{-1} \sup_{u^* \in Q_p^0} p_{\Omega_0}(u^*) = \sup_{u^* \in Q_p^0} \sup_{v \in \Omega_0} |\langle M_p^{-1} v, u^* \rangle| \leq 1, \quad p \in \Gamma.$$

Thus, (2.10) and (2.14) lead to (2.1) and (2.3), noting

$$f(\lambda) = \lambda \int_0^\infty e^{-\lambda t} \tilde{F}_r(t) dt, \quad \lambda > a,$$

which implies by the uniqueness theorem for Laplace transforms that $\tilde{F}_r(\cdot)$ coincides with $F_1(t)$.

(ii)$\Rightarrow$(i). Fix $p \in \Gamma$. Let $u^* \in Q_p^0$ and set $g(t) = \langle \tilde{F}_r(t), u^* \rangle$ for $t \geq 0$. Then

$$\langle f(\lambda), u^* \rangle = \lambda \int_0^\infty e^{-\lambda t} g(t) dt, \quad \lambda > a. \tag{2.16}$$

Using (2.1) and (2.14) we obtain that for $t, h \geq 0$,

$$|g(t + h) - g(t)| \ \leq \ p\left(\tilde{F}_r(t + h) - \tilde{F}_r(t)\right)$$

$$\leq \ M_p e^{\omega t} \max\{e^{\omega h}, \ 1\} h.$$

Hence $g(\cdot)$ is differentiable almost everywhere with

$$|g\prime(t)| \leq M_p e^{\omega t}, \quad \text{for } t \in R^+ \text{ a.e.}$$

Thus, it follows from (2.16) that for $\lambda > a$,

$$\langle f(\lambda),\ u^* \rangle \ =\ -e^{-\lambda t}g(t)\Big|_0^\infty + \int_0^\infty e^{-\lambda t}g'(t)dt$$

$$=\ \int_0^\infty e^{-\lambda t}g'(t)dt,$$

and therefore

$$\left|\left\langle f^{(k)}(\lambda),\ u^* \right\rangle\right| \le M_p k! (\lambda - \omega)^{-k-1}, \quad \lambda > a,\ k \in N_0.$$

Consequently, we obtain by (2.14) again that for $p \in \Gamma$,

$$p(f^{(k)}(\lambda)) \ =\ \sup_{u^* \in Q_p^0} \left|\left\langle f^{(k)}(\lambda),\ u^* \right\rangle\right|$$

$$\le\ M_p k! (\lambda - \omega)^{-k-1}, \quad \lambda > a,\ k \in N_0.$$

The proof is then complete.

1.3   Integrated, regularized semigroups

Throughout this section, we assume that C is a continuous, injective operator on E, and A is a closed linear operator in E such that $CA \subset AC$. The C-resolvent set of A is defined as

$$\rho_C(A) = \{\lambda \in \mathbf{C};\ (\lambda - A) \text{ is injective,}$$

$$\mathcal{R}(C) \subset \mathcal{R}(\lambda - A) \text{ and } (\lambda - A)^{-1}C \in \mathbf{L}(E)\}.$$

Definition 3.1. Let $\omega \in R$, $r \in R^+$. If $(\omega, \infty) \subset \rho_C(A)$ and there exists $S_r(\cdot) : R^+ \to \mathbf{L}(E)$ satisfying that $t \mapsto S_r(t)u \in C(R^+, E)$ for each $u \in E$ such that $\{e^{-\omega t}S_r(t);\ t \ge 0\}$ is equicontinuous and

$$(\lambda - A)^{-1}Cu = \lambda^r \int_0^\infty e^{-\lambda t}S_r(t)u\,dt, \quad \lambda > \omega,\ u \in E,$$

then we say that A is a subgenerator of an r-times integrated, C-regularized semigroup $\{S_r(t)\}_{t\ge 0}$. If $r = 0$ (resp. $C = I$), then $\{S_r(t)\}_{t\ge 0}$ is called a C-regularized (resp. r-times integrated) semigroup.

Remark. There could be more than one subgenerator A for an integrated C-regularized semigroup $\{S_r(t)\}_{t\ge 0}$, unless $C = I$. On the other hand, it is easy to see that the extension $C^{-1}AC$ of A is also a subgenerator of $\{S_r(t)\}_{t\ge 0}$; moreover, if both A_1 and A_2 are subgenerators of $\{S_r(t)\}_{t\ge 0}$, then $C^{-1}A_1C = C^{-1}A_2C$.

The operator $\tilde{A} := C^{-1}AC$ will be called the generator of $\{S_r(t)\}_{t \geq 0}$. Evidently, an r-times integrated semigroup has a unique subgenerator, i.e., its generator. Moreover, 0-times integrated semigroups coincide with (exponentially equicontinuous) strongly continuous semigroups, so do their generators.

Lemma 3.2. *Assume that A is a subgenerator of an r-times integrated, C-regularized semigroup $\{S_r(t)\}_{t \geq 0}$. Then*

(i) $S_r(t)C = CS_r(t)$ $(t \geq 0)$,

(ii) $S_r(t)u \in \mathcal{D}(A)$, *and* $AS_r(t)u = S_r(t)Au$ $(t \geq 0, u \in \mathcal{D}(A))$,

(iii) $S_r(t)u = \dfrac{t^r}{\Gamma(r+1)}Cu + A\displaystyle\int_0^t S_r(s)u\,ds$ $(t \geq 0, u \in E)$,

(iv) *when* $r = 0$,

$$S_0(0) = C, \quad S_0(s)S_0(t) = CS_0(t+s), \quad t,\ s \geq 0,$$

$$\mathcal{D}\left(\tilde{A}\right) = \left\{ u \in E;\ \lim_{t \to 0^+} \frac{S_0(t)u - Cu}{t} \in \mathcal{R}(C) \right\}$$

and

$$\tilde{A}u = C^{-1}\lim_{t \to 0^+}\frac{S_0(t)u - Cu}{t}, \quad \text{for all } u \in \mathcal{D}\left(\tilde{A}\right).$$

Proof. By the uniqueness of Laplace transforms, the properties (i) and (ii) are shown immediately. Next, observe that for each $u \in E$, λ sufficiently large,

$$\int_0^\infty e^{-\lambda t}\left(S_r(t)u - \frac{t^r}{\Gamma(r+1)}Cu\right)dt = \lambda^{-r}(\lambda - A)^{-1}Cu - \lambda^{-r-1}Cu$$

$$= A\left(\lambda^{-r-1}(\lambda - A)^{-1}Cu\right)$$

$$= A\int_0^\infty e^{-\lambda t}\left(\int_0^t S_r(s)u\,ds\right)dt.$$

Then using Theorem 1.10, we deduce that for each $t \geq 0$, $u \in E$, $\int_0^t S_r(s)u\,ds \in \mathcal{D}(A)$ and (iii) is satisfied. Moreover, it follows from (iii) that

$$\frac{d}{ds}\left(S_0(\sigma - s)S_0(s)u\right) = 0, \quad u \in \mathcal{D}(A),\ \sigma \geq s \geq 0;$$

whence

$$S_0(\sigma - s)S_0(s)u = CS_0(\sigma)u.$$

Letting $t = \sigma - s$ gives

$$S_0(t)S_0(s)u = CS_0(t+s)u, \quad t,\ s \geq 0;$$

this equality holds actually for any $u \in E$, since $(\omega + 1 - A)^{-1}u$ (ω as in Definition 3.1) is in $\mathcal{D}(A)$, commutes with $S_0(t)$ for each $t \geq 0$ (by (ii)), and $(\omega + 1 - A)^{-1}$

is injective. The remaining part of assertion (iv) can be derived easily from (ii) and (iii) by taking $r = 0$, $A = \tilde{A}$. The proof is then complete.

Theorem 3.3. *Let $\omega \in R$, $r \in R^+$. Then the following assertions are equivalent.*
(i) *There exists $a \geq \omega$ such that $(a, \infty) \subset \rho_C(A)$ and the family*

$$\left\{ \frac{1}{j!}(\lambda - \omega)^{j+1} \left(\lambda^{-r}(\lambda - A)^{-1}C \right)^{(j)}; \quad \lambda > a, \; j \in N_0 \right\}$$

is equicontinuous (resp. $\Gamma(M_p)$-equicontinuous, $\Gamma(M)$-equicontinuous).
(ii) *For each $\bar{r} \in (r, \; r + 1]$, A is a subgenerator of an $\bar{r}$-times integrated, C-regularized semigroup $\{S_{\bar{r}}(t)\}_{t \geq 0}$ on E, such that the family*

$$\left\{ (S_{r+1}(t + h) - S_{r+1}(t))h^{-1}e^{-\omega t} \min \left\{ e^{-\omega h}, \; 1 \right\}; \quad t, \; h \geq 0 \right\}$$

$$\text{is equicontinuous} \tag{3.1}$$

$$(\text{resp. } \Gamma(M_p) - \text{equicontinuous}, \; \Gamma(M) - \text{equicontinuous}).$$

Proof. Apply Theorem 2.1.

Theorem 3.4. *Assume that $\mathcal{D}(A)$ is dense in E. Then A is a subgenerator of an r-times integrated, C-regularized semigroup $\{S_r(t)\}_{t \geq 0}$ such that the family*

$$\{e^{-\omega t}S_r(t); \; t \geq 0\} \; \text{is equicontinuous}$$

$$\tag{3.2}$$

$$(\text{resp. } \Gamma(M_p) - \text{equicontinuous}, \; \Gamma(M) - \text{equicontinuous}),$$

if and only if condition (i) *of Theorem 3.3 holds.*

Proof. The "only if" part is immediate, since for some $a \geq \omega$,

$$\left(\lambda^{-r}(\lambda - A)^{-1}C \right)^{(j)} u = \int_0^\infty e^{-\lambda t}(-t)^j S_r(t)u dt, \quad \lambda > a, \; j \in N_0, \; u \in E.$$

The "if" part. Let $\{S_{r+1}(t)\}_{t \geq 0}$ be as in Theorem 3.3 (ii). Then for λ sufficiently large,

$$(\lambda - A)^{-1}Cu = \lambda^{r+1} \int_0^\infty e^{-\lambda t}S_{r+1}(t)u dt, \quad \lambda > a, \; u \in E. \tag{3.3}$$

We have from Lemma 3.2 that for $t \geq 0$, $u \in \mathcal{D}(A)$,

$$S_{r+1}(t)u = \frac{t^{r+1}}{\Gamma(r + 2)}Cu + \int_0^t S_{r+1}(s)Au ds.$$

This implies that for each $u \in \mathcal{D}(A)$, $t \mapsto S_{r+1}(t)u$ is differentiable in R^+. Fix $t \geq 0$, $\varepsilon > 0$, $p \in \Gamma$. For $h \in (-t, \; 1)$, $u \in E$, set

$$Q_h u = \frac{1}{h} \left(S_{r+1}(t + h)u - S_{r+1}(t)u \right).$$

(3.1) assures the existence of a continuous seminorm q such that

$$p(Q_h u) \le q(u) \quad (h \in (-t, 1), \ u \in E).$$

Fix $u \in E$. There exists $u_0 \in \mathcal{D}(A)$ such that $q(u - u_0) < \frac{\epsilon}{3}$ since $\mathcal{D}(A)$ is dense in E. But $t \mapsto S_{r+1}(t)u_0$ is differentiable; we have that there is a $\delta \in (0, 1)$ such that

$$p(Q_{h_1} u_0 - Q_{h_2} u_0) < \frac{\epsilon}{3}, \quad \text{for every } h_1, \ h_2 \text{ with } -\min\{t, \ \delta\} < h_1, \ h_2 < \delta.$$

Therefore, for h_1, h_2 as above,

$$p(Q_{h_1} u - Q_{h_2} u)$$

$$\le \quad p(Q_{h_1} u_0 - Q_{h_2} u_0) + p(Q_{h_1}(u - u_0)) + p(Q_{h_2}(u - u_0))$$

$$< \quad \frac{\epsilon}{3} + 2q(u - u_0) < \epsilon.$$

Accordingly, the sequential completeness of E implies that $t \mapsto S_{r+1}(t)u$ (for all $u \in E$) is differentiable in R^+. Set

$$S_r(t)u = \frac{d}{dt}(S_{r+1}(t)u), \quad t \ge 0, \ u \in E.$$

Using (3.1) again gives (3.2). Finally, integrating by parts, we obtain from (3.3) that for λ sufficiently large,

$$(\lambda - A)^{-1}Cu = \lambda^r \int_0^\infty e^{-\lambda t} S_r(t)u\,dt, \quad u \in E.$$

This ends the proof.

The following is a perturbation result for generators of integrated semigroups in SCLCS, whose Banach space relatives can be found in van Neerven-Straub [1] (for the case of $K = aI$, $a \in \mathbf{C}$) or Kellermann-Hieber [1], Neubrander [5], Nicaise [1] for the case of $r \in N$.

Theorem 3.5. *Let $r \in R^+$, M, $\omega > 0$, and let A be the generator of an r-times integrated semigroup $\{S(t)\}_{t \ge 0}$ satisfying $\|S(t)\|_\Gamma \le Me^{\omega t}$ $(t \ge 0)$. Assume $K \in B_\Gamma(E)$ such that $KA \subset AK$. Then $A + K$ is the generator of an r-times integrated semigroup $\{S_K(t)\}_{t \ge 0}$, given by*

$$S_K(t)u \quad = \quad e^{Kt}S(t)u + \sum_{j=1}^\infty \binom{r}{j}(-K)^j \int_0^t \frac{(t-s)^{j-1}}{(j-1)!} e^{Ks} S(s)u\,ds,$$

$$\text{(3.4)}$$

$$t \ge 0, \ u \in E,$$

where $\binom{r}{j} = \dfrac{r(r-1)\cdots(r-j+1)}{j!}$.

Proof. Set $L_0 = \sup\limits_{j \in N} \left| \binom{r}{j} \right|$. Clearly $L_0 < \infty$. Thus, it is easy to see that the series in (3.4) converges absolutely with respect to $\|\cdot\|_{\mathbf{r}}$ and

$$\|S_K(t)\|_{\mathbf{r}} \le M(L_0+1)e^{t(2\|K\|_{\mathbf{r}}+\omega)}, \quad t \ge 0.$$

By hypothesis,

$$R(\lambda; A)u = \lambda^r \int_0^\infty e^{-\lambda t} S(t)u\,dt, \quad \lambda > \omega, \ u \in E.$$

From this, we obtain that for $u \in E$, $\lambda > \omega$, $k \in N$,

$$R(\lambda; A)^{k+1}u$$

$$= (-1)^k \frac{1}{k!} \left(\frac{d}{d\lambda} \right)^k R(\lambda; A)u$$

$$= (-1)^k \frac{1}{k!} \sum_{j=0}^k \binom{k}{j} r(r-1)\cdots(r-j+1)\lambda^{r-j} \int_0^\infty e^{-\lambda t}(-t)^{k-j} S(t)u\,dt$$

$$= \lambda^r \sum_{j=0}^k \binom{r}{j} \frac{(-\lambda)^{-j}}{(k-j)!} \int_0^\infty e^{-\lambda t} t^{k-j} S(t)u\,dt$$

$$= \lambda^r \int_0^\infty e^{-\lambda t} \left(\frac{1}{k!} t^k S(t)u + \sum_{j=1}^k \binom{r}{j} \int_0^t \frac{(t-s)^{j-1}}{(j-1)!} \frac{(-1)^j s^{k-j}}{(k-j)!} S(s)u\,ds \right) dt.$$

Thus for $u \in E$, $\lambda > 2\|K\|_{\mathbf{r}} + \omega$,

$$\sum_{k=0}^\infty K^k R(\lambda; A)^{k+1}u$$

$$= \lambda^r \int_0^\infty e^{-\lambda t} \left\{ \left[\sum_{k=0}^\infty \frac{(tK)^k}{k!} S(t)u \right] \right.$$

$$\left. + \left[\sum_{j=1}^\infty \binom{r}{j} (-K)^j \int_0^t \frac{(t-s)^{j-1}}{(j-1)!} \left(\sum_{k=j}^\infty \frac{(sK)^{k-j}}{(k-j)!} \right) S(s)u\,ds \right] \right\} dt$$

$$= \lambda^r \int_0^\infty e^{-\lambda t} S_K(t)u\,dt,$$

and therefore $\lambda \in \rho(A + K)$ and

$$R(\lambda; \; A + K)u \;\; = \;\; \sum_{k=0}^{\infty} K^k R(\lambda; \; A)^{k+1} u$$

$$= \;\; \lambda^r \int_0^{\infty} e^{-\lambda t} S_K(t) u \, dt.$$

This ends the proof.

Similarly, we have

Theorem 3.6. *Let A be a subgenerator (resp. the generator) of a C-regularized semigroup $\{W(t)\}_{t\geq 0}$ satisfying $\|W(t)\|_{\Gamma} \leq M e^{\omega t}$ $(t \geq 0)$ for some constants $M, \omega > 0$. Assume $K \in B_{\Gamma}(E)$ such that $KA \subset AK$, $KC = CK$. Then $A + K$ is a subgenerator (resp. the generator) of a C-regularized semigroup $\{W_K(t)\}_{t\geq 0}$ given by*

$$W_K(t) = e^{Kt} W(t), \quad t \geq 0.$$

Proof. Obviously,

$$C(A + K) \subset (A + K)C \quad (\text{resp. } A + K = C^{-1}(A + K)C).$$

Observe that for $u \in E$, $k \in N$, λ sufficiently large,

$$(\lambda - A)^{-k-1} Cu \;\; = \;\; (-1)^k \frac{1}{k!} \left(\frac{d}{d\lambda} \right)^k [(\lambda - A)^{-1} Cu]$$

$$= \;\; \int_0^{\infty} e^{-\lambda t} \frac{t^k}{k!} W(t) u \, dt.$$

We have that for $u \in E$, λ sufficiently large,

$$\sum_{k=0}^{\infty} K^k (\lambda - A)^{-k-1} Cu \;\; = \;\; \int_0^{\infty} e^{-\lambda t} \left(\sum_{k=0}^{\infty} \frac{t^k}{k!} K^k W(t) u \right) dt$$

$$= \;\; \int_0^{\infty} e^{-\lambda t} e^{tK} W(t) u \, dt$$

and therefore $\lambda \in \rho_C(A + K)$ and

$$(\lambda - A - K)^{-1} Cu \;\; = \;\; \sum_{k=0}^{\infty} K^k (\lambda - A)^{-k-1} Cu$$

$$= \;\; \int_0^{\infty} e^{-\lambda t} W_K(t) u \, dt.$$

The proof is then complete.

1.4   Integrated, regularized cosine functions

Let C, A be as in Section 1.3.

Definition 4.1. Let $\omega \in R$, $r \in R^+$. If $(\omega^2, \infty) \subset \rho_C(A)$ and there exists $C_r(\cdot) : R^+ \to L(E)$ satisfying that $t \mapsto C_r(t)u \in C(R^+, E)$ for each $u \in E$ such that $\{e^{-\omega t} C_r(t); \ t \geq 0\}$ is equicontinuous and

$$\lambda(\lambda^2 - A)^{-1} Cu = \lambda^r \int_0^\infty e^{-\lambda t} C_r(t) u \, dt, \quad \lambda > \omega, \ u \in E,$$

then we say that A is a subgenerator of an r-times integrated, C-regularized cosine function $\{C_r(t)\}_{t \geq 0}$. If $r = 0$ (resp. $C = I$), then $\{C_r(t)\}_{t \geq 0}$ is called a C-regularized (resp. r-times integrated) cosine function. The operator $\tilde{A} := C^{-1} A C$ will be called the generator of $\{C_r(t)\}_{t \geq 0}$.

Remark. As in the case of semigroups (in Section 1.3), we can see that an r-times integrated cosine function has a unique subgenerator, i.e., its generator, and that 0-times integrated cosine functions coincide with strongly continuous cosine functions, so do their generators.

Using the similar arguments as in Section 1.3 establishes the following results.

Lemma 4.2. *Assume that A is a subgenerator of an r-times integrated, C-regularized cosine function $\{C_r(t)\}_{t \geq 0}$. Then*
  (i) $C_r(t)C = CC_r(t)$ $(t \geq 0)$,
  (ii) $C_r(t)u \in \mathcal{D}(A)$, and $AC_r(t)u = C_r(t)Au$ $(t \geq 0, \ u \in \mathcal{D}(A))$,
  (iii) $C_r(t)u = \dfrac{t^r}{\Gamma(r+1)} Cu + A \displaystyle\int_0^t (t-s) C_r(s) u \, ds$ $(t \geq 0, \ u \in E)$,
  (iv) *when $r = 0$,*

$$C_0(0) = C, \quad 2C_0(t)C_0(s) = CC_0(s+t) + CC_0(|s-t|), \quad t, \ s \geq 0,$$

$$\mathcal{D}\left(\tilde{A}\right) = \left\{ u \in E; \ \lim_{t \to 0+} \frac{2}{t^2}(C_0(t)u - Cu) \in \mathcal{R}(C) \right\}$$

and

$$\tilde{A}u = C^{-1} \lim_{t \to 0+} \frac{2}{t^2}(C_0(t)u - Cu), \quad \text{for all } u \in \mathcal{D}\left(\tilde{A}\right).$$

Theorem 4.3. *Let $\omega \in R$, $r \in R^+$. Then the following assertions are equivalent.*
  (i) *There exists $a \geq \omega$ such that $(a^2, \infty) \subset \rho_C(A)$ and the family*

$$\left\{ \frac{1}{j!}(\lambda - \omega)^{j+1} \left(\lambda^{1-r}(\lambda^2 - A)^{-1}C \right)^{(j)}; \ \lambda > a, \ j \in N_0 \right\}$$

is equicontinuous (resp. $\Gamma(M_p)$-equicontinuous, $\Gamma(M)$-equicontinuous).

(ii) *For each $\bar{r} \in (r,\ r+1]$, A is a subgenerator of an $\bar{r}$-times integrated, C-regularized cosine function $\{C_{\bar{r}}(t)\}_{t\geq 0}$ on E, such that the family*

$$\left\{ (C_{r+1}(t+h) - C_{r+1}(t))h^{-1}e^{-\omega t} \min\left\{e^{-\omega h},\ 1\right\};\ \ t,\ h \geq 0 \right\}$$

is equicontinuous (resp. $\Gamma(M_p)$-equicontinuous, $\Gamma(M)$-equicontinuous).

Theorem 4.4. *Assume that $\mathcal{D}(A)$ is dense in E. Then A is a subgenerator of an r-times integrated, C-regularized cosine function $\{C_r(t)\}_{t\geq 0}$ such that the family*

$$\left\{ e^{-\omega t}C_r(t);\ t \geq 0 \right\}\ \ \text{is equicontinuous}$$

$$(\text{resp. } \Gamma(M_p) - \text{equicontinuous},\ \Gamma(M) - \text{equicontinuous}),$$

if and only if condition (i) of Theorem 4.3 holds.

1.5 Differential operators as generators

Let R^n be the n-dimensional Euclidean space. An n-tuple of nonnegative integers $\alpha = (\alpha_1,\ \alpha_2,\ \cdots, \alpha_n)$ is called a multiindex which we denote by $\alpha \in N_0^n$ and we define

$$|\alpha| = \sum_{i=1}^{n} \alpha_i,$$

$$x^\alpha = x_1^{\alpha_1} x_2^{\alpha_2} \cdots x_n^{\alpha_n} \quad \text{for } x = (x_1, x_2, \cdots, x_n) \in R^n,$$

$$D^\alpha = \left(\frac{\partial}{\partial x_1}\right)^{\alpha_1} \left(\frac{\partial}{\partial x_2}\right)^{\alpha_2} \cdots \left(\frac{\partial}{\partial x_n}\right)^{\alpha_n}.$$

By $\mathcal{S}(R^n)$, we denote the space of all rapidly decreasing functions on R^n with the local convex topology defined by the family of norms

$$\|f\|_m := \sup_{|\beta|\leq m} \sup_{x\in R^n} \left(1 + |x|^2\right)^m \left|(D^\beta f)(x)\right|,\quad m \in N_0.$$

The Fourier transform and its inverse transform are denoted by

$$(\mathcal{F}f)(x) = \hat{f}(x) := \int_{R^n} e^{-i(y,\ x)} f(y) dy$$

and

$$\left(\mathcal{F}^{-1}f\right)(y) := (2\pi)^{-n} \int_{R^n} e^{i(y,\ x)} f(x) dx.$$

A function $u \in L^\infty(R^n)$ is called a Fourier multiplier on $L^p(R^n)$ $(1 \leq p \leq \infty)$, if $\mathcal{F}^{-1}\left(u\hat{\phi}\right) \in L^p(R^n)$ for all $\phi \in \mathcal{S}(R^n)$ and if

$$\|u\|_{\mathcal{M}_p} := \sup\left\{\left\|\mathcal{F}^{-1}\left(u\hat{\phi}\right)\right\|_{L^p};\ \phi \in \mathcal{S}(R^n),\ \|\phi\|_{L^p} \leq 1\right\} < \infty.$$

The space of all Fourier multipliers on $L^p(R^n)$ $(1 \leq p \leq \infty)$ will be denoted by $\mathcal{M}_p$, which is a Banach algebra under pointwise multiplication and addition with the norm $\|\cdot\|_{\mathcal{M}_p}$. We note that

$$\mathcal{M}_p = \mathcal{M}_{p'} \quad \left(\frac{1}{p} + \frac{1}{p'} = 1; \ 1 \leq p \leq \infty \right)$$

with identical norms and

$$\mathcal{F}L^1 \hookrightarrow \mathcal{M}_1 \hookrightarrow \mathcal{M}_p, \quad \text{for all} \ p,$$

where $\mathcal{F}L^1$ denotes the Banach algebra $\{\mathcal{F}f; \ f \in L^1\}$ under pointwise multiplication and addition, with the norm

$$\|u\|_{\mathcal{F}L^1} := \left\| \mathcal{F}^{-1}u \right\|_{L^1};$$

that for a constant $t \in R \setminus \{0\}$ and a function $u : R^n \to \mathbf{C}$, $u \in \mathcal{M}_p$ implies

$$u_t \in \mathcal{M}_p \quad \text{and} \quad \|u\|_{\mathcal{M}_p} = \|u_t\|_{\mathcal{M}_p},$$

where u_t is defined by $u_t(x) = u(tx)$. For details on Fourier multipliers, we refer to Hörmander [1], Stein [1].

Lemma 5.1 (Bernstein's theorem). *Let $j \in N$ with $j > \frac{n}{2}$ and let $f \in C^\infty(R^n) \bigcap H^j(R^n)$. Then $f \in \mathcal{F}L^1$ and there exists a constant M such that*

$$\|f\|_{\mathcal{F}L^1} \leq M\|f\|_{L^2}^{1-\frac{n}{2j}} \left(\sum_{|\alpha|=j} \|D^\alpha f\|_{L^2} \right)^{\frac{n}{2j}}.$$

Lemma 5.2. *Let $1 \leq p \leq \infty$, j, $n \in N$, $j > \frac{n}{2}$ and $\{f_t\}_{t\geq 0}$ be a family of $C^j(R^n)$-functions. Assume that for each $x \in R^n$, $\alpha \in N_0^n$ with $|\alpha| \leq j$, $t \mapsto D^\alpha f_t(x)$ is continuous in R^+, and there exist $a > 0$, $r > n\left|\frac{1}{2} - \frac{1}{p}\right|$, $M_t > 0$ (M_t is bounded on compacts of $t \geq 0$) such that*

$$|D^\alpha f_t(x)| \leq M_t^{|\alpha|}(1 + |x|)^{(a-1)|\alpha|-ar} \quad (|\alpha| \leq j, \ x \in R^n, \ t \geq 0). \tag{5.1}$$

Then, for any $t \geq 0$, $p = 1$, ∞ (resp. $1 < p < \infty$), we have $f_t \in \mathcal{F}L^1$ (resp. $\mathcal{M}_p$), $t \mapsto f_t$ is continuous with respect to $\|\cdot\|_{\mathcal{F}L^1}$ (resp. $\|\cdot\|_{\mathcal{M}_p}$ if $(t,x) \mapsto f_t(x)$ is continuous in $R^+ \times R^n$), and there is a constant C independent of t such that

$$\|f_t\|_{\mathcal{F}L^1} \ (\text{resp.} \ \|f_t\|_{\mathcal{M}_p}) \leq CM_t^{n\left|\frac{1}{2}-\frac{1}{p}\right|}.$$

Proof. We may and do assume $1 \leq p \leq 2$, since $\mathcal{M}_q = \mathcal{M}_p$ with identical norms if $\frac{1}{q} + \frac{1}{p} = 1$. According to a known fact stated in Hörmander [1, p. 36], we can take two $C_c^\infty(R^n)$-functions ϕ, ψ such that

$$\phi(x) = \begin{cases} 1, & |x| \leq 1, \\ \\ 0, & |x| \geq 2, \end{cases}$$

$$\operatorname{supp}\psi \subset \left\{ x;\ \frac{1}{2} < |x| < 2 \right\} \quad \text{and} \quad \sum_{l=-\infty}^{\infty} \psi\left(2^{-l}x\right) = 1 \text{ for } x \neq 0.$$

Defining

$$\psi_l(x) = \psi\left(2^{-l}x\right), \quad x \in R^n,\ l = 0, \pm1, \pm2, \cdots,$$

we have

$$f_t = f_t\phi + f_t\psi_0(1-\phi) + f_t\psi_1(1-\phi) + \sum_{l=2}^{\infty} f_t\psi_l. \tag{5.2}$$

Using Leibniz's formula, we get by (5.1) that, there exists $C' > 0$ such that for $t \geq 0$, $l \geq 2$, $\alpha \in N_0^n$ with $|\alpha| \leq j$,

$$|D^\alpha(f_t(x)\psi_l(x))| \leq C' M_t^{|\alpha|} 2^{l((a-1)|\alpha|-ar)} \quad (x \in R^n);$$

therefore,

$$\|D^\alpha(f_t\psi_l)\|_{L^2} \leq \text{ const } M_t^{|\alpha|} 2^{l((a-1)|\alpha|-ar)} 2^{\frac{ln}{2}}, \tag{5.3}$$

$$\|(f_t\psi_l)\|_{\mathcal{M}_2} = \|(f_t\psi_l)\|_{L^\infty} \leq C' 2^{-lar}. \tag{5.4}$$

Making use of Bernstein's theorem, we obtain by (5.3) that

$$\|(f_t\psi_l)\|_{\mathcal{M}_1} \leq \|(f_t\psi_l)\|_{\mathcal{F}L^1} \leq \text{ const } M_t^{\frac{n}{2}} 2^{la(\frac{n}{2}-r)}; \tag{5.5}$$

moreover, the dominated convergence theorem implies

$$\|f_{t+h}\psi_l - f_t\psi_l\|_{\mathcal{F}L^1}$$

$$\leq \quad \text{const } \|(f_{t+h} - f_t)\psi_l\|_{L^2}^{1-\frac{n}{2j}} \left(\sum_{|\alpha|=j} \|D^\alpha(f_{t+h} - f_t)\psi_l\|_{L^2} \right)^{\frac{n}{2j}}$$

$$\longrightarrow 0, \quad \text{as } h \longrightarrow 0,$$

noting that for each $x \in R^n$, $|\alpha| \leq j$, $t \mapsto D^\alpha f_t(x)$ is assumed to be continuous. Whence, $t \mapsto f_t\psi_l$ is continuous with respect to $\|\cdot\|_{\mathcal{F}L^1}$. Applying the Riesz-Thorin convexity theorem (cf., e.g., Bergh-Löfström [1]) gives by (5.4) and (5.5) that, if $1 < p < 2$,

$$\|(f_t\psi_l)\|_{\mathcal{M}_p} \leq \|(f_t\psi_l)\|_{\mathcal{M}_1}^{2p^{-1}-1} \|(f_t\psi_l)\|_{\mathcal{M}_2}^{2(1-p^{-1})}$$

$$\leq \text{ const } M_t^{n(\frac{1}{p}-\frac{1}{2})} 2^{la(n(\frac{1}{p}-\frac{1}{2})-r)}; \tag{5.6}$$

moreover,

$$\|f_{t+h}\psi_l - f_t\psi_l\|_{\mathcal{M}_p} \leq \|(f_{t+h} - f_t)\psi_l\|_{\mathcal{F}L^1}^{2p^{-1}-1} \|(f_{t+h} - f_t)\psi_l\|_{L^\infty}^{2(1-p^{-1})}$$

$$\longrightarrow 0, \quad \text{as } h \longrightarrow 0,$$

namely $t \mapsto f_t \psi_l$ is continuous with respect to $\| \cdot \|_{\mathcal{M}_p}$. From (5.4) – (5.6), we see that for $p = 1$ (resp. $1 < p \leq 2$) the series $\sum\limits_{l=2}^{\infty} f_t \psi_l$ converges with respect to $\| \cdot \|_{\mathcal{F}L^1}$ (resp. $\| \cdot \|_{\mathcal{M}_p}$) uniformly on compacts of $t \geq 0$, and

$$\left\| \sum_{l=2}^{\infty} f_t \psi_l \right\|_{\mathcal{F}L^1} \left(\text{resp. } \left\| \sum_{l=2}^{\infty} f_t \psi_l \right\|_{\mathcal{M}_p} \right) \leq \text{const } M_t^{n(\frac{1}{p} - \frac{1}{2})} \quad (t \geq 0),$$

since $r > n \left(\frac{1}{p} - \frac{1}{2} \right)$ and M_t is bounded on compacts of $t \geq 0$. The continuity of $t \mapsto \sum\limits_{l=2}^{\infty} f_t \psi_l$ with respect to $\| \cdot \|_{\mathcal{F}L^1}$ (resp. $\| \cdot \|_{\mathcal{M}_p}$) follows. The remainder of the proof is now clear by (5.2).

Lemma 5.3. *Let j, $n \in N$, $j > \frac{n}{2}$ and $\{f_t\}_{t \geq 0}$ be a family of $C^j(R^n)$-functions. Assume that for each $x \in R^n$, $\alpha \in N_0^n$ with $|\alpha| \leq j$, $t \mapsto D^\alpha f_t(x)$ is continuous in R^+, and there exist $b > 0$, and positive, nondecreasing function $M(t) > 0$ such that for all $\alpha \in N_0^n$ with $|\alpha| \leq j$, $x \in R^n$, $t \geq 0$,*

$$|D^\alpha f_t(x)| \leq M(t)(1 + |x|)^{-b - |\alpha|}.$$

Then, for any $t \geq 0$, $f_t \in \mathcal{F}L^1$, $t \mapsto f_t$ is continuous with respect to $\| \cdot \|_{\mathcal{F}L^1}$, and there is a constant C such that $\|f_t\|_{\mathcal{F}L^1} \leq CM(t)$, $t \geq 0$.

Proof. Let ϕ, ψ, ψ_l be $C_c^\infty(R^n)$-functions as in the proof of Lemma 5.2, and so (5.2) holds. Proceeding similarly as in the proof of Lemma 5.2, we can prove that

(i) The series $\sum\limits_{l=2}^{\infty} f_t \psi_l$ converges with respect to $\| \cdot \|_{\mathcal{F}L^1}$ uniformly on compacts of $t \geq 0$, and

$$\left\| \sum_{l=2}^{\infty} f_t \psi_l \right\|_{\mathcal{F}L^1} \leq \text{const } M(t) \quad (t \geq 0);$$

(ii) As a $C_c^\infty(R^n)$-function, each term on the right-hand side of (5.2) satisfies the conclusion (in place of f_t) of Lemma 5.3 by applying Bernstein's theorem, combined with the dominated convergence theorem.

Consequently, the conclusion of the lemma holds. The proof is then complete.

Lemma 5.4. *Let $1 < p < \infty$, j, $n \in N$, $j > \frac{n}{2}$ and $f \in C^j(R^n)$. Assume that there are $a \geq 0$, $r \geq n \left| \frac{1}{2} - \frac{1}{p} \right|$, $M_f \geq 1$, $L_f > 0$ such that for each multiindex α with $|\alpha| \leq j$, $x \in R^n$,*

$$|D^\alpha f(x)| \leq L_f M_f^{|\alpha|}(1 + |x|)^{(a-1)|\alpha| - ar}.$$

Then $f \in \mathcal{M}_p$ and there is a constant C independent of f such that

$$\|f\|_{\mathcal{M}_p} \leq CL_f M_f^{n \left| \frac{1}{2} - \frac{1}{p} \right|}.$$

Proof. Let ϕ be as in the proof of Lemma 5.2. Then using Leibniz's rule gives that for $|\alpha| \leq j$, $x \in R^n$,

$$\left| D^\alpha \left[L_f^{-1}(1 - \phi(x))f(x) \right] \right| \leq \text{const } M_f^{|\alpha|}(1 + |x|)^{(a-1)|\alpha| - ar}.$$

It follows by virtue of Miyachi [1, Theorem 1] that $(1 - \phi(x))f(x) \in \mathcal{M}_p$ and

$$L_f^{-1} \|(1 - \phi)f\|_{\mathcal{M}_p} \leq \text{const } M_f^{n \left| \frac{1}{2} - \frac{1}{p} \right|}.$$

On the other hand, we have that for $|\alpha| \leq j$, $x \in R^n$,

$$|D^\alpha[\phi(x)f(x)]| \begin{cases} \leq \text{const } L_f M_f^{|\alpha|}, & \text{if } |x| \leq 2, \\ \\ = 0, & \text{if } |x| \geq 2. \end{cases}$$

Applying Bernstein's theorem and then the Riesz-Thorin convexity theorem, we obtain that $\phi f \in \mathcal{M}_p$ and

$$\|\phi f\|_{\mathcal{M}_p} \leq \text{const } L_f M_f^{n \left| \frac{1}{2} - \frac{1}{p} \right|}.$$

This completes the proof.

Lemma 5.5. *Let $r > 0$. For $t > 0$, $z \in \mathbf{C} \setminus \{0\}$, define*

$$u_{t,r}(z) = \int_0^t \frac{(t - s)^{r-1}}{\Gamma(r)} e^{sz} ds, \quad v_{t,r}(z) = u_{t,r}(z) - z^{-r} e^{tz},$$

Here z^a $(a \in R)$ denotes the branch of the power function which is analytic in $\mathbf{C}$ slit along $(-\infty, 0]$ such that $1^a = 1$. Then for each $j \in N_0$, there exists $C_j > 0$ such that for all $t > 0$, $z \in \mathbf{C} \setminus (-\infty, 0]$,

$$|D^j v_{t,r}(z)| \leq C_j \begin{cases} (t+1)^{r-1}|z|^{-j-1}\left(1 + |z|^{1-r}\right), & \text{if } r \geq 1, \\ \\ t^{r-1}|z|^{-j-1}, & \text{if } r \in (0, 1). \end{cases} \tag{5.7}$$

Proof. Observe that for $z \in \mathbf{C} \setminus \{0\}$ with $\text{Re} z > 0$,

$$z^{-r} e^{tz} = e^{tz} z^{-r} (\Gamma(r))^{-1} \int_0^\infty s^{r-1} e^{-s} ds$$

$$= \frac{e^{tz}}{\Gamma(r)} z^{-r} \int \mu^{r-1} e^{-\mu} d\mu$$

by Cauchy's theorem, where the integral is taken on the ray with direction z,

$$= e^{tz} \int_0^\infty \frac{s^{r-1}}{\Gamma(r)} e^{-sz} ds = \int_{-\infty}^t \frac{(t - s)^{r-1}}{\Gamma(r)} e^{sz} ds.$$

We have that for z as above,

$$v_{t,r}(z) = -\int_{-\infty}^{0} \frac{(t-s)^{r-1}}{\Gamma(r)} e^{sz} ds$$

$$= -\int_{0}^{\infty} \frac{(t+s)^{r-1}}{\Gamma(r)} e^{-sz} ds. \tag{5.8}$$

Noting that for $j \in N_0$ and z as above,

$$\int_{0}^{\infty} (tz+sz)^{r-1}(sz)^j e^{-sz} z \, ds = \int_{0}^{\infty} (tz+\sigma)^{r-1}\sigma^j e^{-\sigma} d\sigma$$

by Cauchy's theorem, we obtain by (5.8) that for j, z as above,

$$D^j v_{t,r}(z) = \frac{(-1)^{j+1} z^{-r-j}}{\Gamma(r)} \int_{0}^{\infty} (tz+\sigma)^{r-1}\sigma^j e^{-\sigma} d\sigma. \tag{5.9}$$

Furthermore, we find that (5.9) holds for all $z \in \mathbf{C} \setminus (-\infty, 0]$, since the two sides of (5.9) are all analytic in $z \in \mathbf{C} \setminus (-\infty, 0]$.

Let $j_0 \in N_0$ fixed.

When $r \geq 1$, we have that for $t > 0$, $z \in \mathbf{C} \setminus (-\infty, 0]$,

$$\left| D^j v_{t,r}(z) \right| \leq \int_{0}^{\infty} \frac{|z|^{-r-j}}{\Gamma(r)} (t|z|+s)^{r-1} s^j e^{-s} ds,$$

so that

$$\left| D^j v_{t,r}(z) \right| \leq \text{ const } |z|^{-r-j}, \quad \text{whenever } t|z| < 1,$$

$$\left| D^j v_{t,r}(z) \right| \leq \frac{|z|^{-r-j}}{\Gamma(r)} (t|z|)^{r-\lceil r \rceil -1} \int_{0}^{\infty} (t|z|+s)^{\lceil r \rceil} s^j e^{-s} ds$$

$$\leq \text{ const } \left(t^{r-1} |z|^{-j-1} \right), \quad \text{whenever } t|z| \geq 1.$$

When $r \in (0,1)$, we have that for $t > 0$, $z \in \mathbf{C} \setminus (-\infty, 0]$,

$$\left| D^j v_{t,r}(z) \right| \leq \int_{0}^{\infty} \frac{|z|^{-r-j}}{\Gamma(r)} |s-t|z||^{r-1} s^j e^{-s} ds. \tag{5.10}$$

Then we break the integral on the right-hand side of (5.10) into three integrals I_1, I_2, I_3 on the intervals $\left[0, \frac{t|z|}{2}\right]$, $\left[\frac{t|z|}{2}, t|z|\right]$, $[t|z|, \infty)$ respectively. We see

$$I_1 \leq \frac{|z|^{-r-j}}{\Gamma(r)} \left(\frac{t|z|}{2} \right)^{r-1} \int_{0}^{\infty} s^j e^{-s} ds$$

$$\leq \text{ const } t^{r-1} |z|^{-j-1},$$

$$I_2 \leq \frac{|z|^{-r-j}}{\Gamma(r)} e^{-\frac{t|z|}{2}} (t|z|)^j \int_{\frac{t|z|}{2}}^{t|z|} (t|z| - s)^{r-1} ds$$

$$\leq \text{const } t^{r-1} |z|^{-j-1},$$

$$I_3 \leq \frac{|z|^{-r-j}}{\Gamma(r)} e^{-t|z|} \int_0^\infty \sigma^{r-1} (\sigma + t|z|)^j e^{-\sigma} d\sigma$$

$$\leq \text{const } t^{r-1} |z|^{-j-1}.$$

The proof is then complete.

Let $l \in N_0$ with $0 \leq l \leq n$. A multiindex $\alpha = (\alpha_1, \cdots, \alpha_n)$ will be also denoted by $\alpha \in N_0^l$ if $\alpha_{l+1}, \cdots, \alpha_n = 0$. We introduce the following space

$$L_l^p(R^n) := \{ f \in L^p(R^n); \ D^\beta f \in L^p(R^n) \text{ for each } \beta \in N_0^l \}, \quad 1 < p < \infty;$$

for each multiindex $\beta \in N_0^l$, a seminorm q_β is defined on $L_l^p(R^n)$ by

$$q_\beta(f) = \| D^\beta f \|_{L^p} = \left[\int_{R^n} |D^\beta f(x)|^p \, dx \right]^{\frac{1}{p}}, \quad f \in L_l^p(R^n).$$

We show in the following that the totality Γ of these seminorms q_β corresponding to all multiindices $\beta \in N_0^l$ induces a Fréchet topology τ for $L_l^p(R^n)$.

Lemma 5.6. $(L_l^p(R^n), \tau)$ *is a Fréchet space.*

Proof. Clearly, τ is metrizable and locally convex. It remains to show the completeness. To this end, take any Cauchy sequence $\{f_k\}_{k \in N}$ in $(L_l^p(R^n), \tau)$. Then for each $\beta \in N_0^l$, $\varepsilon > 0$, there exists $k_0 > 0$ such that

$$q_\beta(f_k - f_j) < \varepsilon, \quad \text{for } k, j > k_0.$$

It is thus easy to see that for each $\beta \in N_0^l$, $\{D^\beta f_k\}_{k \in N}$ is a Cauchy sequence in $L^p(R^n)$. Therefore the completeness of $L^p(R^n)$ assures the existence of a $g_\beta \in L^p(R^n)$ such that

$$\| D^\beta f_k - g_\beta \|_{L^p} \longrightarrow 0, \quad \text{as } k \longrightarrow \infty. \tag{5.11}$$

Next, for each test function $\psi \in C_c^\infty(R^n)$, we have

$$\int_{R^n} (D^\beta f_k(x)) \, \psi(x) dx = (-1)^{|\beta|} \int_{R^n} f_k(x) D^\beta \psi(x) dx, \quad \beta \in N_0^l.$$

Letting $k \to \infty$ gives

$$\int_{R^n} g_\beta(x) \psi(x) dx = (-1)^{|\beta|} \int_{R^n} g_{(0,\cdots,0)}(x) D^\beta \psi(x) dx, \quad \beta \in N_0^l.$$

Accordingly, for each $\beta \in N_0^l$,

$$D^\beta g_{(0,\cdots,0)} = g_\beta.$$

This in combination with (5.11) indicates that $f_k \to g_{(0,\cdots,0)}$ in $(L_l^p(R^n),\ \tau)$, as $k \to \infty$, which ends the proof.

Remark. $L_l^p(R^n)$ contains the functions taking the form of

$$\sum_{i=1}^k f_i(x_1,\cdots,x_l)g_i(x_{l+1},\cdots,x_n),$$

where $k \in N$, $f_i \in \mathcal{S}(R^l)$ (the space of rapidly decreasing functions on R^l), $g_i \in L^p(R^{n-l})$, $1 \le i \le k$.

Obviously, $L_0^p(R^n)$ coincides with the usual Banach space $L^p(R^n)$. The Sobolev imbedding theorem yields

$$L_n^p(R^n) = \{f \in C^\infty(R^n);\ \ D^\alpha f \in L^p(R^n) \text{ for each multiindex } \alpha\}.$$

Thus $L_n^2(R^n)$, restricted in real-valued functions, is just the space introduced by Miyadera [3, Section 6].

Let $m \in N$, $a_\alpha \in \mathbf{C}$ for each $\alpha \in N_0^n$ with $|\alpha| \le m$. We consider the differential operator $A:\mathcal{D}(A) \to L_l^p(R^n)$, given by

$$Af := \sum_{|\alpha| \le m} a_\alpha D^\alpha f,$$

$$\mathcal{D}(A) := \left\{ f \in L_l^p(R^n);\ \ \sum_{|\alpha| \le m} a_\alpha D^\alpha f \in L_l^p(R^n) \text{ distributionally} \right\}.$$

Clearly, A is a closed linear operator in $L_l^p(R^n)$. The symbol of A will be denoted by

$$P(x) := \sum_{|\alpha| \le m} i^{|\alpha|} a_\alpha x^\alpha, \quad x \in R^n.$$

$P(x)$ is called elliptic if its principal part vanishes only when $x = 0$. We shall prove that under certain conditions, A generates an integrated or regularized semigroup on $L_l^p(R^n)$. Prior to this, we first present two propositions. In the sequel, we denote

$$r_p = n \left| \frac{1}{2} - \frac{1}{p} \right| \quad (1 < p < \infty).$$

Proposition 5.7. *Assume that $P(x)$ is elliptic with $\omega := \sup_{x \in R^n} \mathrm{Re} P(x) < \infty$. For $x \in R^n$, $t \ge 0$, define*

$$s_t(x) = \begin{cases} \displaystyle\int_0^t \frac{(t-\sigma)^{r_p-1}}{\Gamma(r_p)} e^{\sigma P(x)} d\sigma, & \text{if } p \ne 2, \\[2ex] e^{tP(x)}, & \text{if } p = 2. \end{cases}$$

Then for each $t \geq 0$, $s_t \in M_p$ and there exists $C > 0$ such that

$$\|s_t\|_{\mathcal{M}_p} \leq C a_p(t), \quad \text{for all } t \geq 0,$$

where

$$a_p(t) := \begin{cases} (1+t)^{r_p} e^{\omega t}, & \text{when } \omega > 0 \text{ or } p = 2, \\ (1+t)^{2r_p}, & \text{when } \omega = 0, \\ (1+t)^{r_p - 1}, & \text{when } \omega < 0. \end{cases}$$

Proof. Since $P(x)$ is elliptic, we have that there exist constants $L_0, C_0 > 0$ such that

$$|P(x)| \geq C_0 |x|^m, \quad |x| \geq L_0. \tag{5.12}$$

Take $L > L_0$ such that $C_0 L^m \geq 1$. When $p = 2$, using Leibniz's rule and (5.12) gives that for each $\alpha \in N_0^n$ with $|\alpha| \leq [\frac{n}{2}] + 1$,

$$|D^\alpha s_t(x)| \leq \text{const} (1+t)^{|\alpha|} e^{\omega t} (1+|x|)^{(m-1)|\alpha|}, \quad t \geq 0, \ x \in R^n;$$

this leads to the desired result by virtue of Lemma 5.4. We next take care of the case $p \neq 2$. Let ϕ_L be a $C^\infty(R^n)$-function such that

$$\phi_L(x) = \begin{cases} 1, & \text{if } |x| \geq L+1, \\ 0, & \text{if } |x| \leq L. \end{cases}$$

For $t \geq 1$, $x \in R^n$, setting

$$s_{t,1}(x) = (1 - \phi_L(x)) s_t(x),$$

$$s_{t,2}(x) = -\phi_L(x) \sum_{i=1}^{[r_p]} \frac{t^{r_p - i}}{\Gamma(r_p - i + 1)} [P(x)]^{-i},$$

$$s_{t,3}(x) = \begin{cases} [P(x)]^{-r_p} e^{P(x)t}, & \text{if } r_p \in N, \\ \phi_L(x) [P(x)]^{-[r_p]} \int_0^t \frac{(t-s)^{r_p - [r_p] - 1}}{\Gamma(r_p - [r_p])} e^{P(x)s} ds, & \text{otherwise,} \end{cases}$$

then

$$s_t(x) = s_{t,1}(x) + s_{t,2}(x) + s_{t,3}(x).$$

Using Leibniz's rule and recalling (5.12), we have that for any $\alpha \in N_0^n$ with $|\alpha| \leq [\frac{n}{2}] + 1$,

$$|D^\alpha s_{t,2}(x)| \leq \text{const} (1+t)^{r_p - 1} (1+|x|)^{-|\alpha| - 1}, \quad x \in R^n, \ t \geq 1.$$

Consequently, we have by Lemma 5.3 that for each $t \geq 1$,

$$s_{t,2} \in \mathcal{M}_p, \quad \|s_{t,2}\|_{\mathcal{M}_p} \leq \text{const } (1+t)^{r_p-1} \quad (t \geq 1).$$

When $r_p \notin N$, we have from Lemma 5.5 that for $t \geq 1$, $j \in N_0$, $z \in C$ with Re$z \leq \omega$, $|z| \geq 1$,

$$\left| \frac{d^j}{dz^j} \int_0^t \frac{(t-s)^{r_p-[r_p]-1}}{\Gamma(r_p-[r_p])} e^{zs} ds \right| \leq C_j \left(t^{r_p-1}|z|^{-j-1} + t^j z^{[r_p]-r_p} e^{\omega t} \right)$$

and therefore

$$\left| \frac{d^j}{dz^j} \left(z^{-[r_p]} \int_0^t \frac{(t-s)^{r_p-[r_p]-1}}{\Gamma(r_p-[r_p])} e^{zs} ds \right) \right| \leq C_j |z|^{-r_p} \begin{cases} t^j e^{\omega t}, & \text{if } \omega \geq 0, \\ t^{r_p-1}, & \text{if } \omega < 0. \end{cases}$$

Accordingly, we obtain by (5.12) that for any $\alpha \in N_0^n$ with $|\alpha| \leq [\frac{n}{2}]+1$, $x \in R^n$, $t \geq 1$,

$$|D^\alpha s_{t,3}(x)| \leq \text{const } (1+|x|)^{(m-1)|\alpha|-mr_p} \begin{cases} (1+t)^{|\alpha|} e^{\omega t}, & \text{if } \omega \geq 0, \\ (1+t)^{r_p-1}, & \text{if } \omega < 0. \end{cases}$$

It follows by virtue of Lemma 5.4 that for $t \geq 1$, $s_{t,3} \in \mathcal{M}_p$ and

$$\|s_{t,3}\|_{\mathcal{M}_p} \leq \text{const} \begin{cases} (1+t)^{r_p} e^{\omega t}, & \text{if } \omega \geq 0, \\ (1+t)^{r_p-1}, & \text{if } \omega < 0. \end{cases}$$

Now, observe that for every $t \geq 0$ and $\alpha \in N_0^n$ with $|\alpha| \leq [\frac{n}{2}]+1$,

$$\left| D^\alpha \left[(1 - \phi_L(x)) e^{tP(x)} \right] \right| \begin{cases} \leq \text{const } (1+t)^{|\alpha|} e^{\omega t}, & \text{if } |x| \leq L+1, \\ = 0, & \text{if } |x| \geq L+1. \end{cases}$$

Then Lemma 5.4 gives that for $t \geq 0$, $(1-\phi_L)e^{tP} \in \mathcal{M}_p$ and

$$\left\| (1-\phi_L)e^{tP} \right\|_{\mathcal{M}_p} \leq \text{const } (1+t)^{r_p} e^{\omega t}.$$

Therefore

$$\|s_{t,1}\|_{\mathcal{M}_p} \leq \text{const} \int_0^t (t-s)^{r_p-1}(1+s)^{r_p} e^{\omega s} ds$$

$$\leq \text{const } a_p(t), \quad t \geq 0.$$

Here we used Lemma 5.5 again when $\omega > 0$:

$$\int_0^t (t-s)^{r_p-1} s^{r_p} e^{\omega s} ds \leq t^{r_p-[r_p]} \int_0^t (t-s)^{r_p-1} s^{[r_p]} e^{\omega s} ds$$

$$= \Gamma(r_p)t^{r_p-[t_p]}\frac{d^{[r_p]}}{dz^{[r_p]}}\left[z^{-r_p}e^{tz}+v_{t,r_p}(z)\right]\Big|_{z=\omega}$$

$$\leq \text{const}\left[t^{r_p-[r_p]}(1+t)^{[r_p]}e^{\omega t}\right]$$

$$\leq \text{const }(1+t)^{r_p}e^{\omega t}, \quad t\geq 0.$$

Thus, we have proved the conclusion in the case of $t\geq 1$. When $t\in (0,\ 1]$, we consider

$$P_t(x):=tP\left(t^{-\frac{1}{m}}x\right), \quad x\in R^n.$$

Obviously $P_t(x)$ is elliptic and

$$\sup_{x\in R^n}\text{Re}P_t(x)\leq \max\{0,\ \omega\}, \quad |P_t(x)|\geq C_0|x|^m \quad (|x|\geq L_0),$$

$$\sup_{|x|\leq L_0,\ t\in (0,\ 1]}|P_t(x)|<\infty.$$

Proceeding similarly as above, we can show that for each $t\in (0,\ 1]$,

$$\tilde{s}(x;\ t):=t^{r_p}\int_0^1\frac{(1-s)^{r_p-1}}{\Gamma(r_p)}e^{P_t(x)s}ds\in \mathcal{M}_p, \quad \|\tilde{s}(x;\ t)\|_{\mathcal{M}_p}\leq \text{const.}$$

But

$$s_t(x)=t^{r_p}\int_0^1\frac{(1-s)^{r_p-1}}{\Gamma(r_p)}e^{tP(x)s}ds=\tilde{s}\left(t^{\frac{1}{m}}x;\ t\right).$$

So for each $t\in (0,\ 1]$,

$$s_t(x)\in \mathcal{M}_p, \quad \|s_t(x)\|_{\mathcal{M}_p}=\left\|\tilde{s}\left(t^{\frac{1}{m}}x;\ t\right)\right\|_{\mathcal{M}_p}=\|\tilde{s}(x;\ t)\|_{\mathcal{M}_p}\leq \text{const.}$$

This completes the proof.

Proposition 5.8. *Assume* $\omega:=\sup_{x\in R^n}\text{Re}P(x)<\infty$. *For* $r\geq 0$, $t\geq 0$, *define*

$$b_r(x)=\left(1+|x|^2\right)^{-\frac{mr}{2}}, \quad w_{t,r}(x)=b_r(x)e^{tP(x)}, \quad x\in R^n.$$

Then $w_{t,r}\in \mathcal{M}_p$ *whenever* $t\geq 0$, $r\geq r_p$ *and there exists a constant* C *such that for all* $t\geq 0$,

$$\|w_{t,r}(x)\|_{\mathcal{M}_p}\leq C(1+t)^{r_p}e^{\omega t}.$$

Moreover, for each $r>0$, $b_r(x)\in \mathcal{FL}^1$ *and for each* $t\geq 0$, $w_{t,r}\in \mathcal{FL}^1$ *whenever* $r\geq [\frac{n}{2}]+1$.

Proof. It is easy to see that for each $\alpha\in N_0^n$ with $|\alpha|\leq [\frac{n}{2}]+1$,

$$|D^\alpha b_r(x)|\leq \text{const }(1+|x|)^{-|\alpha|-mr}, \quad x\in R^n. \tag{5.13}$$

Thus, applying Leibniz's rule shows the existence of a constant C such that for $t \geq 0$, $x \in R^n$, $\alpha \in N_0^n$ with $|\alpha| \leq [\frac{n}{2}] + 1$,

$$|D^\alpha w_{t,r}(x)| \leq C(1+t)^{|\alpha|} e^{\omega t} (1+|x|)^{(m-1)|\alpha|-mr}. \tag{5.14}$$

This leads to the first statement of our conclusions with the aid of Lemma 5.4. From (5.13), we have by Lemma 5.3 that $b_r(x) \in \mathcal{F}L^1$ for each $r > 0$. Also, (5.14) implies that for each $t \geq 0$, $w_{t,r} \in H^{[\frac{n}{2}]+1}(R^n)$ and therefore $w_{t,r} \in \mathcal{F}L^1$ (by Lemma 5.1) whenever $r \geq [\frac{n}{2}] + 1$. This ends the proof.

With a given $u \in \mathcal{M}_p$, we associate a bounded linear operator $\mathbf{T}\langle u \rangle$ on $L^p(R^n)$ such that for $\phi \in \mathcal{S}(R^n)$, $\mathbf{T}\langle u \rangle \phi = \mathcal{F}^{-1}\left(u\hat{\phi} \right)$. Observe that for each $f \in \mathcal{S}(R^n)$,

$$D^\beta(\mathbf{T}\langle u \rangle f) = \mathbf{T}\langle u \rangle \left(D^\beta f \right), \quad \beta \in N_0^l. \tag{5.15}$$

It is clear that for each $f \in L_l^p(R^n)$, there exists a sequence $\{\phi_k\}_{k \in N} \subset \mathcal{S}(R^n)$ such that

$$\left\| D^\beta(\phi_k - f) \right\|_{L^p} \longrightarrow 0, \quad \text{as} \quad k \longrightarrow \infty \quad (\beta \in N_0^l).$$

This implies that (5.15) holds for each $f \in L_l^p(R^n)$. In conclusion,

$$\mathbf{T}\langle u \rangle L_l^p(R^n) \subset L_l^p(R^n).$$

Let $\mathbf{T}_l \langle u \rangle$ be the part of $\mathbf{T}\langle u \rangle$ in $L_l^p(R^n)$. Then (5.15) gives that

$$q_\beta(\mathbf{T}_l \langle u \rangle f) \leq \|u\|_{\mathcal{M}_p} q_\beta(f), \quad \text{for all} \quad q_\beta \in \Gamma, \ f \in L_l^p(R^n);$$

hence

$$\mathbf{T}_l \langle u \rangle \in \mathcal{B}_\Gamma \left(L_l^p(R^n) \right), \quad \|\mathbf{T}_l \langle u \rangle\|_\Gamma = \|u\|_{\mathcal{M}_p}. \tag{5.16}$$

Theorem 5.9. *Let $P(x)$, $a_p(t)$ be as in Proposition 5.7. Then A is the generator of an r_p-times integrated semigroup $\{S(t)\}_{t \geq 0}$ on $L_l^p(R^n)$ satisfying*

$$\|S(t)\|_\Gamma = O(a_p(t)) \quad (t \to \infty).$$

Proof. For each $t \geq 0$, define

$$S(t) = \mathbf{T}_l \langle s_t \rangle,$$

where s_t is given in Proposition 5.7. Then we have by (5.16) that for all $t \geq 0$, $S(t) \in \mathcal{B}_\Gamma(L_l^p(R^n))$ and there exists $C > 0$ such that

$$\|S(t)\|_\Gamma \leq C a_p(t). \tag{5.17}$$

Observing

$$s_t(x) = \int_0^t s_\sigma(x) P(x) d\sigma + \frac{t^{r_p}}{\Gamma(r_p + 1)} \quad (t \geq 0, \ x \in R^n),$$

we get by (5.15) and Fubini's theorem that for $t \geq 0$, $\phi \in \mathcal{S}(R^n)$, $\beta \in N_0^l$,

$$
\begin{aligned}
D^\beta (S(t)\phi)(x) &= S(t)D^\beta \phi \\
&= \int_0^t S(\sigma)\widetilde{\phi}d\sigma + \frac{t^{r_p}}{\Gamma(r_p + 1)}D^\beta \phi.
\end{aligned}
$$

(5.18)

Here $\widetilde{\phi} = \mathcal{F}^{-1}\left(P(D^\beta \phi)^\wedge\right)$, which still belongs to $\mathcal{S}(R^n)$. Note that for each $\psi \in \mathcal{S}(R^n)$, the $L^p(R^n)$-valued function $S(\cdot)\psi$ is weakly continuous and therefore strongly measurable. We have by (5.16), (5.17) and (5.18) that for ϕ, $\widetilde{\phi}$ as above and for $q_\beta \in \Gamma$, $f \in L_l^p(R^n)$, t, $h \geq 0$,

$$
q_\beta(S(t+h)f - S(t)f)
$$

$$
\leq C(a_p(t+h) + a_p(t))q_\beta(f - \phi) + hC \max\{a_p(t+h),\ 1\}\left\|\widetilde{\phi}\right\|_{L^p}
$$

$$
+ \frac{(t+h)^{r_p} - t^{r_p}}{\Gamma(r_p + 1)}\left\|D^\beta \phi\right\|_{L^p}.
$$

This observation implies, recalling the statement below (5.15), that for each $f \in L_l^p(R^n)$,

$$
S(\cdot)f \in C\left(R^+, L_l^p(R^n)\right).
$$

Next, for each $f \in L_l^p(R^n)$, $\lambda > \widetilde{\omega}$ ($\widetilde{\omega} := \max\{0,\ \omega + 1\}$), setting

$$
K_\lambda f = \int_0^\infty e^{-\lambda t}S(t)f dt,
$$

then $K_\lambda \in \mathcal{B}r\left(L_l^p(R^n)\right)$. Applying Fubini's theorem yields that for $\lambda > \widetilde{\omega}$, f, $\phi \in \mathcal{S}(R^n)$,

$$
\left\langle \lambda^{r_p} K_\lambda f,\ \left(\lambda - \sum_{|\alpha| \leq m}(-1)^{|\alpha|}a_\alpha D^\alpha\right)\phi \right\rangle
$$

$$
= \int_0^\infty \lambda^{r_p}e^{-\lambda t}\left\langle \mathcal{F}^{-1}\left(s_t\widehat{f}\right),\ \left(\lambda - \sum_{|\alpha| \leq m}(-1)^{|\alpha|}a_\alpha D^\alpha\right)\phi \right\rangle dt
$$

$$
= \int_0^\infty \lambda^{r_p}e^{-\lambda t}\mathcal{F}^{-1}\left(s_t\widehat{f}\right) * \left(\lambda - \sum_{|\alpha| \leq m}a_\alpha D^\alpha\right)\phi_-(0)dt
$$

$$
= \int_0^\infty \lambda^{r_p}e^{-\lambda t}\mathcal{F}^{-1}\left(s_t\widehat{f}(\lambda - P)\widehat{\phi_-}\right)(0)dt
$$

$$
= (f * \phi_-)(0)
$$

$$
= \langle f,\ \phi \rangle,
$$

where and in the sequel $\phi_-(x) := \phi(-x)$. Also for $f \in \mathcal{S}(R^n)$,

$$
\begin{aligned}
K_\lambda(\lambda - A)f &= \int_0^\infty e^{-\lambda t} \mathcal{F}^{-1}\left(s_t \mathcal{F}\left(\lambda f - \sum_{|\alpha| \le m} a_\alpha D^\alpha f \right) \right) dt \\
&= \int_0^\infty e^{-\lambda t} \mathcal{F}^{-1}\left(s_t (\lambda - P)\widehat{f} \right) dt \\
&= \left(\lambda - \sum_{|\alpha| \le m} a_\alpha D^\alpha \right) K_\lambda f.
\end{aligned}
$$

Accordingly, we have for each $f \in \mathcal{S}(R^n)$,

$$
\lambda^{r_p} K_\lambda(\lambda - A)f = f, \quad \lambda^{r_p}(\lambda - A)K_\lambda f = f, \quad \lambda > \widetilde{\omega}. \tag{5.19}
$$

Let $f \in \mathcal{D}(A)$. Then the ellipticity of $P(x)$ assures the existence of a sequence $\{\phi_k\}_{k \in N} \subset \mathcal{S}(R^n)$ such that

$$
\lim_{k \to \infty} \|\phi_k - f\|_{L^p} = 0, \quad \lim_{k \to \infty} \|A\phi_k - Af\|_{L^p} = 0.
$$

Therefore,

$$
\lim_{k \to \infty} \|K_\lambda(\lambda - A)(\phi_k - f)\|_{L^p} = 0,
$$

recalling that $K_\lambda \in \mathcal{B_r}\left(L^p_l(R^n) \right)$. Consequently, the first equality in (5.19) holds for all $f \in \mathcal{D}(A)$. Similarly, we can show that the second equality in (5.19) holds for all $f \in L^p_l(R^n)$. Thus we have proved that for $\lambda > \widetilde{\omega}$, $(\lambda - A)^{-1} = \lambda^{r_p} K_\lambda$. This completes the proof.

Theorem 5.10. *Let* $r \ge r_p$ *and let* $P(x)$, $b_r(x)$ *be as in Proposition 5.8. Define* $C_{r,l} = T_l \langle b_r(x) \rangle$. *Then* A *is the generator of a* $C_{r,l}$-*regularized semigroup* $\{W_{r,l}(t)\}_{t \ge 0}$ *on* $L^p_l(R^n)$ *satisfying*

$$
\|W_{r,l}(t)\|_{\mathbf{r}} = O\left(t^{r_p} e^{\omega t} \right) \quad (t \to \infty).
$$

Proof. Since $b_r(x) \in \mathcal{F}L^1$ (in case of $r > 0$), we have that for each $f \in L^p_l(R^n)$, $C_{r,l} f = \mathcal{F}^{-1}\left(b_r \widehat{f} \right)$; therefore $C_{r,l} f = 0$ implies $f = 0$. In conclusion, $C_{r,l}$ is an injective, continuous operator on $L^p_l(R^n)$. Moreover, we see that $A = C_{r,l}^{-1} A C_{r,l}$ by the following observation:

$$
f \in \mathcal{D}\left(C_{r,l}^{-1} A C_{r,l} \right)
$$

if and only if

$$
f \in L^p_l(R^n) \quad \text{and} \quad \mathcal{F}^{-1}\left[b_r^{-1}(x) P(x) b_r(x) \widehat{f} \right] \in L^p_l(R^n)
$$

if and only if

$$f \in L_l^p(R^n) \quad \text{and} \quad \mathcal{F}^{-1}\left(P(x)\hat{f}\right) \in L_l^p(R^n)$$

if and only if

$$f \in \mathcal{D}(A).$$

For each $t \geq 0$, define

$$W_{r,l}(t) = \mathbf{T}_l \langle w_{t,r} \rangle,$$

where $w_{t,r}$ is given in Proposition 5.8. Then we have by (5.16) again that for $t \geq 0$, $W_{r,l}(t) \in \mathcal{B}_{\mathbf{r}}\left(L_l^p(R^n)\right)$ and there is $M > 0$ such that

$$\|W_{r,l}(t)\|_{\mathbf{r}} \leq M(1+t)^{r_r}e^{\omega t} \quad (t \geq 0).$$

Set

$$J_{\lambda,r,l}f = \int_0^\infty e^{-\lambda t}W_{r,l}(t)f\,dt \quad (f \in L_l^p(R^n),\ \lambda > \tilde{\omega}).$$

Proceeding similarly as in the treatment of $\{S(t)\}_{t\geq 0}$, we can show that for each $f \in L_l^p(R^n)$, $W_{r,l}(\cdot)f \in C\left(R^+, L_l^p(R^n)\right)$ using the equality

$$w_{t,r}(x) = \int_0^t w_{\sigma,r}(x)P(x)\,d\sigma + \left(1 + |x|^2\right)^{-\frac{mr}{2}} \quad (t \geq 0,\ x \in R^n);$$

that for $\lambda > \tilde{\omega}$,

$$(\lambda - A)J_{\lambda,r,l}f = C_{r,l}f, \quad \text{for } f \in \mathcal{S}(R^n), \tag{5.20}$$

$$J_{\lambda,n+r,l}(\lambda - A)f = C_{n+r,l}f, \quad \text{for } f \in \mathcal{D}(A). \tag{5.21}$$

In order to obtain (5.21), we used the fact that for each $t \geq 0$, $w_{t,n+r} \in \mathcal{F}L^1$ (see Proposition 5.8), which implies that

$$W_{n+r,l}(t)f = \mathcal{F}^{-1}\left(w_{t,n+r}\hat{f}\right), \quad \text{for all } f \in L_l^p(R^n).$$

It is clear that

$$J_{\lambda,n+r,l} = C_{n,l}J_{\lambda,r,l}, \quad C_{n+r,l} = C_{n,l}C_{r,l}, \quad \lambda > \tilde{\omega}.$$

Whence, (5.21) together with the injectivity of $C_{n,l}$ yields that

$$J_{\lambda,r,l}(\lambda - A)f = C_{r,l}f, \quad \lambda > \tilde{\omega},\ f \in \mathcal{D}(A).$$

On the other hand, taking a sequence $\{\phi_k\}_{k\in N} \subset \mathcal{S}(R^n)$ such that

$$\lim_{k\to\infty} \|\phi_k - f\|_{L^p} = 0, \quad \text{for a fixed } f \in L_l^p(R^n)$$

will show that (5.20) holds for all $f \in L_l^p(R^n)$. Consequently, for $\lambda > \tilde{\omega}$, $\mathcal{R}\left(C_{r,l}\right) \subset \mathcal{R}(\lambda - A)$ and

$$(\lambda - A)^{-1}C_{r,l}f = J_{\lambda,r,l}f \quad (f \in L_l^p(R^n)). \tag{5.22}$$

The proof is then complete.

In the sequel, for $0 \leq l \leq n$, $a > 0$, we write

$$W_l^{a,p}(R^n) = \left\{ f \in L_l^p(R^n); \ D^\beta f \in W^{a,p}(R^n) \text{ for each } \beta \in N_0^l \right\}.$$

It is easy to verify that $C_{r,l}$ in Theorem 5.10 satisfies

$$\mathcal{R}(C_{r,l}) = W_l^{mr,p}(R^n), \tag{5.23}$$

$$C_{r,l}(W_l^{a,p}(R^n)) = W_l^{a+mr,p}(R^n) \tag{5.24}$$

and for each $\beta \in N_0^l$, $f \in \mathcal{R}(C_{r,l})$,

$$q_\beta\left(C_{r,l}^{-1} f\right) = \left\| D^\beta f \right\|_{W^{mr,p}(R^n)}, \tag{5.25}$$

where

$$\|g\|_{W^{a,p}(R^n)} := \left\| \mathcal{F}^{-1}\left((1+|x|^2)^{\frac{a}{2}} \hat{g} \right) \right\|_{L^p(R^n)}. \tag{5.26}$$

Corollary 5.11. *Let $P(x)$ be as in Proposition 5.8. Then, in the space $L_n^p(R^n)$, A generates a (exponentially equicontinuous) strongly continuous semigroup (i.e. 0-times integrated semigroup) $\{W_n(t)\}_{t \geq 0}$.*

Proof. Let $C_{r,n}$, $\{W_{r,n}(t)\}_{t\geq 0}$ be as in Theorem 5.10 and fix $r = [\frac{n}{2}] + 1$. It is easy to see by (5.23) that $\mathcal{R}(C_{r,n}) = L_n^p(R^n)$. Define

$$W_n(t) = W_{r,n}(t)C_{r,n}^{-1}, \quad t \geq 0.$$

Then for each $\alpha \in N_0^n$,

$$q_\alpha\left(t^{-r_p} e^{-\omega t} W_n(t) f\right) \leq \text{const } q_\alpha\left(C_{r,n}^{-1} f\right)$$

$$\leq \text{const } \sum_{|\gamma| = |\alpha|, |\alpha| + mr} q_\gamma(f), \quad t \geq 0, \ f \in L_n^p(R^n).$$

This implies that $\{t^{-r_p} e^{-\omega t} W_n(t)\}_{t \geq 0}$ is equicontinuous. Also, (5.22) gives that for λ sufficiently large,

$$(\lambda - A)^{-1} g = \int_0^\infty e^{-\lambda t} W_n(t) g \, dt, \quad \text{for each } g \in L_n^p(R^n),$$

which ends the proof.

Remark. Let A be a partial differential operator of the $2m$th order in R^n defined by

$$A = (-1)^{m+1} \sum_{|\rho|, |\nu| = 0}^m a_{\rho,\nu} D^\rho D^\nu, \quad a_{\rho,\nu} \in R.$$

Assume $a_{\rho,\nu} = a_{\nu,\rho}$ for $|\rho| = |\nu| = m$, and there exists $\varepsilon_0 > 0$ such that

$$\sum_{|\rho|=|\nu|=m} a_{\rho,\nu} x^\rho x^\nu \geq \varepsilon_0 |x|^{2m}, \quad x \in R^n.$$

Miyadera [3] has shown that A generates a (exponentially equicontinuous) strongly continuous semigroup in $L_n^2(R^n)$ (restricted to real-valued functions). This result is also an easy consequence of Corollary 5.11 noting that in this case,

$$P(x) \;\; = \;\; -\sum_{|\rho|=|\nu|=m} a_{\rho,\nu} x^\rho x^\nu + (-1)^{m+1} \sum_{|\rho|=|\nu|=0}^{m-1} i^{|\rho|+|\nu|} x^\rho x^\nu, \quad x \in R^n,$$

$$\overline{P(x)} \;\; = \;\; P(-x), \quad x \in R^n,$$

and therefore for each real-valued $f \in S(R^n)$,

$$\overline{W_n(t)f} = W_n(t)f, \quad t \geq 0.$$

1.6 Relationship to Cauchy problems

In this section, we consider the following Cauchy problems:

$$\begin{cases} u'(t) = Au(t) \;\; (t \geq 0), \\[2mm] u(0) = u_0, \end{cases} \tag{6.1}$$

$$\begin{cases} u''(t) = Au(t) \;\; (t \geq 0), \\[2mm] u(0) = u_0, \;\; u'(0) = u_1, \end{cases} \tag{6.2}$$

where A is a linear operator in E.

Definition 6.1. By a solution of (6.1) we mean a map $u(\cdot) \in C^1(R^+, E)$ satisfying (6.1); by a solution of (6.2) we mean a map $u(\cdot) \in C^2(R^+, E)$ satisfying (6.2).

The following result is easy to verify

Lemma 6.2. *Let* $\lambda_0 \in \rho(A)$, $u_0 \in \mathcal{D}(A)$, $l \in N_0$. *Then*
  (i) $u(\cdot)$ *is a solution of* (6.1) *if and only if* $v(\cdot) := (\lambda_0 - A)u(\cdot) \in C(R^+, E)$ *and satisfies*

$$v(t) = (\lambda_0 - A)u_0 + A \int_0^t v(s)ds, \quad t \geq 0;$$

  (ii) $w(\cdot) \in C(R^+, E)$ *and satisfies*

$$w(t) = \frac{t^l}{l!}u_0 + A \int_0^t w(s)ds, \quad t \geq 0$$

if and only if

$$w_0(t) := (\lambda_0 - A) \int_0^t w(s)ds \in C(R^+, E)$$

and satisfies

$$w_0(t) = \frac{t^{l+1}}{(l+1)!}(\lambda_0 - A)u_0 + A \int_0^t w_0(s)ds, \quad t \geq 0.$$

In the sequel, we assume as in Section 1.3 that C is a continuous, injective operator on E such that $CA \subset AC$.

Theorem 6.3. *Suppose E is a Fréchet space topologized by the seminorms $\{\|\cdot\|_j\}_{j \in N}$, $\rho(A)$ is nonempty, $r \in N_0$, $\omega > 0$. If (6.1) has a unique solution $u(\cdot)$, for each $u_0 \in C\left(\mathcal{D}\left(A^{r+1}\right)\right)$ with*

$$\sup_{t \geq 0} \left\|e^{-\omega t}u'(t)\right\|_j < \infty \quad (j \in N),$$

then A is the generator of an r-times integrated, C-regularized semigroup.

Proof. In view of Lemma 6.2, we obtain that for each $u \in E$,

$$v_r(t) = \frac{t^r}{r!}Cu + A \int_0^t v_r(s)ds, \quad t \geq 0 \tag{6.3}$$

has a unique solution $v_r(t) = v_r(t; \; Cu) \in C(R^+, E)$, with

$$\sup_{t \geq 0} \left\|e^{-\omega t}v_r(t; \; Cu)\right\|_j < \infty \quad (j \in N); \tag{6.4}$$

$v_r(t; \; Cu) \in \mathcal{D}(A)$ for any $t \geq 0$, if $u \in \mathcal{D}(A)$. As a consequence,

$$v_r(t; \; CAu) = Av_r(t; \; Cu), \quad t \geq 0, \; u \in \mathcal{D}(A).$$

Define $S_r : E \to C(R^+, E)$ by

$$(S_r u)(t) = v_r(t; \; Cu),$$

where $C(R^+, E)$ denotes the Fréchet space topologized by the seminorms

$$\|\Phi\|_{a,b,j} := \sup_{t \in [a, \; b]} \|\Phi(t)\|_j, \quad j \in N, \; a, \; b \in Q^+$$

(Q^+ is the set of nonnegative rational numbers). We see easily that S_r is closed. Hence S_r is continuous according to the closed graph theorem.

Now for each $t \geq 0$, define $S_r(t) : E \to E$ by $S_r(t)u = (S_r u)(t)$. Then for any $t \geq 0$, $u \in \mathcal{D}(A)$,

$$S_r(t) \in L(E), \quad S_r(t)Au = AS_r(t)u,$$

and for all $u \in E$, $t \mapsto S_r(t)u \in C(R^+, E)$; moreover $\{e^{-\omega t} S_r(t); t \geq 0\}$ is equicontinuous by (6.4) and an application of the Banach-Steinhause theorem. Also, we get from (6.3) that

$$S_r(t)u = \frac{t^r}{r!} Cu + A \int_0^t S_r(s)u \, ds, \quad t \geq 0, \ u \in E.$$

Taking Laplace transforms yields that for $u \in E$, $\lambda > \omega$,

$$\mathcal{L}[S_r(t)u](\lambda) = \lambda^{-r-1} Cu + \lambda^{-1} A \mathcal{L}[S_r(t)u](\lambda);$$

that is, $(\lambda - A)\mathcal{L}[S_r(t)u](\lambda) = \lambda^{-r-1} Cu$. Noting

$$\mathcal{L}[S_r(t)Au](\lambda) = A\mathcal{L}[S_r(t)u](\lambda), \quad u \in \mathcal{D}(A), \ \lambda > \omega,$$

we deduce that for each $\lambda > \omega$, $\lambda - A$ is injective by the same property of C. Consequently, $(\omega, \infty) \subset \rho_C(A)$ and

$$(\lambda - A)^{-1} Cu = \lambda^r \int_0^\infty e^{-\lambda t} S_r(t)u \, dt, \quad \lambda > \omega, \ u \in E,$$

as desired.

Corollary 6.4. *Let the hypothesis of Theorem 6.3 hold, except that the estimate on $u'(t)$ is replaced by*

$$\sup_{t \geq 0} \left\| e^{-\omega t} \int_0^t \frac{1}{p!} (t-s)^p u(s) ds \right\|_j < \infty \qquad (j \in N), \tag{6.5}$$

for some $p \in N_0$. Then A is the generator of an $(r + p + 2)$-times integrated, C-regularized semigroup.

Proof. Take $\lambda_0 \in \rho(A)$. For each $t \geq 0$, define $T(t) : C(\mathcal{D}(A^{r+1})) \to E$ by $T(t)u_0 = u(t; u_0)$, $(u(t; u_0)$ is the solution of (6.1)). The uniqueness of solutions yields that for $t \geq 0$,

$$(\lambda_0 - A)^{-1} T(t)u = T(t)(\lambda_0 - A)^{-1}u, \quad u \in \mathcal{D}(A^{r+1}). \tag{6.6}$$

Now, for any $t \geq 0$, set $S(t) : C(\mathcal{D}(A^{r+1})) \to E$ by

$$
\begin{aligned}
S(t)u &= \int_0^t \frac{1}{p!} (t-s)^p T(s)u \, ds \\
&= \int_0^t \frac{1}{p!} s^p T(t-s)u \, ds, \quad u \in C(\mathcal{D}(A^{r+1})).
\end{aligned}
\tag{6.7}
$$

Then, for any $t \geq 0$, $u \in C(\mathcal{D}(A^{r+2}))$,

$$
\begin{aligned}
S'(t)u &= \frac{1}{p!} t^p u + \int_0^t \frac{1}{p!} s^p T(t-s) Au \, ds \\
&= \frac{1}{p!} t^p u + S(t) Au,
\end{aligned}
\tag{6.8}
$$

noting that (6.6) implies that

$$AT(t)u = T(t)Au, \quad t \geq 0, \ u \in C\left(\mathcal{D}\left(A^{r+2}\right)\right). \tag{6.9}$$

Thus, by (6.7), (6.8) and (6.9), we have that for $t \geq 0$, $u \in C\left(\mathcal{D}\left(A^{r+p+3}\right)\right)$,

$$T(t)u = S^{(p+1)}(t)u = \sum_{k=0}^{p} \frac{t^k}{k!} A^k u + S(t) A^{p+1} u. \tag{6.10}$$

It follows from (6.5), (6.8) and (6.10) that for every $u \in C\left(\mathcal{D}\left(A^{r+p+3}\right)\right)$,

$$\sup_{t \geq 0} \left\| e^{-\omega t} \frac{d}{dt}(T(t)u) \right\|_j < \infty \quad (j \in N).$$

Now an application of Theorem 6.3 concludes the proof.

For a general SCLCS, we have

Theorem 6.5. *Assume that $\rho(A)$ is nonempty. Let $r \in N_0$, $\omega > 0$. If for each $u_0 \in C\left(\mathcal{D}\left(A^{r+1}\right)\right)$, (6.1) has a unique solution $u(t; u_0)$ such that, to each $p \in \Gamma$ corresponds a continuous seminorm q (independent of u_0) satisfying*

$$p(u'(t; u_0)) \leq e^{\omega t} \sum_{i=0}^{r} q\left(A^{i+1} C^{-1} u_0\right), \quad t \geq 0, \tag{6.11}$$

then A is the generator of an r-times integrated, C-regularized semigroup.

Proof. By Lemma 6.2 again, we obtain that for each $u \in E$, (6.3) has a unique solution $v_r(t; Cu) \in C(R^+, E)$ such that, to each $p \in \Gamma$ corresponds a continuous seminorm $\bar{q}$ (independent of u) satisfying

$$p(v_r(t; Cu)) \leq e^{\omega t} \bar{q}(u), \quad t \geq 0.$$

The remaining argument is similar to the latter part of the proof of Theorem 6.3 and so is omitted.

Conversely, we obtain

Theorem 6.6. *Let $r \in N_0$. If A is the generator of an r-times integrated, C-regularized semigroup $\{S_r(t)\}_{t \geq 0}$, then for each $u_0 \in C\left(\mathcal{D}\left(A^{r+1}\right)\right)$, (6.1) has a unique solution $u(t; u_0)$ satisfying (6.11) and*

$$p(u(t; u_0)) \leq e^{\omega t} \sum_{i=0}^{r} q\left(A^i C^{-1} u_0\right), \quad t \geq 0, \tag{6.12}$$

for some $\omega \in R$.

Proof. Let $u_0 \in C\left(\mathcal{D}\left(A^{r+1}\right)\right)$. For $t \geq 0$, put

$$u(t;\ u_0) := \sum_{i=0}^{r-1} \left(\frac{t^i}{i!}A^i u_0\right) + S_r(t)A^r C^{-1} u_0. \tag{6.13}$$

A simple computation shows by (ii) and (iii) of Lemma 3.2 that $u(t;\ u_0)$ is a solution of (6.1) satisfying (6.11) and (6.12). For uniqueness, let $u(\cdot)$ be a solution of (6.1) with $u_0 = 0$. Noting $u(s) \in \mathcal{D}(A)$, we have by (ii) and (iii) of Lemma 3.2 that for $0 \leq s \leq t$,

$$
\begin{aligned}
\frac{d}{ds}\left[S_r(t-s)u(s)\right] &= -\frac{(t-s)^{r-1}}{(r-1)!}u(s) - S_r(t-s)Au(s) + S_r(t-s)u'(s) \\
&= -\frac{(t-s)^{r-1}}{(r-1)!}u(s).
\end{aligned}
$$

Consequently,

$$
\begin{aligned}
-\int_0^t \frac{(t-s)^{r-1}}{(r-1)!}u(s)ds &= S_r(0)u(t) - S_r(t)u(0) \\
&= 0, \qquad t \geq 0.
\end{aligned}
$$

So $u(t) = 0$ for all $t \geq 0$. This ends the proof.

Combining Theorem 6.5 and Theorem 6.6 yields immediately

Theorem 6.7. *Let* $r \in N_0$, $\lambda_0 \in \rho(A)$. *Then* A *is the generator of an r-times integrated semigroup if and only if A is the generator of a $(\lambda_0 - A)^{-r}$-regularized semigroup.*

The following results concerning (6.2) are shown analogically as in the case of (6.1).

Lemma 6.8. *Let* $\lambda_0 \in \rho(A)$, u_0, $u_1 \in \mathcal{D}(A)$, $l \in N_0$. *Then*
(i) $u(\cdot)$ *is a solution of* (6.2) *with* $u_1 = 0$ *if and only if* $v(\cdot) := (\lambda_0 - A)u(\cdot) \in C(R^+, E)$ *and satisfies*

$$v(t) = (\lambda_0 - A)u_0 + A\int_0^t (t-s)v(s)ds, \qquad t \geq 0;$$

(ii) $u(\cdot)$ *is a solution of* (6.2) *with* $u_0 = 0$ *if and only if* $v_1(\cdot) := (\lambda_0 - A)u(\cdot) \in C(R^+, E)$ *and satisfies*

$$v_1(t) = t(\lambda_0 - A)u_1 + A\int_0^t (t-s)v_1(s)ds, \qquad t \geq 0;$$

(iii) $w(\cdot) \in C(R^+, E)$ *and satisfies*

$$w(t) = \frac{t^l}{l!}u_0 + A \int_0^t (t-s)w(s)ds, \quad t \geq 0$$

if and only if

$$w_0(t) := (\lambda_0 - A) \int_0^t (t-s)w(s)ds \in C(R^+, E)$$

and satisfies

$$w_0(t) = \frac{t^{l+2}}{(l+2)!}(\lambda_0 - A)u_0 + A \int_0^t (t-s)w_0(s)ds, \quad t \geq 0.$$

Theorem 6.9. *Let E, $\rho(A)$, r, ω be as in Theorem 6.3.*
    (i) *If (6.2) with $u_1 = 0$ has a unique solution $u(\cdot)$, for each $u_0 \in C\left(\mathcal{D}\left(A^{r+1}\right)\right)$, satisfying*

$$\sup_{t \geq 0} \left\| e^{-\omega t} u''(t) \right\|_j < \infty \quad (j \in N), \tag{6.14}$$

then A is the generator of a $(2r)$-times integrated, C-regularized cosine function.
    (ii) *If (6.2) with $u_0 = 0$ has a unique solution $u(\cdot)$, for each $u_1 \in C\left(\mathcal{D}\left(A^{r+1}\right)\right)$, satisfying (6.14), then A is the generator of a $(2r + 1)$-times integrated, C-regularized cosine function.*

Theorem 6.10. *Let $r \in N_0$.*
    (i) *If A is the generator of a $(2r)$-times integrated, C-regularized cosine function $\{C_{2r}(t)\}_{t \geq 0}$, then for each u_0, $u_1 \in C\left(\mathcal{D}\left(A^{r+1}\right)\right)$, (6.2) has a unique solution*

$$u(t; u_0, u_1) := \left[\sum_{i=0}^{r-1} \frac{t^{2i}}{(2i)!}A^i u_0 + C_{2r}(t)A^r C^{-1} u_0\right]$$
$$+ \left[\sum_{i=0}^{r-1} \frac{t^{2i+1}}{(2i+1)!}A^i u_1 + \int_0^t C_{2r}(s)A^r C^{-1} u_1 ds\right].$$

    (ii) *If A is the generator of a $(2r + 1)$-times integrated, C-regularized cosine function $\{C_{2r+1}(t)\}_{t \geq 0}$, then for $u_0 \in C\left(\mathcal{D}\left(A^{r+2}\right)\right)$, $u_1 \in C\left(\mathcal{D}\left(A^{r+1}\right)\right)$, (6.2) has a unique solution*

$$u(t; u_0, u_1) := \left[\sum_{i=0}^{r} \frac{t^{2i}}{(2i)!}A^i u_0 + \int_0^t C_{2r+1}(s)A^{r+1} C^{-1} u_0 ds\right]$$
$$+ \left[\sum_{i=0}^{r-1} \frac{t^{2i+1}}{(2i+1)!}A^i u_1 + C_{2r+1}(t)A^r C^{-1} u_1\right].$$

Proof. Use (ii) and (iii) of Lemma 4.2.

Example 6.11. Consider the initial value problem on $L^p(R^n)$ $(1 < p < \infty)$

$$
\begin{cases}
\dfrac{\partial u(t, x)}{\partial t} = \displaystyle\sum_{|\alpha| \le m} a_\alpha D^\alpha u(t, x), & (t, x) \in R^+ \times R^n, \\[4mm]
u(0, x) = f(x), & x \in R^n.
\end{cases}
\tag{6.15}
$$

Assume that

$$
\sup_{x \in R^n} \mathrm{Re} \left(\sum_{|\alpha| \le m} a_\alpha (ix)^\alpha \right) < \infty.
$$

Then applying Theorems 5.10 and 6.6 gives by (5.24), (5.25) that for every

$$
f \in W_l^{m(n|\frac{1}{2} - \frac{1}{p}| + 1), p}(R^n),
$$

(6.15) admits a unique solution $u(\cdot, x) \in C\left(R^+, L_l^p(R^n)\right)$ and there are constants M, ω such that for each $\beta \in N_0^l$,

$$
\left\| D^\beta u(t, x) \right\|_{L^p(R^n)} \le M e^{\omega t} \left\| D^\beta f \right\|_{W^{mn|\frac{1}{2} - \frac{1}{p}|, p}(R^n)}, \qquad t \ge 0.
$$

Example 6.12. Consider the initial value problem on $L^p(R^3)$ $(1 < p < \infty)$

$$
\begin{cases}
\dfrac{\partial u(t, x)}{\partial t} = \Delta u(t, x) + \displaystyle\sum_{i=1}^{3} c_i \dfrac{\partial}{\partial x_i} u(t, x) \\[4mm]
\qquad + c_4 u(t, x) + \displaystyle\int_{R^3} h(x - y) u(t, y) dy, \quad (t, x) \in R^+ \times R^3, \\[4mm]
u(0, x) = f(x), \quad x \in R^3,
\end{cases}
\tag{6.16}
$$

where Δ denotes the Laplacian, c_i $(i = 1, 2, 3, 4) \in C$ and $h(x) \in L^1(R^3)$.

Let $E = L_l^p(R^3)$, $r = r_p = 3 \left| \frac{1}{2} - \frac{1}{p} \right|$, and let

$$
Ag = \Delta g + \sum_{i=1}^{3} c_i \frac{\partial}{\partial x_i} g + c_4\, g, \quad g \in \mathcal{D}(A) := W_1^{2,p}(R^3),
$$

$$
(Kg)(x) = \int_{R^3} h(x - y) g(y) dy, \quad g \in \mathcal{D}(K) := E,
$$

$$
C_{r,1} = \mathbf{T}_1 \left\langle (1 + |x|^2)^{-r} \right\rangle.
$$

Observe that for each $\beta \in N_0^1$, $g \in E$,

$$
D^\beta (Kg) = \mathcal{F}^{-1} \left(\widehat{h} \cdot \widehat{D^\beta g} \right) = K \left(D^\beta g \right);
$$

therefore, $q_\beta(Kg) \leq \|h\|_{L_1} q_\beta(g)$, which indicates $K \in \mathcal{B}_\Gamma(E)$. The same reasoning gives that $KA \subset AK$, $KC_{r,1} = C_{r,1}K$. Denote by $P(x)$ the symbol of A. Then $\omega := \sup_{x \in R^3} \mathrm{Re}P(x) < \infty$. Therefore, we can apply Theorems 5.10, 3.6 and 6.6 to obtain by (5.24), (5.25) that for each $f \in W_1^{5,p}(R^3)$, (6.16) admits a unique solution $u(\cdot, x) \in C(R^+, L_1^p(R^3))$ and there are constants M, ω_0 such that for all $k \in N_0$, $t \geq 0$,

$$\left\|\frac{\partial^k}{\partial x_1^k} u(t, x)\right\|_{L^p(R^3)} \leq M e^{\omega_0 t} \left\|\frac{\partial^k}{\partial x_1^k} f(x)\right\|_{W^{3,p}(R^3)}.$$

1.7 Notes

The results of Section 1.1 are mainly from Liang-Xiao [10] and Xiao-Liang [2]. The complex representation theorem for Laplace transforms, Theorem 1.12, is essentially due to Arendt-Kellermann [1, Proposition 3.1], where the Hölder continuity of $h_\alpha(t)$ is also shown.

The theory of operator families (including operator semigroups and cosine operator functions) and abstract differential equations is closely related to the method of Laplace transforms. How to extend the classical Widder's representation theorem of Laplace transforms to vector-valued functions has all along been a notable problem. Following the work of Miyadera [2], Zaidman [1] proved in 1960 that Widder's theorem holds in a Banach space X if and only if X has the Radon-Nikodym property. In 1987, Arendt [2] present a significant integrated version of Widder's theorem in an arbitrary Banach space X:

Given $F \in C^\infty(R^+, X)$, $\lambda^{-1}F(\lambda)$ is representable as the Laplace transform of a Lipschitz continuous function if and only if

$$\sup_{\lambda > 0} \left\|\frac{1}{j!}\lambda^{j+1}F^{(j)}(\lambda)\right\| \leq \text{const} \quad (j = 0,\ 1,\ 2,\ \cdots).$$

Later in 1989, Hieber [1, 2] further obtained that the growth assumption as above implies that for each $r \in (0, 1]$, $\lambda^{-r}F(\lambda)$ is a Laplace transform. The integrated version of Widder's theorem provides a concise and heuristic proof (see Arendt [2]) for the well-known Hille-Yosida theorem and it has particularly stimulated the establishment and development of the theory of many new types of operator families in Banach spaces such as integrated semigroups, regularized semigroups, integrated cosine functions, existence families, so that a great deal of wellposed and illposed problems for abstract differential equations can be unified to deal with (cf., e.g., Arendt [1, 2], Arendt-Kellermann [1], deLaubenfels [9], deLaubenfels-Kantorovitz [1], deLaubenfels-Sun-Wang [1], Grimmer-Liu [1], Hieber [1-5], Kellermann-Hieber [1], Kuo-Shaw [1, 2], Y.-C. Li [1], Li-Shaw [1], Liang-Xiao [10, 12], Neubrander [4, 5], Nicaise [1], Xiao-Liang [15, 19, 23].)

Theorem 2.1, which extends the integrated version of Widder's theorem to a SCLCS for r-times integrated Laplace transforms (for any $r > 0$), as well as its proof comes from Xiao-Liang [21]. The special case $r = 1$ of Theorem 2.1 can be found in Y.-C. Li [1] and Xiao-Liang [16].

Integrated semigroups were introduced by Arendt [1, 2] in 1987; earlier on, similar concepts had appeared in Sova [2, 3], Oharu [1]. Regularized semigroups were introduced by Da Prato [1]; they were also introduced independently by Davies-Pang [1], where they were called C-semigroups. Now these operator families and other new types of operator families have been obtaining extensive developments and wide applications. The reader is referred to deLaubenfels [9] for a thorough presentation on this topic. Sections 1.3, 1.4 and 1.6 are intended to give a brief introduction about the basic properties of integrated, regularized semigroups or cosine functions, as well as their relationship to the Cauchy problem in SCLCS. Usually the given object is the operator, as in Section 1.5. Therefore we employ Laplace transforms to define the generators of these operator families. Thus these operator families are required to be exponentially equicontinuous. For material intimately related to the content within these sections, we refer to Hieber [2] (introducing α-times integrated semigroups, with fractional values α), Miyadera [5] (about once integrated regularized semigroups), Hieber-Holderrieth-Neubrander [1] (n-times integrated regularized semigroups, with integer n), deLaubenfels [5] (more general n-times integrated C-existence families), Kuo-Shaw [1, 2] (α-times integrated regularized semigroups), Y.-C. Li [1] and Li-Shaw [1, 2] (n-times integrated regularized semigroups and cosine functions in SCLCS); the operator families therein may not be exponentially equicontinuous. The results similar to Theorems 6.3, 6.5, 6.7 and 6.9 can be found in deLaubenfels [9, Theorems 4.15 and 18.3], Neubrander [4, Theorem 3.1], Tanaka-Miyadera [3]. Corollary 6.4 is from Xiao-Liang [17].

The study of the operator families in locally convex spaces dates back to 1958 when Schwartz [1] initiated the study of equicontinuous C_0-semigroups (i.e. strongly continuous semigroups), and this theory were further extended to quasi-equicontinuous (i.e. exponentially equicontinuous) C_0-semigroups (see Yosida [3, 4], Miyadera [3], Fattorini [1], Choe [1]). The characterizations for integrated, regularized semigroups in Section 1.3 due to Xiao-Liang [21] are generalizations of the corresponding ones in the aforementioned references. All of these results originate from the celebrated work of Hille [2] and Yosida [1] in 1948. They therein established independently a theorem giving the first complete characterization of the generator of a strongly continuous semigroup of contractions in Banach spaces. The significant extension of the theorem to arbitrary strongly continuous semigroups were given in the early fifties independently by Feller [1], Miyadera [1] and Phillips [4]. This theorem is exactly the well known Hille-Yosida-Feller-Miyadera-Phillips theorem, or in short Hille-Yosida-Phillips theorem or Hille-Yosida theorem. The characterizations for integrated, regularized cosine functions in Section 1.4 are generalizations of the characterization for strongly continuous cosine functions in Banach spaces, which was proved by Sova [1]. Other proofs were given independently by Da Prato-Giusti [1] and Fattorini [2] (in barrelled complete locally convex spaces).

Section 1.5 is from Xiao-Liang [21]. Lemmas 5.2 and 5.3 are based on Lemmas

3.3 and 3.2 in Hieber [3]. Specialized to the Banach space $L^p(R^n)$ $(1 < p < \infty)$, Theorem 5.9 improves Theorem 4.2 (b) in Hieber [3] which required that $\mathrm{Re}P(x) = 0$ for all $x \in R^n$. Theorem 5.10 perfects Theorem 2.3 in Lei-Zheng [1] by allowing r to take the critical value $n|\frac{1}{2} - \frac{1}{p}|$. Other related references are Arendt-Kellermann [1], Hieber [1-5], Kellermann-Hieber [1], Xiao-Liang [19], deLaubenfels-Lei [1].

Chapter 2

Wellposedness and solvability

Summary

When more than one coefficient operators are involved, the theory of (ACP_n) is considerably more complicated than that of (ACP_1) or incomplete (ACP_2) (i.e. the Cauchy problem for $u''(t) + A_0 u(t) = 0$). As made clear by Fattorini [7], the usual wellposedness of (ACP_n) can not ensure the exponential growth of its solutions. This motivates the introduction of the notion of strong wellposedness for general (ACP_n), which makes it possible to evolve a rich theory for the propagators of (ACP_n), just as that for the classical strongly continuous semigroups.

We start in Section 2.1 with the explicit definitions of wellposedness and strong wellposedness of (ACP_n). One see easily that, in the case of (ACP_1) or incomplete (ACP_2), strong wellposedness is equivalent to wellposedness. Moreover, some basic facts regarding a strongly wellposed (ACP_n) are also discussed.

Section 2.2 is devoted to the proof of the characterization (Theorem 2.2) for (ACP_n) to be strongly wellposed. This is a Hille-Yosida-Feller-Miyadera-Phillips type theorem.

In Theorem 2.2, the estimates of the derivatives of the three terms

$$\lambda^{n-1} R_\lambda, \quad \lambda^{k-1} A_k R_\lambda, \quad \lambda^{k-1} \overline{R_\lambda A_k} \ (1 \le k \le n-1)$$

together lead to the strong wellposedness of (ACP_n). The removal of the condition on the last term then suggests a solvability (i.e. existence and uniqueness without continuous dependence) result in Section 2.3. Its proof is also based on a general Ljubic type uniqueness theorem.

In Section 2.4, we are concerned with perturbation problems for both strong wellposedness and solvability. Interesting cases, when the generators of strongly continuous semigroups or cosine functions act as 'principal' coefficient operators, can be found there.

Finally in Section 2.5, we look at complete (ACP_2) (i.e. the Cauchy problem for $u''(t) + A_1 u'(t) + A_0 u(t) = 0$) and take care of the situation when either one of the two coefficient operators is bounded. Concise equivalent conditions are presented.

2.1   Basic properties

In this section, as well as in Sections 2.2 and 2.3, E and Γ will be as in Chapter 1.

Definition 1.1. A function $u(\cdot) \in C^n(R^+, E)$ is said to be a solution of (ACP_n) if for $0 \leq i \leq n-1$, $t \geq 0$, $u^{(i)}(t) \in \mathcal{D}(A_i)$, $A_i u^{(i)}(\cdot) \in C(R^+, E)$, and (ACP_n) is satisfied.

Definition 1.2. (ACP_n) is said to be wellposed if
  (i) There exist dense subspaces $D_0, \cdots, D_{n-1}$ of E such that, for any $u_0 \in D_0$, $\cdots$, $u_{n-1} \in D_{n-1}$, (ACP_n) has a solution.
  (ii) There exists a nondecreasing, positive function $M(t)$ defined in R^+ such that, to each $p \in \Gamma$ corresponds a continuous seminorm q satisfying

$$p(u(t)) \leq M(t) \sum_{k=0}^{n-1} q\left(u^{(k)}(0)\right), \quad t \geq 0, \tag{1.1}$$

for any solution $u(t)$ of (ACP_n).

For each $0 \leq k \leq n-1$, we define the operator $S_k(\cdot)$ by

$$S_k(t)u = u_k(t),$$

where $u_k(\cdot)$ is the solution of (ACP_n) with $u_k^{(l)}(0) = \delta_{kl} u$, δ_{kl} the Kronecker delta $(0 \leq l \leq n-1)$; the definition of $S_k(\cdot)$ makes sense for $u \in D_k$. We call $S_0, \cdots, S_{n-1}$ the propagators or solution operators of (ACP_n) if they can be extended to all of E as continuous operators; the extension is obviously unique, and when E is complete (not only sequentially complete), it automatically exists by (1.1) and the density of $D_0, \cdots, D_{n-1}$. It is easy to show that for each $0 \leq k \leq n-1$,

$$S_k(\cdot)u \in C(R^+, E), \quad u \in E,$$

the operator family $\{S_k(t)\}_{t \geq 0}$ is equicontinuous in compacts of $t \geq 0$, and for any solution $u(\cdot)$, we have

$$u(t) = \sum_{k=0}^{n-1} S_k(t)u^{(k)}(0), \quad t \geq 0. \tag{1.2}$$

Definition 1.3. (ACP_n) is called strongly wellposed if

(i) It is wellposed with the n propagators S_0, $\cdots$, S_{n-1} existing.

(ii) For each $u \in E$, $1 \le k \le n-1$, $S_k(\cdot)u \in C^k(R^+, E)$, $S_{n-1}^{(k-1)}(t)u \in \mathcal{D}(A_k)$ (for all $t \ge 0$) and $A_k S_{n-1}^{(k-1)}(t)u \in C(R^+, E)$.

We now define the linear operators $S_k^{(j)}(t)$ ($t \ge 0$, $1 \le k \le n-1$, $1 \le j \le k$) as follows

$$S_k^{(j)}(t)u = \frac{d^j}{dt^j}\left(S_k(t)u\right), \quad u \in E.$$

It is clear

$$S_k^{(j)}(0) = \delta_{kj}I, \quad 0 \le j \le n-1. \tag{1.3}$$

Theorem 1.4. *Let A_0, $\cdots$, A_{n-1} be closed linear operators in E such that (ACP_n) is strongly wellposed. Then*

(i) P_λ *is closable for all $\lambda \in \mathbf{C}$.*

(ii) $\bigcap_{i=0}^{n-1} \mathcal{D}(A_i)$ *is dense in E, and $\bigcap_{i=0}^{k} \mathcal{D}(A_i) \subset D_k$ for $0 \le k \le n-1$.*

(iii)

$$S_0(t)u = u - \int_0^t S_{n-1}(s)A_0 u\, ds, \quad u \in \mathcal{D}(A_0),\ t \ge 0, \tag{1.4}$$

and for $1 \le k \le n-1$,

$$S_k'(t)u = S_{k-1}(t)u - S_{n-1}(t)A_k u, \quad u \in D_{k-1}\bigcap\mathcal{D}(A_k),\ t \ge 0. \tag{1.5}$$

(iv) *For any $u \in \bigcap_{i=0}^{n-1}\mathcal{D}(A_i)$,*

$$S_{n-1}^{(n)}(t)u + \sum_{j=0}^{n-1} S_{n-1}^{(j)}(t)A_j u = 0, \quad t \ge 0.$$

Proof. First of all, we show that given $\tilde{u} \in E$ and $f \in C^1(R^+, \mathbf{C})$, then

$$u(t) := \int_0^t f(s)S_{n-1}(t-s)\tilde{u}\, ds$$

is a solution of

$$\begin{cases} u^{(n)}(t) + \displaystyle\sum_{j=0}^{n-1} A_j u^{(j)}(t) = f(t)\tilde{u}, & t \ge 0, \\[2mm] u^{(k)}(0) = 0, & 0 \le k \le n-1. \end{cases} \tag{1.6}$$

In fact, it is clear that for $u \in D_{n-1}$,

$$A_0 S_{n-1}(t)u = -\sum_{j=1}^{n-1} A_j S_{n-1}^{(j)}(t)u - S_{n-1}^{(n)}(t)u. \tag{1.7}$$

Integrating it yields that

$$A_0 \int_0^t S_{n-1}(s)u\,ds = -\sum_{j=1}^{n-1} A_j S_{n-1}^{(j-1)}(t)u - S_{n-1}^{(n-1)}(t)u + u. \qquad (1.8)$$

We see that the right-hand side is a continuous operator of u by the strong wellposedness; hence (1.8) holds for all $u \in E$ due to the density of D_{n-1} and the closedness of A_0. Now writing, by integration by parts,

$$u(t) = \int_0^t f(0)S_{n-1}(s)\tilde{u}\,ds + \int_0^t f'(s)\left(\int_0^{t-s} S_{n-1}(\sigma)\tilde{u}\,d\sigma\right)ds,$$

and differentiating, we obtain that for $1 \le k \le n-1$,

$$u^{(k)}(t) = f(0)S_{n-1}^{(k-1)}(t)\tilde{u} + \int_0^t f'(s)S_{n-1}^{(k-1)}(t-s)\tilde{u}\,ds.$$

Thus, (1.6) can be verified easily, using (1.8).

Assume (i) is false for some λ. Then there exists a sequence $\{u_m\} \subset \mathcal{D}(P_\lambda)$ such that

$$u_m \to 0, \quad v_m := P_\lambda u_m \to v \ne 0.$$

A simple computation shows that $u_m(t) := e^{\lambda t}u_m$ satisfies (1.6) with

$$\tilde{u} = v_m, \quad f(t) = e^{\lambda t}.$$

Consequently,

$$e^{\lambda t}u_m - \int_0^t e^{\lambda s}S_{n-1}(t-s)v_m\,ds = \sum_{k=0}^{n-1} \lambda^k S_k(t)u_m, \quad t \ge 0.$$

Letting $m \to \infty$ we obtain

$$\int_0^t e^{\lambda(t-s)}S_{n-1}(s)v\,ds = \int_0^t e^{\lambda s}S_{n-1}(t-s)v\,ds = 0, \quad t \ge 0.$$

So by differentiation,

$$S_{n-1}(t)v + \lambda \int_0^t e^{\lambda s}S_{n-1}(t-s)v\,ds = 0, \quad t \ge 0.$$

Differentiating this expression $n - 1$ times yields

$$S_{n-1}^{(n-1)}(t)v + \lambda \int_0^t e^{\lambda s}S_{n-1}^{(n-1)}(t-s)v\,ds = 0, \quad t \ge 0.$$

Putting $t = 0$, we get $v = 0$. This is a contradiction. So (i) is true.

Next, it is easy to verify by (1.6) that, for $u \in \mathcal{D}(A_0)$,

$$x(t) := u - \int_0^t S_{n-1}(s)A_0 u\, ds$$

satisfies (ACP_n) with initial conditions $u_0 = u$, $u_j = 0$ $(1 \le j \le n-1)$; for $u \in D_{k-1} \bigcap \mathcal{D}(A_k)$, $1 \le k \le n-1$,

$$y_k(t) := \int_0^t [S_{k-1}(s)u - S_{n-1}(s)A_k u]\, ds$$

satisfies (ACP_n) with initial conditions $u_k = u$, $u_j = 0$ $(j \ne k)$.

Accordingly, we find that for each $0 \le k \le n-1$, $\bigcap_{i=0}^k \mathcal{D}(A_i) \subset D_k$, and by (1.2),

$$S_0(t)u = x(t), \qquad u \in \mathcal{D}(A_0),$$

$$S_k(t)u = y_k(t), \qquad u \in D_{k-1} \bigcap \mathcal{D}(A_k), \ 1 \le k \le n-1.$$

Therefore we obtain (iii). Immediately, (iv) follows from (iii).

Finally, we show the denseness of $\bigcap_{i=0}^{n-1} \mathcal{D}(A_i)$. Let $u \in E$ be arbitrary. We take a sequence $\{\phi_m\}$ of nonnegative infinitely differentiable functions, each ϕ_m with support in $\left(0, \frac{1}{m}\right)$ and such that $\int \phi_m\, ds = 1$. We have

$$w_m := \int \phi_m(s) S_{n-1}^{(n-1)}(s) u\, ds \to u$$

as $m \to \infty$. On the other hand, we discover

$$w_m \in \bigcap_{i=0}^{n-1} \mathcal{D}(A_i)$$

by observing

$$w_m = (-1)^{n-k} \int \phi_m^{(n-k)}(s) S_{n-1}^{(k-1)}(s) u\, ds$$

$$= (-1)^n \int \phi_m^{(n)}(s) \left(\int_0^s S_{n-1}(\sigma) u\, d\sigma \right) ds, \qquad 1 \le k \le n-1,$$

and noting (1.8) as well as the statement below it. This justifies our claim and ends the proof.

Remark. Arguing similarly as in the first part of the above proof, one can obtain the following result:

Suppose that for each $1 \leq k \leq n-1$, *both of the operator families* $\left\{ S_k^{(k)}(t) \right\}_{t \geq 0}$, $\left\{ A_k S_{n-1}^{(k-1)}(t) \right\}_{t \geq 0}$ *are equicontinuous in compacts of* $t \geq 0$ *(which is automatically satisfied for the special case of a Fréchet space by the Banach-Steinhause theorem and the closed graph theorem (cf., e.g., Rudin [1, Chapter 2]). Then, given* $g \in C^1(R^+, E)$,

$$v(t) := \int_0^t S_{n-1}(t-s)g(s)ds$$

satisfies

$$
\begin{cases}
v^{(n)}(t) + \displaystyle\sum_{j=0}^{n-1} A_j v^{(j)}(t) = g(t), \quad t \geq 0, \\[2mm]
v^{(k)}(0) = 0, \quad 0 \leq k \leq n-1.
\end{cases}
$$

Theorem 1.5. *Let* E *be a Fréchet space and let* $A_0, \cdots, A_{n-1}$ *be closed linear operators in* E *such that* (ACP_n) *is wellposed. If either*

(i) A_k *is continuous for each* $1 \leq k \leq n-1$,   *or*

(ii) *there is a* $\lambda_0 \in C$ *such that* P_{λ_0} *is closed and densely defined with* $\mathcal{R}(P_{\lambda_0}) = E$, (ACP_n) *has a solution for every initial value* $(u_0, \cdots, u_{n-1}) \in \left(\bigcap_{i=0}^{n-1} D(A_i) \right)^n$, *and for each* $u \in E$, $1 \leq k \leq n-2$, $S_k(\cdot)u \in C^k(R^+, E)$ *(in the case of* $n \geq 3$),

then (ACP_n) *is strongly wellposed.*

Proof. (i) Integrating (1.7) n-times yields that for $u \in D_{n-1}$,

$$A_0 \int_0^t \frac{(t-s)^{n-1}}{(n-1)!} S_{n-1}(s)u\,ds$$

$$= \frac{t^{n-1}}{(n-1)!}u - S_{n-1}(t)u - \sum_{j=1}^{n-1} A_j \int_0^t \frac{(t-s)^{n-j-1}}{(n-j-1)!} S_{n-1}(s)u\,ds, \quad t \geq 0.$$

The equality holds for all $u \in E$, by a similar argument as for (1.8) (but now using the continuity assumption on A_k, $1 \leq k \leq n-1$, instead of the strong wellposedness). Therefore, for each $1 \leq k \leq n-1$, $u \in E$,

$$u(t) := \int_0^t \frac{(t-s)^{n-1}}{(n-1)!} S_{n-1}(s)A_k u\,ds$$

satisfies

$$u^{(n)}(t) + \sum_{j=0}^{n-1} A_j u^{(j)}(t) = \frac{t^{n-1}}{(n-1)!} A_k u, \quad t \geq 0, \qquad (1.9)$$

with $u^{(k)}(0) = 0$, $0 \le k \le n - 1$. Analogously, we can verify that for each $1 \le k \le n - 1$, $u \in E$,

$$v(t) := \int_0^t \frac{(t - s)^{n-1}}{(n - 1)!} S_{k-1}(s) u \, ds$$

satisfies

$$v^{(n)}(t) + \sum_{j=0}^{n-1} A_j v^{(j)}(t) = \sum_{j=k}^{n} \frac{t^{n+k-j-1}}{(n + k - j - 1)!} A_j u, \quad t \ge 0,$$

and if $u \in D_k$,

$$w(t) := \int_0^t \frac{(t - s)^{n-2}}{(n - 2)!} S_k(s) u \, ds$$

satisfies

$$w^{(n)}(t) + \sum_{j=0}^{n-1} A_j w^{(j)}(t) = \sum_{j=k+1}^{n} \frac{t^{n+k-j-1}}{(n + k - j - 1)!} A_j u, \quad t \ge 0,$$

so that $x(t) := v(t) - w(t)$ satisfies (1.9) with $x^{(k)}(0) = 0$, $0 \le k \le n - 1$. In conclusion, $u(\cdot) = x(\cdot)$; differentiating $(n - 1)$-times yields

$$S_k(t) u = \int_0^t [S_{k-1}(s) - S_{n-1}(s) A_k] u \, ds, \quad 1 \le k \le n - 1.$$

This holds for any $u \in E$, since D_k is dense in E and A_k is continuous. Then the assertion follows by an inductive inference.

(ii) Let $u(\cdot)$ be a solution of (ACP_n) with $u^{(k)}(0) \in \bigcap_{i=0}^{n-1} \mathcal{D}(A_i)$, $0 \le k \le n-1$ and set

$$v(t) := e^{-\lambda_0 t} u(t), \quad t \ge 0.$$

Then it is easy to verify

$$v^{(n)}(t) + \sum_{j=0}^{n-1} \tilde{A}_j v^{(j)}(t) = 0, \quad t \ge 0, \tag{1.10}$$

where

$$\tilde{A}_0 := P_{\lambda_0}, \quad \tilde{A}_j := \sum_{i=j}^{n} \binom{i}{j} \lambda_0^{i-j} A_i, \quad 1 \le j \le n - 1.$$

Obviously

$$u^{(k)}(0) = \sum_{i=0}^{k} \binom{k}{i} \lambda_0^i v^{(k-i)}(0), \quad 0 \le k \le n - 1;$$

hence for every initial value

$$(v_0, \cdots, v_{n-1}) \in \left(\mathcal{D}\left(\tilde{A}_0 \right) \right)^n = \left(\bigcap_{i=0}^{n-1} \mathcal{D}(A_i) \right)^n,$$

the equation (1.10) has a solution with $v^{(k)}(0) = v_k$ $(0 \leq k \leq n - 1)$. Moreover, if $v(t)$ is a solution of (1.10) then

$$u(t) := e^{\lambda_0 t} v(t)$$

is a solution of (ACP_n). From these observations we deduce that the Cauchy problem for (1.10) is wellposed. Denoting the propagators of (1.10) by $\widetilde{S}_0(\cdot), \cdots, \widetilde{S}_{n-1}(\cdot)$, we obtain

$$\widetilde{S}_{n-1}(t) = e^{-\lambda_0 t} S_{n-1}(t), \quad t \geq 0. \tag{1.11}$$

By hypothesis, $\widetilde{A}_0$ is closed and $\widetilde{A}_0 \mathcal{D}\left(\widetilde{A}_0\right) = E$. We get that the graph G of $\widetilde{A}_0$ is closed in the product space $E \times E$, and is therefore a Fréchet space. Considering the operator $\mathcal{A}$ from G to E : $\left(u, \widetilde{A}_0 u\right) \rightarrow \widetilde{A}_0 u$, then $\mathcal{A}$ is onto. By virtue of the open mapping theorem (cf., e.g., Rudin [1, Chapter 2]), $\mathcal{A}$ transforms open sets into open sets. This implies that

$$K := \left\{ \left(u, \widetilde{A}_0 u\right) \in G; \ \widetilde{A}_0 u \in D_{n-1} \right\}$$

is dense in G, since D_{n-1} is dense in E. Let $\left(u, \widetilde{A}_0 u\right) \in K$. A simple computation shows that

$$w(t) := u - \int_0^t \widetilde{S}_{n-1}(s) \widetilde{A}_0 u \, ds$$

is a solution of (1.10) with initial values $w(0) = u$, $w^{(j)}(0) = 0$, $1 \leq j \leq n - 1$. Accordingly,

$$\widetilde{S}_0(t) u - u = -\int_0^t \widetilde{S}_{n-1}(s) \widetilde{A}_0 u \, ds, \quad t \geq 0.$$

In fact, the equality holds for all $u \in \mathcal{D}\left(\widetilde{A}_0\right)$ by the density of K. Since $\widetilde{S}_0(\cdot) u$ is a solution of (1.10) for each $u \in \mathcal{D}\left(\widetilde{A}_0\right)$, and any element of E can be written in the form $\widetilde{A}_0 u$, we deduce that the Cauchy problem for (1.10) is strongly wellposed. So is the (ACP_n) by (1.11). This ends the proof.

Theorem 1.6. *Let E be a Banach space. Let $A_0, \cdots, A_{n-1}$ be closed linear operators in E such that (ACP_n) is strongly wellposed. Then there exist constants $C, \omega > 0$ such that for $1 \leq k \leq n - 1$, $t \geq 0$,*

$$\|S_0(t)\|, \ \left\|S_k^{(k)}(t)\right\|, \ \left\|A_k S_{n-1}^{(k-1)}(t)\right\|, \ \left\|\overline{S_{n-1}^{(k-1)}(t) A_k}\right\| \leq C e^{\omega t}.$$

The detailed proof in the case of (ACP_2) can be found in Fattorini [3] and [7, Chapter VIII]. The generalization to (ACP_n) proceeds in the same way and the following proof is also adapted from Fattorini [3].

Outline of proof of Theorem 1.6. Set

$$R(\lambda, \phi)u = \int_0^\infty e^{-\lambda t}\phi(t)S_{n-1}(t)u\,dt$$

where ϕ is an n-times continuously differentiable function with compact support and such that $\phi(0) = 1$. It can be easily seen that $R(\lambda, \phi)E \subset \bigcap_{i=0}^{n-1}\mathcal{D}(A_i)$; moreover, if $u \in E$,

$$
\begin{aligned}
P_\lambda R(\lambda, \phi)u &= u + \int_0^\infty e^{-\lambda t}M(t, \phi)u\,dt \\
&= u + \widetilde{M}(\lambda, \phi)u,
\end{aligned}
$$

where

$$
\begin{aligned}
M(t, \phi) &:= \sum_{j=0}^{n-1}\binom{n}{j}\phi^{(n-j)}(t)S_{n-1}^{(j)}(t) \\
&\quad + \sum_{k=1}^{n-1}\sum_{j=1}^{k-1}\binom{k}{j}\phi^{(k-j)}(t)A_k S_{n-1}^{(j)}(t).
\end{aligned}
\tag{1.12}
$$

Observe that for $j \leq k - 1$,

$$S_{n-1}^{(j)}(t)u = \frac{1}{(k-j-2)!}\int_0^t (t-s)^{k-j-2}S_{n-1}^{(k-1)}(s)u\,ds.$$

It follows by the strong wellposedness that $M(\cdot, \phi)$ is an $L(E)$-valued, strongly continuous function. Define

$$R(\lambda) = R(\lambda, \phi)\left(I + \widetilde{M}(\lambda, \phi)\right)^{-1}$$

for $\mathrm{Re}\lambda \geq \omega_0$, ω_0 such that

$$\int_0^\infty e^{-\omega_0 t}|M(t, \phi)|\,dt < 1.$$

Once proved that $R(\lambda) = P_\lambda^{-1}$ in $\mathrm{Re}\lambda \geq \omega_0$ we construct the "right-handed" representation of $R(\lambda)$, namely

$$R(\lambda) = \left(I + \widetilde{N}(\lambda, \phi)\right)^{-1}R(\lambda, \phi).$$

Here $\widetilde{N}(\lambda, \phi)$ is the Laplace transform of $N(t, \phi)$; $N(t, \phi)$ is defined by

$$N(t, \phi)u := \sum_{j=0}^{n-1}\binom{n}{j}\phi^{(n-j)}(t)S_{n-1}^{(j)}(t)u + \sum_{k=1}^{n-1}\sum_{j=1}^{k-1}\binom{k}{j}\phi^{(k-j)}(t)S_{n-1}^{(j)}(t)A_k u,$$

for $u \in \bigcap_{i=0}^{n-1} \mathcal{D}(A_i)$. It can be seen by (1.5) that $N(\cdot, \phi)$ has an $L(E)$-valued, strongly continuous extension to all of E. Next, we define

$$\mathcal{M}(t, \phi) = \sum_{m=1}^{\infty} (-1)^m M(t, \phi)^{*m},$$

$$\mathcal{N}(t, \phi) = \sum_{m=1}^{\infty} (-1)^m N(t, \phi)^{*m};$$

they are $L(E)$-valued, strongly continuous functions satisfying

$$|\mathcal{M}(t, \phi)|, \quad |\mathcal{N}(t, \phi)| \leq Ce^{\omega t}, \quad t \geq 0 \tag{1.13}$$

for convenient constants $C, \omega > 0$. The propagator S_{n-1} satisfies

$$S_{n-1}(t) = (\delta \otimes I + \mathcal{N}(t, \phi)) * (\phi S_{n-1})(t) \tag{1.14}$$

and

$$S_{n-1}(t) = (\phi S_{n-1})(t) * (\delta \otimes I + \mathcal{M}(t, \phi)). \tag{1.15}$$

From (1.5) we get that

$$S_k(t) = S_{n-1}^{(n-k-1)}(t) + \sum_{j=k+1}^{n-1} \overline{S_{n-1}^{(j-k-1)}(t) A_j}, \quad 0 \leq k \leq n-1.$$

Consequently,

$$S_k(t) = (\delta \otimes I + \mathcal{N}(t, \phi)) * \left((\phi S_{n-1})^{(n-k-1)} \right.$$

$$\left. + \sum_{j=k+1}^{n-1} \overline{(\phi S_{n-1})^{(j-k-1)} A_j} \right), \quad 0 \leq k \leq n-1. \tag{1.16}$$

Equations (1.14)-(1.16) are used to prove that $S_0, \cdots, S_{n-1}$ increase at most exponentially at ∞. Observe that, using (1.14) the same property can be proved of $\overline{S_{n-1}^{(k-1)}(t) A_k}$ (and so of $S_k^{(k)}(t)$ (by (1.5)), $1 \leq k \leq n-1$; on the other hand, (1.15) can be used to show exponential increase of $A_k S_{n-1}^{(k-1)}(t)$, $1 \leq k \leq n-1$.

2.2 Strong wellposedness

Lemma 2.1. *Let* $A_0, \cdots, A_{n-1}$ *be closed linear operators in* E. *Suppose that* R_λ *exists and* $\mathcal{D}(R_\lambda) = E$ *for* $\lambda > \omega$ *(with some* $\omega > 0$*). Let* $u_0, \cdots, u_{n-1}$ *be such that*

$$\lambda^j A_j R_\lambda u_{n-1}, \quad \lambda^j A_j \sum_{i=0}^{k} \lambda^{i-k-1} R_\lambda A_i u_k \in LT - E$$

$$(0 \le k \le n-2, \ 0 \le j \le n-1),$$

as functions of λ. Then (ACP_n) has a solution.

Proof. Observe that for $0 \le k \le n-2$, λ sufficiently large,

$$\sum_{j=0}^{n-1} \lambda^j A_j \sum_{i=0}^{k} \lambda^{i-k-1} R_\lambda A_i u_k \ = \ \sum_{i=0}^{k} \lambda^{i-k-1} (R_\lambda - \lambda^n) R_\lambda A_i u_k$$

$$= \ \sum_{j=0}^{k} \lambda^{j-k-1} A_j u_k - \lambda^n \sum_{i=0}^{k} \lambda^{i-k-1} R_\lambda A_i u_k.$$

We then obtain for $0 \le k \le n-2$,

$$\Phi_k(\lambda; n) u_k = - \sum_{j=0}^{n-1} A_j \Phi_k(\lambda; j) u_k, \tag{2.1}$$

where

$$\Phi_k(\lambda; j) := \begin{cases} \lambda^j \left(\lambda^{-k-1} - \displaystyle\sum_{i=0}^{k} \lambda^{i-k-1} R_\lambda A_i \right), & \text{if} \quad 0 \le j \le k, \\[4mm] -\lambda^j \displaystyle\sum_{i=0}^{k} \lambda^{i-k-1} R_\lambda A_i, & \text{if} \quad k+1 \le j \le n. \end{cases}$$

Setting

$$\Phi_{n-1}(\lambda; j) := \lambda^j R_\lambda \ (0 \le j \le n-1), \qquad \Phi_{n-1}(\lambda; n) := \lambda^n R_\lambda - I,$$

then it is easy to see that (2.1) also holds for $k = n-1$. It follows by hypothesis that

$$A_j \Phi_k(\lambda; j) u_k \in LT - E \quad (0 \le k, \ j \le n-1).$$

Thus (2.1) shows that

$$\Phi_k(\lambda; n) u_k \in LT - E \quad (0 \le k \le n-1).$$

Note that for $0 \le k, \ j \le n-1$,

$$\Phi_k(\lambda; j+1) u_k = \begin{cases} \lambda \Phi_k(\lambda; j) u_k, & \text{if} \quad j \ne k, \\[2mm] \lambda \Phi_k(\lambda; j) u_k - u_k, & \text{if} \quad j = k. \end{cases}$$

Then applying Lemma 1.1.4 and Theorems 1.1.9 and 1.1.10, we obtain that, for every $0 \le k \le n-1$, there is $u_k(\cdot) \in C^n(R^+, E)$ satisfying

$$u_k{}^{(j)}(0) = \delta_{kj} u_k \quad (0 \le j \le n-1), \qquad u_k{}^{(j)}(t) \in \mathcal{D}(A_j) \quad (t \ge 0),$$

and

$$A_j u_k^{(j)}(\cdot) \in C(R^+, E) \quad (0 \le j \le n-1),$$

such that for λ sufficiently large,

$$\Phi_k(\lambda; 0)u_k = \int_0^\infty e^{-\lambda t} u_k(t)dt,$$

$$A_j \Phi_k(\lambda; j)u_k = \int_0^\infty e^{-\lambda t} A_j u_k^{(j)}(t)dt, \quad 0 \le j \le n-1,$$

$$\Phi_k(\lambda; n)u_k = \int_0^\infty e^{-\lambda t} u_k^{(n)}(t)dt.$$

Accordingly, we obtain that

$$\int_0^\infty e^{-\lambda t} \left[u_k^{(n)}(t) + \sum_{j=0}^{n-1} A_j u_k^{(j)}(t) \right] dt = 0,$$

which implies that

$$u_k^{(n)}(t) + \sum_{j=0}^{n-1} A_j u_k^{(j)}(t) = 0 \quad (t \ge 0, \ 0 \le k \le n-1).$$

In conclusion, $u(\cdot) := \sum_{k=0}^{n-1} u_k(\cdot)$ is a solution of (ACP_n).

Theorem 2.2 (Hille-Yosida-Feller-Miyadera-Phillips type theorem). *Let $a > 0$ and let $A_0, \cdots, A_{n-1}$ be closed linear operators in E. Then the following statements are equivalent.*

(i) (ACP_n) *is strongly wellposed. To each $p \in \Gamma$ corresponds a continuous seminorm q such that for any $t \ge 0$, $u \in E$,*

$$p\left(S_k^{(k)}(t)u\right) \le e^{at}q(u), \quad 0 \le k \le n-1, \tag{2.2}$$

$$p\left(A_k S_{n-1}^{(k-1)}(t)u\right) \le e^{at}q(u), \quad 1 \le k \le n-1. \tag{2.3}$$

(ii) $\bigcap_{i=0}^{n-1} \mathcal{D}(A_i)$ *is dense in E. For any $\lambda > a$, $R_\lambda \in L(E)$. For each $0 \le k \le n-1$, $R_\lambda A_k$ is closable, and for every $p \in \Gamma$, there exists a continuous seminorm q such that for any $1 \le k \le n-1$, $u \in E$,*

$$p\left\{ [\lambda^{n-1} R_\lambda u]^{(m)} \right\}, \quad p\left\{ [\lambda^{k-1} A_k R_\lambda u]^{(m)} \right\}, \quad p\left\{ [\lambda^{k-1} \overline{R_\lambda A_k} u]^{(m)} \right\}$$

$$\le m!(\lambda - a)^{-m-1}q(u), \quad \lambda > a, \ m \in N_0. \tag{2.4}$$

Moreover, in this case, the propagators of (ACP_n) have the following expressions:

$$S_k(t)u = \lim_{m \to \infty} \frac{(-1)^m}{m!} \left(\frac{m}{t}\right)^{m+1} \left[\lambda^{-k-1} \sum_{i=k+1}^{n} \lambda^i \overline{R_\lambda A_i} u\right]^{(m)}\Bigg|_{\lambda = \frac{m}{t}},$$

$$u \in E, \quad t > 0, \quad 0 \le k \le n-1;$$

$$S_0(t)u = u - \lim_{\lambda \to \infty} \sum_{m=1}^{\infty} (-1)^{m+1} \frac{1}{m!} e^{m\lambda t} R_{m\lambda} A_0 u, \quad u \in \mathcal{D}(A_0), \quad t \ge 0,$$

$$S_k(t)u = \lim_{\lambda \to \infty} \sum_{m=1}^{\infty} (-1)^{m+1} \frac{1}{m!} e^{m\lambda t} (m\lambda)^{-k} \sum_{i=k+1}^{n} (m\lambda)^i \overline{R_{m\lambda} A_i} u,$$

$$u \in E, \quad t \ge 0, \quad 1 \le k \le n-1;$$

and for every $\bar{a} > a$,

$$S_0(t)u = \frac{1}{2\pi i} \int_{\bar{a}-i\infty}^{\bar{a}+i\infty} e^{\lambda t} \left(\lambda^{-1} - \lambda^{-1} R_\lambda A_0\right) u \, d\lambda, \quad u \in \mathcal{D}(A_0), \quad t > 0,$$

$$S_k(t)u = \frac{1}{2\pi i} \int_{\bar{a}-i\infty}^{\bar{a}+i\infty} e^{\lambda t} \left(\lambda^{-k-1} \sum_{i=k+1}^{n} \lambda^i \overline{R_\lambda A_i} u\right) d\lambda,$$

$$u \in E, \quad t > 0, \quad 1 \le k \le n-1.$$

Remark. If E is a Banach space, then the conditions (2.2) and (2.3) read

$$\left\|S_k^{(k)}(t)\right\| \le C e^{at}, \quad t \ge 0, \ 0 \le k \le n-1, \tag{2.5}$$

$$\left\|A_k S_{n-1}^{(k-1)}(t)\right\| \le C e^{at}, \quad t \ge 0, \ 1 \le k \le n-1, \tag{2.6}$$

for some constant C. But (2.5) and (2.6) can be deduced from the strong wellposedness; see Theorem 1.6. Therefore, as a corollary of Theorem 2.2, we get

Theorem 2.3. *Let E be a Banach space and let $A_0, \cdots, A_{n-1}$ be closed linear operators in E. Then (ACP_n) is strongly wellposed if and only if $\bigcap_{i=0}^{n-1} \mathcal{D}(A_i)$ is dense in E, there exist constants a, $C > 0$ such that for every $\lambda > a$, $R_\lambda \in L(E)$, $R_\lambda A_k$ $(0 \le k \le n-1)$ is closable, and for $1 \le k \le n-1$,*

$$\left\|[\lambda^{n-1} R_\lambda]^{(m)}\right\|, \quad \left\|[\lambda^{k-1} A_k R_\lambda]^{(m)}\right\|, \quad \left\|[\lambda^{k-1} \overline{R_\lambda A_k}]^{(m)}\right\|$$

$$\le C m! (\lambda - a)^{-m-1}, \quad \lambda > a, \ m \in N_0.$$

Next, we give a counterexample to indicate that the strong wellposedness of (ACP_n) does not generally imply (2.2), (2.3).

Example 2.4. Let E be the space of all complex number sequences $u = \{u_k; \ k \in N\} = \{u_k\}$ topologized by the separating family of seminorms

$$p_m(u) = \left(\sum_{k=1}^{m} |u_k|^2 \right)^{\frac{1}{2}} \qquad (u \in E, \ m \in N).$$

Clearly, E is a Fréchet space. Let $n = 2$, and A_0, A_1 be defined as

$$A_0 u = \{(2 + \ln k)u_k\}, \quad u = \{u_k\} \in E,$$

$$A_1 u = \{-(3 + \ln k)u_k\}, \quad u = \{u_k\} \in E.$$

It is easy to verify that for any u_0, $u_1 \in E$, the (ACP_2) has a unique solution $u(\cdot)$ satisfying $u(0) = u_0$, $u'(0) = u_1$, and

$$S_0(t)\{u_k\} = \left\{ \frac{(2 + \ln k)e^t - e^{(2+\ln k)t}}{1 + \ln k} u_k \right\},$$

$$S_1(t)\{u_k\} = \left\{ \frac{e^{(2+\ln k)t} - e^t}{1 + \ln k} u_k \right\}.$$

We see that for each $t \geq 0$, $u \in E$, $m \in N$,

$$p_m\left(S_0(t)u\right), \ p_m(S_1(t)u) \ \leq \ \left(2e^t + e^{2t}m^t\right) p_m(u)$$

$$\leq \ \left(2e^t + e^{2t}\left(t^{t+1} + m^{m+1}\right)\right) p_m(u). \tag{2.7}$$

Set

$$\overline{q}_m = m^{m+1} p_m, \quad \text{for each } m \in N.$$

Then $\overline{q}_m$ is a continuous seminorm, and we get by (2.7)

$$p_m(S_0(t)u), \ p_m(S_1(t)u) \leq 3e^{2t}\overline{q}_m(u), \quad m \in N, \ t \geq 0, \ u \in E.$$

Consequently, the (ACP_2) is wellposed, and furthermore strongly wellposed by an easy observation. However, (2.2) and (2.3) do not hold in this case. In fact, if there exists a constant $a > 0$ such that for each p_m, there exists a continuous seminorm q_m such that

$$p_m\left(e^{-(1+a)t}S_0(t)u\right) \leq q_m(u), \quad t \geq 0, \ u \in E,$$

then taking

$$m_0 = \left[e^{a+1}\right] + 1,$$

$$u_0 = \{u_k; \ u_k = 1 \text{ when } k = m_0 \text{ and } u_k = 0 \text{ when } k \neq m_0\},$$

we have

$$q_{m_0}(u_0) \geq p_{m_0}\left(e^{-(1+a)t}S_0(t)u_0\right)$$

$$\geq e^{-(1+a)t}\left[\frac{m_0^t e^t}{1+\ln m_0} - 2e^t\right].$$

Letting $t \to +\infty$, the function on the right side tends to $+\infty$. It is a contradiction. Hence, (2.2) does not hold for $k = 0$. Similarly, we can see that (2.2), (2.3) does not hold for $k = 1$ either.

Proof of Theorem 2.2. (i)$\Longrightarrow$(ii).
   By Theorem 1.4, $\bigcap_{i=0}^{n-1}\mathcal{D}(A_i) \subset D_{n-1}$. So for any $u \in \bigcap_{i=0}^{n-1}\mathcal{D}(A_i)$,

$$S_{n-1}^{(n)}(s)u + \sum_{i=0}^{n-1} A_i S_{n-1}^{(i)}(s)u = 0, \quad s \geq 0.$$

Integrating from 0 to t and noting that

$$S_{n-1}^{(i)}(0)u = \delta_{n-1,i}u \quad (0 \leq i \leq n-1),$$

we have that for any $t \geq 0$,

$$S_{n-1}^{(n-1)}(t)u + \sum_{i=1}^{n-1} A_i S_{n-1}^{(i-1)}(t)u + A_0\int_0^t S_{n-1}(s)u\,ds = u.$$

The hypotheses (2.2) and (2.3) enable us to take the Laplace transform to each term of the above equality. The result is

$$\int_0^\infty e^{-\lambda t}S_{n-1}^{(n-1)}(t)u\,dt + \sum_{i=1}^{n-1}\int_0^\infty e^{-\lambda t}A_i S_{n-1}^{(i-1)}(t)u\,dt$$

$$+ \int_0^\infty e^{-\lambda t}A_0\left(\int_0^t S_{n-1}(s)u\,ds\right)dt$$

$$= \frac{1}{\lambda}u, \quad \mathrm{Re}\,\lambda > a.$$

Integrating by parts and using the closedness of A_k $(0 \leq k \leq n-1)$, we obtain

$$\left(\lambda^{n-1} + \sum_{i=0}^{n-1}\lambda^{i-1}A_i\right)\int_0^\infty e^{-\lambda t}S_{n-1}(t)u\,dt = \frac{1}{\lambda}u, \quad \mathrm{Re}\,\lambda > a;$$

that is, for each $u \in \bigcap_{i=0}^{n-1}\mathcal{D}(A_i)$,

$$P_\lambda\int_0^\infty e^{-\lambda t}S_{n-1}(t)u\,dt = u, \quad \mathrm{Re}\,\lambda > a. \tag{2.8}$$

By (2.2), for any $\text{Re}\lambda > a$,

$$u \longmapsto \int_0^\infty e^{-\lambda t} S_{n-1}(t)u\,dt, \quad u \in E$$

is a continuous linear operator on E. Moreover, combining (2.3) with the closedness of A_k, we get

$$\int_0^\infty e^{-\lambda t} S_{n-1}(t)u\,dt \in \mathcal{D}(A_k), \quad u \in E, \ 0 \le k \le n-1.$$

Accordingly, (2.8) holds for each $u \in E$, since $\bigcap_{i=0}^{n-1} \mathcal{D}(A_i)$ is dense in E and P_λ is closable (see Theorem 1.4).

Next, we prove that for any $\text{Re}\lambda > a$, P_λ^{-1} exists and $R_\lambda \in \mathbf{L}(E)$.

If this is not true, then there exist $v_0 \ne 0$, $\lambda_0 \in \mathbf{C}$ with $\text{Re}\lambda_0 > a$ such that $P_{\lambda_0} v_0 = 0$. Clearly, $u(t) := e^{\lambda_0 t} v_0$ is the solution of (ACP_n) with $u_k = \lambda_0^k v_0$ $(0 \le k \le n-1)$. So by (1.2),

$$e^{\lambda_0 t} v_0 = \sum_{k=0}^{n-1} \lambda_0^k S_k(t) v_0, \quad t \ge 0.$$

Taking $p_0 \in \Gamma$ such that $p_0(v_0) > 0$. (2.2) implies that there exists a continuous seminorm q_0 such that

$$e^{\text{Re}\lambda_0 t} p_0(v_0) \le \sum_{k=0}^{n-1} |\lambda_0|^k e^{at} q_0(v_0), \quad t \ge 0.$$

Thus,

$$\sum_{k=0}^{n-1} e^{-(\text{Re}\lambda_0 - a)t} |\lambda_0|^k q_0(v_0) \ge p_0(v_0), \quad t \ge 0.$$

Letting $t \to +\infty$, we have $0 \ge p_0(v_0)$. It is a contradiction. Consequently, for any $\text{Re}\lambda > a$, P_λ^{-1} exists and $R_\lambda \in \mathbf{L}(E)$.

It follows from this fact and (2.8) that

$$R_\lambda u = \int_0^\infty e^{-\lambda t} S_{n-1}(t)u\,dt, \quad \text{Re}\lambda > a, \ u \in E. \tag{2.9}$$

Hence,

$$\lambda^{n-1} R_\lambda u = \int_0^\infty e^{-\lambda t} S_{n-1}^{(n-1)}(t)u\,dt, \quad \text{Re}\lambda > a, \ u \in E,$$

$$\lambda^{k-1} A_k R_\lambda u = \int_0^\infty e^{-\lambda t} A_k S_{n-1}^{(k-1)}(t)u\,dt,$$

$$\text{Re}\lambda > a, \ u \in E, \ 1 \le k \le n-1.$$

Thus, the first two estimates in (2.4) are derived from (2.2) and (2.3).

On the other hand, in view of (1.4) and (1.5) we have that for any $\text{Re}\lambda > a$,

$$
\begin{aligned}
R_\lambda A_0 u &= \int_0^\infty e^{-\lambda t} S_{n-1}(t) A_0 u \, dt \\
&= -\int_0^\infty e^{-\lambda t} S_0'(t) u \, dt \\
&= u - \lambda \int_0^\infty e^{-\lambda t} S_0(t) u \, dt, \quad u \in \mathcal{D}(A_0),
\end{aligned}
$$

$$
\begin{aligned}
R_\lambda A_k u &= \int_0^\infty e^{-\lambda t} S_{n-1}(t) A_k u \, dt \\
&= \int_0^\infty e^{-\lambda t} S_{k-1}(t) u \, dt - \lambda \int_0^\infty e^{-\lambda t} S_k(t) u \, dt,
\end{aligned}
$$

$$
u \in \bigcap_{i=0}^k \mathcal{D}(A_i), \ 1 \le k \le n-1.
$$

So, (2.2) implies that for any $p \in \Gamma$, there exists a continuous seminorm q such that for $0 \le k \le n-1$,

$$
p(R_\lambda A_k u) \le (|\lambda| + 1)(\text{Re}\lambda - a)^{-1} q(u), \quad u \in \bigcap_{i=0}^k \mathcal{D}(A_i), \ \text{Re}\lambda > a.
$$

Accordingly, for each $0 \le k \le n-1$, $\text{Re}\lambda > a$, $R_\lambda A_k$ is closable.

Finally, by (1.5),

$$
S_{n-1}^{(k-1)}(t) A_k u = S_{k-1}^{(k-1)}(t) u - S_k^{(k)}(t) u, \quad u \in \bigcap_{i=0}^k \mathcal{D}(A_i), \ 1 \le k \le n-1. \tag{2.10}
$$

Therefore, $S_{n-1}^{(k-1)}(t) A_k$ is closable and for any $1 \le k \le n-1$,

$$
\lambda^{k-1} \overline{R_\lambda A_k} u = \int_0^\infty e^{-\lambda t} \overline{S_{n-1}^{(k-1)}(t) A_k} u \, dt, \quad \text{Re}\lambda > a, \ u \in E.
$$

Thus, using (2.10) and (2.2) we obtain the last estimate in (2.4).

(ii)$\Longrightarrow$ (i). Firstly, we show that for any $1 \le k \le n-1$,

$$
\lambda \longmapsto \lambda^{n-1} R_\lambda, \quad \lambda \longmapsto \lambda^{k-1} A_k R_\lambda \in LT - L(E). \tag{2.11}
$$

It is plain that for $\lambda > a$,

$$
\lambda^{-1} A_0 R_\lambda = \lambda^{-1} - \lambda^{n-1} R_\lambda - \sum_{i=1}^{n-1} \lambda^{i-1} A_i R_\lambda. \tag{2.12}
$$

Then (2.4) shows by Leibniz's formula that for any $p \in \Gamma$, there exists a continuous seminorm $\bar{q}$ such that

$$p\left\{ \left[\lambda^{-j} \lambda^{k-1} A_k R_\lambda u \right]^{(m)} \right\} \leq m! (\lambda - a)^{-m-1} \bar{q}(u),$$

(2.13)

$$\lambda > a, \ u \in E, \ 0 \leq k \leq n, \ j \in N_0, \ m \in N_0.$$

We can see that for each $u \in \bigcap_{i=0}^{n-1} \mathcal{D}(A_i)$, $1 \leq k \leq n$,

$$\lambda \longmapsto \sum_{j=0}^{n-1} \left(\lambda^{k+j-n} A_k R_\lambda \right) A_j u$$

satisfies the condition in Theorem 1.2.1. But

$$\lambda^{k-1} A_k R_\lambda u = \lambda^{-(n-k+1)} A_k u - \lambda^{-1} \sum_{j=0}^{n-1} \left(\lambda^{k+j-n} A_k R_\lambda \right) A_j u,$$

$$u \in \bigcap_{i=0}^{n-1} \mathcal{D}(A_i), \ 1 \leq k \leq n, \ \lambda > a.$$

Applying Theorem 1.2.1 yields that for any $u \in \bigcap_{i=0}^{n-1} \mathcal{D}(A_i)$, $1 \leq k \leq n$, there exist $T_k(\cdot)u \in C(R^+, E)$ $(1 \leq k \leq n)$ with $T_n(0)u = u$, $T_k(0)u = 0$ $(0 \leq k \leq n-1)$ such that for λ sufficiently large,

$$\lambda^{k-1} A_k R_\lambda u = \int_0^\infty e^{-\lambda t} T_k(t) u \, dt, \quad 1 \leq k \leq n,$$

(2.14)

noting

$$\lambda^{-(n-k+1)} A_k u = \int_0^\infty e^{-\lambda t} \frac{t^{n-k}}{(n-k)!} A_k u \, dt.$$

Thus, an application of Theorem 1.1.5 gives that for any $u \in \bigcap_{i=0}^{n-1} \mathcal{D}(A_i)$, $1 \leq k \leq n$,

$$T_k(t)u = \lim_{m \to \infty} \frac{(-1)^{m-1}}{(m-1)!} \left(\frac{m}{t} \right)^m \left[\lambda^{k-1} A_k R_\lambda u \right]^{(m-1)} \Bigg|_{\lambda = \frac{m}{t}}, \quad t > 0.$$

(2.15)

Observe by (2.4) that for each $1 \leq k \leq n$, $t > 0$, $u \in E$,

$$p \left(\frac{(-1)^{m-1}}{(m-1)!} \left(\frac{m}{t} \right)^m \left[\lambda^{k-1} A_k R_\lambda u \right]^{(m-1)} \Bigg|_{\lambda = \frac{m}{t}} \right)$$

$$\leq \quad \frac{1}{(m-1)!} \left(\frac{m}{t} \right)^m (m-1)! \left(\frac{m}{t} - a \right)^{-m} \bar{q}(u)$$

$$= \quad \left(1 - \frac{at}{m} \right)^{-m} \bar{q}(u)$$

$$\longrightarrow \quad e^{at} \bar{q}(u), \quad \text{as } m \to \infty.$$

We discover by the denseness of $\bigcap_{i=0}^{n-1} \mathcal{D}(A_i)$ and the sequential completeness that for each $1 \le k \le n$, $t \ge 0$, the limit in the right side of (2.15) exists for each $u \in E$. Therefore $T_k(t)$ can be extended continuously to all of E (which we denote by the same symbol), so that (2.15) holds for each $u \in E$. It is clear that $T_k(\cdot)u \in C(R^+, E)$ for any $u \in E$; to each $p \in \Gamma$ corresponds a continuous seminorm $\bar{q}$ such that

$$p(T_k(t)u) \le e^{at}\bar{q}(u), \quad 1 \le k \le n, \ u \in E, \ t \ge 0,$$

and (2.14) hold for all $u \in E$. Hence, (2.11) follows for every $1 \le k \le n-1$. Moreover, it is easy to see by (2.12) that (2.11) holds also for $k = 0$.

It follows from (2.11) that for any $0 \le j \le n-1$,

$$\lambda \longmapsto \lambda^{j+i-k-1} A_j R_\lambda A_i u \in LT - E,$$

$$0 \le k \le n-2, \ 0 \le i \le k, \ u \in \bigcap_{i=0}^{k} \mathcal{D}(A_i),$$

$$\lambda \longmapsto \lambda^j A_j R_\lambda u = \lambda^{-(n-j)} A_j u + \sum_{l=0}^{n-1} \left(\lambda^{j+l-n} A_j R_\lambda \right) A_l u$$

$$\in LT - E, \quad u \in \bigcap_{i=0}^{n-1} \mathcal{D}(A_i).$$

Thus, Lemma 2.1 implies that for any $u \in \bigcap_{i=0}^{n-1} \mathcal{D}(A_i)$, (ACP_n) has a solution $S_k(\cdot)u$ satisfying

$$S_k^{(j)}(0)u = \delta_{kj}u \quad (0 \le k, \ j \le n-1).$$

In addition, from the proof of Lemma 2.1 we see that for λ sufficiently large,

$$\Phi_k(\lambda; j)u = \int_0^\infty e^{-\lambda t} S_k^{(j)}(t)u\, dt,$$

$$(2.16)$$

$$0 \le k \le n-1, \ 0 \le j \le k, \ u \in \bigcap_{i=0}^{k} \mathcal{D}(A_i).$$

According to Theorem 1.1.5, for every $0 \le k \le n-1$, $0 \le j \le k$, $u \in \bigcap_{i=0}^{k} \mathcal{D}(A_i)$,

$$S_k^{(j)}(t)u = \lim_{m \to \infty} \frac{(-1)^{m-1}}{(m-1)!} \left(\frac{m}{t} \right)^m \Phi_k^{(m-1)} \left(\frac{m}{t}; j \right) u, \quad t > 0. \qquad (2.17)$$

By (2.4) and the Leibniz formula, we obtain that for any $p \in \Gamma$, there exists a continuous seminorm $\bar{q}$ such that

$$p\left\{ \left[\lambda^{-j} \lambda^{k-1} \overline{R_\lambda A_k} u \right]^{(m)} \right\} \le m!(\lambda - a)^{-m-1} \bar{q}(u),$$

$$(2.18)$$

$$\lambda > a, \ u \in E, \ 0 \le k \le n, \ j \in N_0, \ m \in N_0.$$

Observing

$$R_\lambda A_k u = \overline{R_\lambda A_k} u, \quad u \in \bigcap_{i=0}^{k} \mathcal{D}(A_i), \ 0 \le k \le n - 1$$

and estimating (2.17) by (2.18), we get that for each $0 \le k \le n-1$, $t \ge 0$, $S_k(t)$ can be extended to all of E as a continuous operator (which we denote by the same symbol); that for any $u \in E$, $S_k(\cdot)u \in C^k(R^+, E)$ and (2.2) holds. On the other hand, taking $k = n - 1$ in (2.16), we have

$$\lambda^j R_\lambda u = \int_0^\infty e^{-\lambda t} S_{n-1}^{(j)}(t) u \, dt, \quad 0 \le j \le n - 1, \ u \in E.$$

This together with (2.11) yields by Theorem 1.1.10 that for any $1 \le k \le n - 1$, $u \in E$,

$$S_{n-1}^{(k-1)}(t)u \in \mathcal{D}(A_k) \quad (t \ge 0)$$

and

$$A_k S_{n-1}^{(k-1)}(\cdot)u \in C(R^+, E),$$

and (2.3) holds.

Finally, it is easy to verify by (2.16) that for $\lambda > a$, $1 \le k \le n - 1$,

$$\int_0^\infty e^{-\lambda t} S_0(t) u \, dt = \int_0^\infty e^{-\lambda t} \left(u - \int_0^t S_{n-1}(s) A_0 u \, ds \right) dt, \quad u \in \mathcal{D}(A_0),$$

$$\int_0^\infty e^{-\lambda t} S_k'(t) u \, dt = \int_0^\infty e^{-\lambda t} \left[S_{k-1}(t) - S_{n-1}(t) A_k \right] u \, dt, \quad u \in \bigcap_{i=0}^{k} \mathcal{D}(A_i).$$

Therefore, for $1 \le k \le n - 1$,

$$S_0'(t)u = -S_{n-1}(t)A_0 u, \quad t \ge 0, \ u \in \mathcal{D}(A_0). \tag{2.19}$$

$$S_k'(t)u = S_{k-1}(t)u - S_{n-1}(t)A_k u, \quad t \ge 0, \ u \in \bigcap_{i=0}^{k} \mathcal{D}(A_i). \tag{2.20}$$

Let $w(t)$ be an arbitrary solution of (ACP_n). We take a sequence $\{b_m(\cdot)\}_{m \in N}$ of nonnegative and infinitely differentiable functions such that the support of $b_m(\cdot)$ is contained in $(0, \frac{1}{m})$ and

$$\int_0^{\frac{1}{m}} b_m(s) ds = 1.$$

Fixing $k \in \{1, 2, \cdots, n-1\}$, define

$$\phi_m(t) = \int_0^{\frac{1}{m}} b_m(s) w^{(k)}(t + s) ds, \quad m \in N, \ t \ge 0.$$

Then $\phi_m(t) \in \mathcal{D}(A_k)$ $(m \in N, \ t \ge 0)$, and for any $t \ge 0$,

$$\phi_m(t) \to w^{(k)}(t), \quad A_k \phi_m(t) \to A_k w^{(k)}(t), \quad \text{as } m \to \infty. \tag{2.21}$$

Integrating by parts, we obtain that for every $t \geq 0$, $m \in N$,

$$\phi_m(t) = (-1)^{k-j} \int_0^{\frac{1}{m}} b_m^{(k-j)}(s) w^{(j)}(t+s) ds \in \mathcal{D}(A_j), \quad 0 \leq j \leq k;$$

hence, $\phi_m(t) \in \bigcap_{i=0}^k \mathcal{D}(A_i)$. Thus using (2.21) shows that (2.20) holds for $u = w^{(k)}(t)$ $(t \geq 0)$. From this and (2.19), we get that for $0 \leq s \leq t$,

$$\frac{d}{ds}\left[\sum_{k=0}^{n-1} S_k(t-s) w^{(k)}(s)\right]$$

$$= -\sum_{k=1}^{n-1} S_{k-1}(t-s) w^{(k)}(s) + \sum_{k=0}^{n-1} S_{n-1}(t-s) A_k w^{(k)}(s)$$

$$+ \sum_{k=0}^{n-1} S_k(t-s) w^{(k+1)}(s)$$

$$= \sum_{k=0}^{n-1} S_{n-1}(t-s) A_k w^{(k)}(s) + S_{n-1}(t-s) w^{(n)}(s)$$

$$= 0.$$

Accordingly,

$$w(t) = \sum_{k=0}^{n-1} S_k(t) w^{(k)}(0), \quad t \geq 0.$$

Consequently, (ACP_n) is strongly wellposed. The proof of sufficiency is then complete.

By (2.16) and Theorems 1.1.5, 1.1.7 and 1.1.8, we obtain the explicit expressions for the propagators of (ACP_n) as stated in the theorem. This ends the proof of Theorem 2.2.

Remark 2.5. From the above proof we know that, if condition (i) in Theorem 2.2 holds, then for any $\lambda \in \mathbb{C}$ with $\mathrm{Re}\lambda > a$, $R_\lambda \in L(E)$ and for $u \in E$

$$\lambda^{k-1} A_k R_\lambda u = \int_0^\infty e^{-\lambda t} A_k S_{n-1}^{(k-1)}(t) u \, dt, \quad 1 \leq k \leq n,$$

$$\overline{R_\lambda A_0} u = u - \lambda \int_0^\infty e^{-\lambda t} S_0(t) u \, dt,$$

$$\lambda^{k-1} \overline{R_\lambda A_k} u = \int_0^\infty e^{-\lambda t} \left[S_{k-1}^{(k-1)}(t) - S_k^{(k)}(t) \right] u \, dt, \quad 1 \leq k \leq n-1.$$

Remark 2.6. In general the premise that the operator $R_\lambda A_k$ has a continuous extension $\overline{R_\lambda A_k} \in L(E)$ does not imply that

$$\overline{R_\lambda A_k} u = R_\lambda A_k u, \quad \text{for all } u \in \mathcal{D}(R_\lambda A_k) = \mathcal{D}(A_k),$$

unless $R_\lambda A_k$ is assumed closable.

The following is a counterexample for the case of $n = 2$.

Example 2.7. Let E be a Hilbert space and A an unbounded, strictly positive self-adjoint operator in E. Let $\lambda > 0$. Take $u_0 \notin \mathcal{D}(A)$, $u_1 \in E$ such that

$$\left(\lambda^2 + \lambda A + \frac{1}{4}A^2\right)^{-1} u_1 \neq \overline{\left(\lambda^2 + \lambda A + \frac{1}{4}A^2\right)^{-1}} Au_0. \qquad (2.22)$$

Define a linear operator B by

$$\mathcal{D}(B) = \{au_0 + v; \quad a \in \mathbf{C}, \ v \in \mathcal{D}(A)\},$$

$$B(au_0 + v) = au_1 + Av, \quad a \in \mathbf{C}, \ v \in \mathcal{D}(A).$$

Clearly, $Bu = Au$ for all $u \in \mathcal{D}(A)$, $Bu_0 = u_1$ and

$$\left(\lambda^2 + \lambda B + \frac{1}{4}A^2\right)^{-1} = \left(\lambda^2 + \lambda A + \frac{1}{4}A^2\right)^{-1}.$$

Hence we get by (2.22) that

$$\overline{\left(\lambda^2 + \lambda B + \frac{1}{4}A^2\right)^{-1}} Au_0 \neq \left(\lambda^2 + \lambda B + \frac{1}{4}A^2\right)^{-1} Bu_0.$$

It remains to show that B is a closed operator. To this end, let

$$a_m u_0 + v_m \to z_0, \quad a_m u_1 + Av_m \to z.$$

Then

$$a_m \left(A^{-1}u_1 - u_0\right) \to A^{-1}z - z_0.$$

So $\sup\{|a_m|, \ m \geq 1\} < \infty$. Thus $\{a_m\}$ has a subsequence $\{a_{m_k}\}$ such that $a_{m_k} \to a$. It follows that

$$v_{m_k} \to z_0 - au_0, \quad Av_{m_k} \to z - au_1.$$

By the closedness of A, we obtain

$$z_0 - au_0 \in \mathcal{D}(A), \quad A(z_0 - au_0) = z - au_1.$$

That is $z_0 \in \mathcal{D}(B)$ and $Bz_0 = z$. Accordingly, B is a closed operator.

2.3 Solvability

Let D be a dense subspace of E^n. The (ACP_n) is called solvable for D if it has a unique solution for every initial value $(u_0, \cdots, u_{n-1}) \in D$.

Lemma 3.1. *Let $A_0, \cdots, A_{n-1}$ be closed linear operators in E. Let R_λ exist and $\mathcal{D}(R_\lambda) = E$ for λ sufficiently large. Suppose that there exists $\delta > 0$ such that*

$$\lim_{\lambda \to \infty} e^{-\delta\lambda} p(R_\lambda u) = 0, \quad u \in E, \ p \in \Gamma.$$

Then (ACP_n) has at most one solution for every initial value $(u_0, \cdots, u_{n-1})$.

Proof. Let $u(t)$ be a solution of (ACP_n) with $u_k = 0$, $0 \le k \le n - 1$. Then integrating by parts, we have that for $t \ge 0$, λ sufficiently large,

$$\lambda^k \int_0^t e^{\lambda(t-s)} u(s)ds$$

$$= -\sum_{j=0}^{k-1} \lambda^{k-j-1} u^{(j)}(t) + \int_0^t e^{\lambda(t-s)} u^{(k)}(s)ds, \quad 1 \le k \le n.$$

Therefore

$$P_\lambda \int_0^t e^{\lambda(t-s)} u(s)ds$$

$$= \sum_{k=1}^{n} A_k \lambda^k \int_0^t e^{\lambda(t-s)} u(s)ds + A_0 \int_0^t e^{\lambda(t-s)} u(s)ds$$

$$= -\sum_{k=1}^{n}\sum_{j=0}^{k-1} \lambda^{k-j-1} A_k u^{(j)}(t) + \int_0^t e^{\lambda(t-s)} \left(u^{(n)}(s) + \sum_{i=0}^{n-1} A_i u^{(i)}(s) \right) ds$$

$$= -\sum_{k=1}^{n}\sum_{j=0}^{k-1} \lambda^{k-j-1} A_k u^{(j)}(t).$$

Thus, taking $\sigma > \delta$ we have that for every $t \ge 0$, $p \in \Gamma$,

$$\lim_{\lambda \to \infty} e^{-\sigma\lambda} p\left(\int_0^t e^{\lambda(t-s)} u(s)ds \right)$$

$$= -\lim_{\lambda \to \infty} \sum_{k=1}^{n}\sum_{j=0}^{k-1} \left(\lambda^{k-j-1} e^{-(\sigma-\delta)\lambda} \right) \left[e^{-\delta\lambda} p \left(R_\lambda A_k u^{(j)}(t) \right) \right]$$

$$= 0.$$

Consequently, noting

$$\int_{t-\sigma}^t e^{\lambda(t-\sigma-s)} ds \longrightarrow 0$$

as $\lambda \to \infty$, we obtain

$$\lim_{\lambda \to \infty} \int_0^{t-\sigma} e^{\lambda(t-\sigma-s)} u(s) ds = 0, \quad t \geq \sigma. \tag{3.1}$$

Now recall the following result (cf. Pazy [2, Lemma 4.1.1]) which states:

Let $u(\cdot) \in C([0, T_0], E)$ *for some* $T_0 > 0$, *then*

$$\left\| \int_0^{T_0} e^{ms} u(s) ds \right\| \leq \text{const}, \quad m \in N$$

implies $u(t) \equiv 0$ *on* $[0, T_0]$.

In view of this result, it follows from (3.1) that $u(\tau) = 0$, for $0 \leq \tau \leq t - \sigma$. Since t and σ were arbitrary, $u(t) \equiv 0$ for $t \geq 0$. This leads to the conclusion.

Theorem 3.2. *Let* $A_0, \cdots, A_{n-1}$ *be closed linear operators in* E *such that* $\bigcap_{i=0}^{n-1} \mathcal{D}(A_i)$ *is dense in* E. *Assume that there exists a constant* $a > 0$ *such that* $R_\lambda \in L(E)$ *for* $\lambda > a$, *and for every* $p \in \Gamma$ *there exists a continuous seminorm* q *such that for any* $1 \leq k \leq n-1$, $u \in E$, $\lambda > a$, $m \in N_0$,

$$p\left\{ \left[\lambda^{n-1} R_\lambda u \right]^{(m)} \right\}, \quad p\left\{ \left[\lambda^{k-1} A_k R_\lambda u \right]^{(m)} \right\} \leq m!(\lambda - a)^{-m-1} q(u).$$

Then for every $u_k \in \bigcap_{i=0}^k \mathcal{D}(A_i)$ $(0 \leq k \leq n-1)$, (ACP_n) *has a unique solution.*

Proof. From the proof (the first part prior to the equality (2.16)) of Theorem 2.2, we get the existence of solutions. The uniqueness is guaranteed by Lemma 3.1.

2.4   Perturbation

Throughout this section, we assume that E is a Banach space. Letting $B_0, \cdots, B_{n-1}$ be linear operators in E, consider the perturbed (ACP_n)

$$\begin{cases} u^{(n)}(t) + \displaystyle\sum_{i=0}^{n-1}(A_i + B_i)u^{(i)}(t) = 0, \quad t \geq 0, \\[2mm] u^{(k)}(0) = u_k, \quad 0 \leq k \leq n-1. \end{cases} \qquad (ACP_n)_{[B_{n-1}, \cdots, B_0]}$$

Definition 4.1. By a solution of $(ACP_n)_{[B_{n-1}, \cdots, B_0]}$ we mean a function $u(\cdot) \in C^n(R^+, E)$ such that for $0 \leq i \leq n-1$, $t \geq 0$, $u^{(i)}(t) \in \mathcal{D}(A_i) \bigcap \mathcal{D}(B_i)$, and $A_i u^{(i)}(\cdot)$, $B_i u^{(i)}(\cdot) \in C(R^+, E)$, satisfying $(ACP_n)_{[B_{n-1}, \cdots, B_0]}$.

The wellposedness and propagators $S_k(\cdot)$ ($0 \le k \le n-1$) of $(ACP_n)_{[B_{n-1},\cdots,B_0]}$ are defined in the same way as in Section 2.1.

$(ACP_n)_{[B_{n-1},\,\cdots,\,B_0]}$ is called strongly wellposed if it is wellposed and for each $u \in E$, $1 \le k \le n-1$, $S_k(\cdot)u \in C^k(R^+, E)$, $S_{n-1}^{(k-1)}(t)u \in \mathcal{D}(A_k) \bigcap \mathcal{D}(B_k)$ (for all $t \ge 0$) and $A_k S_{n-1}^{(k-1)}(\cdot)u$, $B_k S_{n-1}^{(k-1)}(\cdot)u \in C(R^+, E)$.

Clearly, the above definition of solution is slightly stronger compared with Definition 1.1 if the A_i there is replaced by $A_i + B_i$; the two definitions coincide in the case when for each $0 \le k \le n-1$, either A_k or B_k is bounded. The same comment applies to the definition of strong wellposedness.

The arguments similar to those in the proof of Theorem 1.6 (with $\overline{A_k} + \overline{B_k}$ in place of A_k, $0 \le k \le n-1$, everywhere) show that, letting A_i, B_i ($0 \le i \le n-1$) be closable operators in E such that $(ACP_n)_{[B_{n-1},\,\cdots,\,B_0]}$ is strongly wellposed, then for $t \ge 0$,

$$\left\| S_k^{(k)}(t) \right\| \le C e^{\omega t}, \quad 0 \le k \le n-1,$$

$$\left\| A_k S_{n-1}^{(k-1)}(t) \right\|, \ \left\| B_k S_{n-1}^{(k-1)}(t) \right\| \le C e^{\omega t}, \quad 1 \le k \le n-1,$$

for suitable constants C, ω.

Theorem 4.2. *Let* $A_0, \cdots, A_{n-1}$ *be closed linear operators in* E *and let* $\bigcap_{i=0}^{n-1} \mathcal{D}(A_i)$ *be dense in* E. *Assume*
  (i) *There exist* C, $a > 0$ *such that* $R_\lambda \in \mathbf{L}(E)$ *for* $\lambda > a$, *and*

$$\left\| [\lambda^{n-1} R_\lambda]^{(m)} \right\|, \ \left\| [\lambda^{k-1} A_k R_\lambda]^{(m)} \right\| \le C m! (\lambda - a)^{-m-1},$$

$$1 \le k \le n-1, \quad m \in N_0, \quad \lambda > a.$$

  (ii) $B_0, \cdots, B_{n-1}$ *are closable linear operators in* E *satisfying that for each* $0 \le k \le n-1$ *there is an* i_k *with* $k+1 \le i_k \le n$ *such that* $\mathcal{D}(B_k) \supset \mathcal{D}(A_{i_k})$ *and* $\rho(A_{i_k})$ *is nonempty.*
Then for any

$$u_0 \in \mathcal{D}(A_0 + B_0), \quad \cdots, \quad u_{n-2} \in \bigcap_{i=0}^{n-2} \mathcal{D}(A_i + B_i), \quad u_{n-1} \in \bigcap_{i=0}^{n-1} \mathcal{D}(A_i),$$

the Cauchy problem $(ACP_n)_{[B_{n-1},\,\cdots,\,B_0]}$ *has a unique solution.*

Proof. From the proof of Theorem 2.2, we know that condition (i) implies

$$\lambda \longmapsto \lambda^{n-1} R_\lambda, \quad \lambda \longmapsto \lambda^{k-1} A_k R_\lambda \in LT - \mathbf{L}(E), \quad 0 \le k \le n-1. \quad (4.1)$$

By condition (ii), we get that

$$A_{i_{n-1}} = A_n = I, \quad \text{i.e. } B_{n-1} \in \mathbf{L}(E),$$

$$B_k \left(\lambda_k - A_{i_k}\right)^{-1} \in L(E), \quad 0 \le k \le n-2, \tag{4.2}$$

where $\lambda_k \in \rho(A_{i_k})$. Hence, for each $0 \le k \le n-1$, $0 \le j \le k$,

$$\lambda \longmapsto \lambda^j B_k R_\lambda \in LT - L(E),$$

noting

$$B_k R_\lambda = B_k \left(\lambda_k - A_{i_k}\right)^{-1} \left(\lambda_k R_\lambda - A_{i_k} R_\lambda\right), \quad 0 \le k \le n-1.$$

Consequently, for any $0 \le k \le n-1$,

$$\lambda \longmapsto \lambda^{k-1} A_k R_\lambda, \quad \lambda \longmapsto \lambda^{k-1} B_k R_\lambda \in LT - L(E),$$

$$\lambda \longmapsto \lambda^{n-1} R_\lambda \in LT - L(E),$$

$$\lambda \longmapsto \sum_{k=0}^{n-1} \lambda^k B_k R_\lambda \in LT - L(E). \tag{4.3}$$

Define

$$\widetilde{P}_\lambda = \lambda^n + \sum_{i=0}^{n-1} (A_i + B_i)\lambda^i, \quad \lambda \in \mathbf{C}, \tag{4.4}$$

then for λ sufficiently large, $\widetilde{P}_\lambda^{-1}$ exists and

$$\widetilde{R}_\lambda := \widetilde{P}_\lambda^{-1} = R_\lambda \left(I + \sum_{k=0}^{n-1} \lambda^k B_k R_\lambda\right)^{-1} \in L(E). \tag{4.5}$$

From (4.3), we can apply Theorem 1.1.11 to conclude that

$$\begin{cases} \lambda \longmapsto \lambda^{k-1} A_k \widetilde{R}_\lambda, \quad \lambda \longmapsto \lambda^{k-1} B_k \widetilde{R}_\lambda \in LT - L(E), \quad 0 \le k \le n-1, \\ \lambda \longmapsto \lambda^{n-1} \widetilde{R}_\lambda \in LT - L(E). \end{cases} \tag{4.6}$$

Accordingly, for $0 \le j \le n-1$, $0 \le k \le n-2$,

$$\lambda \longmapsto \lambda^{j+i-k-1} A_j \widetilde{R}_\lambda (A_i + B_i)u \in LT - E,$$

$$0 \le i \le k, \; u \in \bigcap_{i=0}^{k} \mathcal{D}(A_i + B_i),$$

$$\lambda \longmapsto \lambda^j A_j \widetilde{R}_\lambda u \in LT - E, \quad u \in \bigcap_{i=0}^{n-1} \mathcal{D}(A_i + B_i) = \bigcap_{i=0}^{n-1} \mathcal{D}(A_i).$$

Thus, using (4.2) we obtain the result as claimed by an application of the following Lemmas 4.3 and 4.4.

Lemma 4.3. *Let $A_0, \cdots, A_{n-1}$ be closed linear operators in E, $B_0, \cdots, B_{n-1}$ closable linear operators in E such that*

$$\mathcal{D}(B_k) \supset \bigcap_{i=0}^{n-1} \mathcal{D}(A_i), \quad 0 \leq k \leq n-1.$$

Define $\widetilde{P}_\lambda$, $\widetilde{R}_\lambda$ as in (4.4) and (4.5). Suppose that $\widetilde{R}_\lambda$ exists and $\mathcal{D}\left(\widetilde{R}_\lambda\right) = E$ for λ sufficiently large, and there is $\delta > 0$ such that

$$\lim_{\lambda \to \infty} e^{-\delta\lambda} \widetilde{R}_\lambda u = 0,$$

for each $u \in E$. Then $(ACP_n)_{[B_{n-1}, \cdots, B_0]}$ has at most one solution for every initial value $(u_0, \cdots, u_{n-1})$.

Proof. Argue similarly as in the proof of Lemma 3.1 and note

$$\widetilde{P}_\lambda := \lambda^n + \sum_{i=0}^{n-1} (A_i + \overline{B}_i).$$

Lemma 4.4. *Let A_k, $B_{0,k}$ $(0 \leq k \leq n-1)$ be closed linear operators in E with $0 \in \rho(B_{0,k})$ $(0 \leq k \leq n-1)$. Let $B_0, \cdots, B_{n-1}$ be closable linear operators in E such that*

$$\mathcal{D}(B_k) \supset \mathcal{D}(B_{0,k}) \quad (0 \leq k \leq n-1).$$

If $u_0, \cdots, u_{n-1}$ satisfy that for $0 \leq j \leq n-1$, $0 \leq k \leq n-2$,

$$\lambda \longmapsto \lambda^j A_j \sum_{i=0}^{k} \lambda^{i-k-1} \widetilde{R}_\lambda (A_i + B_i) u_k \in LT - E,$$

$$\lambda \longmapsto \lambda^j B_{0,j} \sum_{i=0}^{k} \lambda^{i-k-1} \widetilde{R}_\lambda (A_i + B_i) u_k \in LT - E,$$

$$\lambda \longmapsto \lambda^j A_j \widetilde{R}_\lambda u_{n-1}, \quad \lambda \longmapsto \lambda^j B_{0,j} \widetilde{R}_\lambda u_{n-1} \in LT - E,$$

then $(ACP_n)_{[B_{n-1}, \cdots, B_0]}$ has a solution.

Proof. We proceed in the same way as in the proof of Lemma 2.1, and make the observation as below. Let $v_j(\cdot)$ be the determining function of $\lambda \longmapsto \lambda^j \widetilde{R}_\lambda u_{n-1}$. Since

$$\lambda \longmapsto \lambda^j B_{0,j} \widetilde{R}_\lambda u_{n-1} \in LT - E$$

by hypothesis, we have by Theorem 1.1.10 that $B_{0,j} v_j(\cdot) \in C(R^+, E)$. Therefore $B_j v_j(\cdot) \in C(R^+, E)$, due to

$$B_j = \left(B_j B_{0,j}^{-1}\right) B_{0,j} \quad \text{and} \quad B_j B_{0,j}^{-1} \in L(E).$$

This ends the proof.

An examination of the steps of the proof of Theorem 4.2 shows the following result.

Theorem 4.5. *Let $A_0, \cdots, A_{n-1}$ be as in Theorem 4.2. Let $B_0, \cdots, B_{n-1}$ be closed linear operators in E. Assume that for each $0 \leq k \leq n-1$, $\mathcal{D}(B_k) \supset \bigcap_{i=0}^{n-1} \mathcal{D}(A_i)$ and*

$$\lambda \longmapsto \lambda^k B_k R_\lambda \in LT - \mathbf{L}(E).$$

Then the conclusion of Theorem 4.2 holds.

Theorem 4.6. *Let $A_0, \cdots, A_{n-1}$ be closed linear operators in E such that (ACP_n) is strongly wellposed and $R_\lambda A_k$ is closable for $1 \leq k \leq n-1$ and λ large enough. Assume that $B_0, \cdots, B_{n-1}$ are closable linear operators in E satisfying that for each $0 \leq k \leq n-1$, there is an i_k with $k+1 \leq i_k \leq n$ such that $\mathcal{D}(B_k) \supset \mathcal{D}(A_{i_k})$; moreover there exists $\lambda_k \in \rho(A_{i_k})$ such that $(\lambda_k - A_{i_k})^{-1} B_k$ has a bounded extension to E. Then $(ACP_n)_{[B_{n-1}, \cdots, B_0]}$ is strongly wellposed.*

Proof. From the proof of Theorem 2.2, we see

$$
\begin{cases}
\lambda \longmapsto \lambda^{n-1} R_\lambda \in LT - \mathbf{L}(E), \\[2mm]
\lambda \longmapsto \lambda^{k-1} A_k R_\lambda, \quad \lambda \longmapsto \lambda^{k-1} \overline{R_\lambda A_k} \in LT - \mathbf{L}(E), \\[2mm]
\hspace{4cm} 0 \leq k \leq n-1.
\end{cases}
\tag{4.7}
$$

Accordingly, defining $\widetilde{P}_\lambda$, $\widetilde{R}_\lambda$ as in (4.4), (4.5), we then get (4.6).

On the other hand, we have by hypothesis that for each $0 \leq k \leq n-1$, λ sufficiently large,

$$R_\lambda B_k = \left(\lambda_k R_\lambda - \overline{R_\lambda A_{i_k}}\right)\left(\lambda_k - A_{i_k}\right)^{-1} B_k$$

is closable, and so by (4.7)

$$\lambda \longmapsto \lambda^j \overline{R_\lambda B_k} \in LT - \mathbf{L}(E), \quad 0 \leq j \leq k.$$

Therefore for any $0 \leq k \leq n-1$,

$$\lambda \longmapsto \lambda^{k-1} \overline{R_\lambda(A_k + B_k)}, \quad \lambda \longmapsto \sum_{k=0}^{n-1} \lambda^k \overline{R_\lambda B_k} \in LT - \mathbf{L}(E). \tag{4.8}$$

Observe

$$\widetilde{R}_\lambda = \left(I + \sum_{k=0}^{n-1} \lambda^k \overline{R_\lambda B_k}\right)^{-1} R_\lambda.$$

We discover that for each $0 \leq k \leq n - 1$, λ sufficiently large, $\widetilde{R}_\lambda(A_k + B_k)$ is closable and by (4.8) and Theorem 1.1.11,

$$\lambda \longmapsto \lambda^{k-1}\overline{\widetilde{R}_\lambda(A_k + B_k)} \in LT - \mathbf{L}(E).$$

This together with (4.6) leads to the desired result by Lemma 4.4 and by an argument similar to the latter part of the proof of Theorem 2.2.

The arguments in the proof of Theorem 4.6 also implies the following theorem.

Theorem 4.7. *Let $A_0, \cdots, A_{n-1}$ be closed linear operators in E such that (ACP_n) is strongly wellposed. Suppose that $B_0, \cdots, B_{n-1}$ are closed linear operators in E such that for each $0 \leq k \leq n - 1$, $\mathcal{D}(B_k) \supset \bigcap_{i=0}^{n-1} \mathcal{D}(A_i)$ and*

$$\lambda \longmapsto \lambda^k B_k R_\lambda \in LT - \mathbf{L}(E),$$

$$\lambda \longmapsto \lambda^k \overline{R_\lambda \mathcal{B}_k} \in LT - \mathbf{L}(E).$$

where $\mathcal{B}_k$ denotes the restriction of B_k on $\bigcap_{i=0}^k \mathcal{D}(B_i)$. Then $(ACP_n)_{[B_{n-1},\cdots,B_0]}$ is strongly wellposed.

Corollary 4.8. *Let $n \geq 2$, $-A_{n-1}$ be the generator of a strongly continuous operator semigroup on E. Let $A_0, \cdots, A_{n-2}$ be closable linear operators in E such that $\mathcal{D}(A_k) \supset \mathcal{D}(A_{n-1})$ $(0 \leq k \leq n - 2)$. Then for every*

$$u_{n-1} \in \mathcal{D}(A_{n-1}), \quad u_k \in \bigcap_{i=0}^k \mathcal{D}(A_i) \ (0 \leq k \leq n - 2),$$

(ACP_n) has a unique solution. If in addition, there exists $\lambda_k \in \rho(A_{n-1})$ such that $(\lambda_k - A_{n-1})^{-1}A_k$ has a bounded extension to E for any $0 \leq k \leq n-2$, then (ACP_n) is strongly wellposed.

Proof. By hypothesis, we have

$$\lambda \longmapsto (\lambda + A_{n-1})^{-1} \in LT - \mathbf{L}(E).$$

Noting that

$$(\lambda + A_{n-1})^{-1} = \lambda^{-1} - \lambda^{-1}A_{n-1}(\lambda + A_{n-1})^{-1},$$

we get that, as functions of λ,

$$\lambda^{n-1}\left(\lambda^n + \lambda^{n-1}A_{n-1}\right)^{-1} \in LT - \mathbf{L}(E),$$

$$\lambda^{n-2}A_{n-1}\left(\lambda^n + \lambda^{n-1}A_{n-1}\right)^{-1} \in LT - \mathbf{L}(E).$$

Thus, the Cauchy problem

$$\begin{cases} u^{(n)}(t) + A_{n-1}u^{(n-1)}(t) = 0, \quad t \geq 0 \\ u^{(k)}(0) = u_k, \quad 0 \leq k \leq n-1, \end{cases}$$

is strongly wellposed. Making use of Theorems 4.2 and 4.6, we obtain the conclusion.

Corollary 4.9. *Let $n \geq 3$, $A_{n-1} \in L(E)$ and $-A_{n-2}$ be the generator of a strongly continuous cosine operator function on E. Let $A_0, \cdots, A_{n-3}$ be closable linear operators in E such that*

$$\mathcal{D}(A_k) \supset \mathcal{D}(A_{n-2}) \quad (0 \leq k \leq n-3).$$

Then for any

$$u_{n-1}, \quad u_{n-2} \in \mathcal{D}(A_{n-2}), \quad u_k \in \bigcap_{i=0}^{k} \mathcal{D}(A_i) \quad (0 \leq k \leq n-3),$$

(ACP_n) has a unique solution. If in addition, there exists $\lambda_k \in \rho(A_{n-2})$ such that $(\lambda_k - A_{n-2})^{-1}A_k$ has a bounded extension to E for any $0 \leq k \leq n-3$, then (ACP_n) is strongly wellposed.

Proof. By hypothesis, we have

$$\lambda \longmapsto \lambda \left(\lambda^2 + A_{n-2}\right)^{-1} \in LT - L(E).$$

Thus, it is easy to verify that the Cauchy problem

$$\begin{cases} u^{(n)}(t) + A_{n-2}u^{(n-2)}(t) = 0, \quad t \geq 0, \\ u^{(k)}(0) = u_k, \quad 0 \leq k \leq n-1, \end{cases}$$

is strongly wellposed. According to Theorems 4.2 and 4.6, we obtain the conclusion.

2.5 Two typical cases

Let A_0, A_1 be closed, densely defined linear operators in a Banach space E. We look at the complete second order Cauchy problem

$$\begin{cases} u''(t) + A_1 u'(t) + A_0 u(t) = 0, \quad t \geq 0, \\ u(0) = u_0, \quad u'(0) = u_1, \end{cases} \qquad (ACP_2)$$

and focus our attention on two typical cases: $A_1 \in L(E)$ and $A_0 \in L(E)$.

Clearly, P_λ, R_λ take the form

$$P_\lambda = \lambda^2 + \lambda A_1 + A_0, \quad R_\lambda = (\lambda^2 + \lambda A_1 + A_0)^{-1}. \tag{5.1}$$

Theorem 5.1. *Assume $A_1 \in \mathbf{L}(E)$. Then the following statements are equivalent.*

(i) (ACP_2) *is strongly wellposed.*
(ii) (ACP_2) *is wellposed.*
(iii) $-A_0$ *is the generator of a strongly continuous cosine function.*

Proof. By virtue of Theorem 1.5, (i) and (ii) are equivalent. The implication (iii) $\implies$ (i) is given by Corollary 4.9. To see that (i) implies (iii), we appeal to Theorem 4.6 and infer that the Cauchy problem for

$$u''(t) + A_0 u(t) = 0, \quad t \geq 0$$

is strongly wellposed; this establishes the result as desired by applying Theorem 2.3 and Theorem 1.4.4 (with $r = 0$, $C = I$).

Theorem 5.2. *Assume $A_0 \in \mathbf{L}(E)$. Then the following statements are equivalent.*

(i) (ACP_2) *is strongly wellposed.*
(ii) (ACP_2) *is wellposed; there exist constants M, $\omega > 0$ and a norm $\| \cdot \|_0$ on E, which is equivalent to the original norm $\| \cdot \|$ on E, such that*

$$\|S_0(t)\| \leq M e^{\omega t}, \quad \|S_1(t)\|_0 \leq t e^{\omega t}, \quad t \geq 0. \tag{5.2}$$

(iii) $-A_1$ *is the generator of a strongly continuous semigroup.*

Proof. The implication (iii) $\implies$ (i) follows immediately from Corollary 4.8.

For the converse, we make use of Theorem 4.6 again and obtain that, assuming (i) then the Cauchy problem for

$$u''(t) + A_1 u'(t) = 0, \quad t \geq 0$$

is wellposed, and so is the Cauchy problem for

$$u'(t) + A_1 u(t) = 0, \quad t \geq 0,$$

which gives (iii).

(iii) $\implies$ (ii). By the equivalence of (i) and (iii) we just proved, (ACP_2) is wellposed and there are M_0, $\omega_0 > 0$ such that

$$\|S_0(t)\|, \quad \|S_1'(t)\| \leq M_0 e^{\omega_0 t}, \quad t \geq 0,$$

$$\lambda R_\lambda u = \int_0^\infty e^{-\lambda t} S_1'(t) u \, dt, \quad \lambda > \omega_0, \, u \in E. \tag{5.3}$$

On the other hand, it is known (cf. Pazy [2, Lemma 5.1, p. 17 and Theorem 8.3, p. 33]) that there is a constant $\omega_1 > 0$ and a norm $\| \cdot \|_0$ on E, which is equivalent to the original norm $\| \cdot \|$ on E, such that

$$\|T(t)\|_0 \leq e^{\omega_1 t}, \quad t \geq 0, \tag{5.4}$$

where $T(t)$ denotes the semigroup generated by $-A_1$. Observe that for λ large enough,

$$\lambda R_\lambda = \left[I + \lambda^{-1} (\lambda + A_1)^{-1} A_0 \right]^{-1} (\lambda + A_1)^{-1}$$

and for any $u \in E$,

$$(\lambda + A_1)^{-1} u = \int_0^\infty e^{-\lambda t} T(t) u \, dt,$$

$$\lambda^{-1} (\lambda + A_1)^{-1} A_0 u = \int_0^\infty e^{-\lambda t} \left(\int_0^t T(s) A_0 u \, ds \right) dt,$$

$$\left\| \int_0^t T(s) A_0 u \, ds \right\|_0 \leq \omega_1^{-1} e^{\omega_1 t} \|A_0\|_0 \|u\|_0, \quad t \geq 0.$$

We obtain, using (1.1.2) with $C_1 = 1$, $C_0 = \omega_1^{-1} \|A_0\|_0$, $a = \omega_1$ and noting (5.3), (5.4), that

$$\|S_1'(t)\|_0 \leq e^{(\omega_1^{-1} \|A_0\|_0 + \omega_1) t}, \quad t \geq 0.$$

Accordingly

$$\begin{aligned} \|S_1(t)\|_0 &\leq \int_0^t e^{(\omega_1^{-1} \|A_0\|_0 + \omega_1) s} \, ds \\ &= t e^{(\omega_1^{-1} \|A_0\|_0 + \omega_1) t}, \quad t \geq 0. \end{aligned}$$

as desired.

(ii) $\Longrightarrow$ (iii). Obviously, P_λ is closed and densely defined for each $\lambda \in \mathbb{C}$. It is easy to verify that, if $u \in D_1$ (see Definition 1.2),

$$P_\lambda \int_0^\infty e^{-\lambda t} S_1(t) u \, dt = u,$$

for $\lambda > \omega$. This equality holds actually for all $u \in E$, since D_1 is dense and P_λ is closed. Thus we see that P_λ is surjective. To show that it is one-to-one, let $P_{\lambda_0} u_0 = 0$, $\lambda_0 > \omega$, $u_0 \in \mathcal{D}(A_0) \bigcap \mathcal{D}(A_1)$. Then

$$e^{\lambda_0 t} u_0 = S_0(t) u_0 + \lambda_0 S_1(t) u_0, \quad t \geq 0.$$

Whence we obtain that

$$e^{\lambda_0 t} \|u_0\| \leq \text{const } e^{\omega t} (\|u_0\| + |\lambda_0| t \|u_0\|_0), \quad t \geq 0,$$

which implies that $u_0 = 0$. Consequently,

$$R_\lambda u = \int_0^\infty e^{-\lambda t} S_1(t) u dt, \quad u \in E, \ \lambda > \omega. \tag{5.5}$$

Observe by (5.2) and (5.5) that for $u \in E$, $\lambda > 2\omega$,

$$\|A_0 R_\lambda u\|_0 \ \leq \ \int_0^\infty \|A_0\|_0 e^{-\lambda t} t e^{\omega t} \|u\|_0 dt$$

$$\leq \ \int_0^\infty \omega^{-1} \|A_0\|_0 e^{-(\lambda - 2\omega)t} \|u\|_0 dt$$

$$\leq \ \omega^{-1} \|A_0\|_0 (\lambda - 2\omega)^{-1} \|u\|_0,$$

$$\|\lambda R_\lambda u\|_0 \ \leq \ \int_0^\infty \lambda e^{-\lambda t} t e^{\omega t} \|u\|_0 dt$$

$$= \ \int_0^\infty e^{-\lambda t} \left(e^{\omega t} + \omega t e^{\omega t} \right) \|u\|_0 dt$$

$$\leq \ \int_0^\infty e^{-(\lambda - 2\omega)t} \|u\|_0 dt$$

$$= \ (\lambda - 2\omega)^{-1} \|u\|_0.$$

We find that for $\lambda > 2\omega + \omega^{-1} \|A_0\|_0$, $[I - A_0 R_\lambda]^{-1} \in L(E)$ and

$$\left\| [I - A_0 R_\lambda]^{-1} \right\|_0 \ \leq \ \left[1 - \omega^{-1} \|A_0\|_0 (\lambda - 2\omega)^{-1} \right]^{-1}$$

$$= \ (\lambda - 2\omega) \left(\lambda - 2\omega - \omega^{-1} \|A_0\|_0 \right)^{-1},$$

and therefore

$$\left\| (\lambda I + A_1)^{-1} \right\|_0 \ = \ \left\| \lambda R_\lambda \left[I - A_0 R_\lambda \right]^{-1} \right\|_0$$

$$\leq \ (\lambda - 2\omega)^{-1} (\lambda - 2\omega) \left(\lambda - 2\omega - \omega^{-1} \|A_0\|_0 \right)^{-1}$$

$$= \ \left(\lambda - 2\omega - \omega^{-1} \|A_0\|_0 \right)^{-1}.$$

In conclusion, $-A_1$ generates a strongly continuous semigroup by Theorem 1.3.4 (with $r = 0$, $C = I$). This completes the proof of the theorem.

The following example shows that in the case of $A_0 \in L(E)$, the wellposedness of (ACP_2) is not equivalent to its strong wellposedness in general.

Example 5.3. Let $E = l^2$, the set of all sequences $v = \{v_m\}_{m \in N}$ of complex numbers with

$$\|\{v_m\}\| = \sum_{m=1}^{\infty} |v_m|^2 < \infty.$$

Let $A_0 = 0$ and A_1 the operator given by

$$A_1\{v_m\} = \{(-\ln(1 + \ln m) + mi)v_m\}.$$

It is not difficult to verify that the Cauchy problem for

$$u''(t) + A_1 u'(t) = 0 \quad (t \geq 0)$$

is wellposed, and the two propagators take the form

$$S_0(t) = I, \quad S_1(t)\{v_m\} = \left\{ \frac{(1 + \ln m)^t e^{-mit} - 1}{\ln(1 + \ln m) - mi} v_m \right\},$$

with

$$\|S_1(t)\| = \sup_{m \in N} \left| \frac{(1 + \ln m)^t e^{-mit} - 1}{\ln(1 + \ln m) - mi} \right| < \infty, \quad t \geq 0.$$

But $-A_1$ does not generate a strongly continuous semigroup, since

$$\sigma(-A_1) \supset \{\ln(1 + \ln m) + mi; \quad m \in N\}.$$

Thus in view of Theorem 5.2, the strong wellposedness fails to hold.

2.6 Notes

The notion of abstract Cauchy problem is due to Hille [5]. The notion of wellposed (ACP_n) in Banach spaces was introduced by Fattorini [3] (see also Chapter VIII of Fattorini [7]), along the line of Lax's definition for (uniformly) wellposed (ACP_1) (see Lax-Richtmyer [1]). This notion for certain special (ACP_n) in barrelled complete locally convex spaces also appeared in an earlier paper of Fattorini [1]. The formulation of the notion goes back to Hadamard [1].

The concept of strongly wellposed (ACP_n) was introduced by Xiao-Liang [2, 5, 12], motivated by the work of Fattorini [3, 7].

The characterization of the strong wellposedness for complete (ACP_2) in Banach spaces was discovered by Xiao-Liang [2]. The extension to (ACP_n) and to SCLCS further (Theorems 2.2 and 2.3) were given by Xiao-Liang [5, 12] and Liang-Xiao [7].

It is well known that the famous Hille-Yosida-Feller-Miyadera-Phillips theorem (as well as its versions in SCLCS, see the notes to Chapter 1) characterizes also the wellposed (ACP_1), since the wellposedness of (ACP_1) is equivalent to the coefficient operator $-A_0$ generating a strongly continuous semigroup. On the other hand, it is easy to see that strong wellposedness is equivalent to wellposedness for (ACP_1) or incomplete (ACP_2). Therefore, Theorems 2.2 and 2.3

are generalizations of the Hille-Yosida-Feller-Miyadera-Phillips theorem (as well as its versions in SCLCS). Moreover, Theorems 2.2 and 2.3 generalize the theorem for characterization of wellposed incomplete (ACP_2), or equivalently, the characterization of generators of strongly continuous cosine functions, which was proved independently by Sova [1], Da Prato-Giusti [1] and Fattorini [2] as mentioned in the notes to Chapter 1. The first paper showing the relation between strongly continuous cosine functions and the propagators of incomplete (ACP_2) is Fattorini [1].

Other types of sufficient and necessary conditions for (ACP_n) to be wellposed, in some sense or another, can be found in Sova [2] (for the case of $n = 2$ and the coefficient operators A_0, A_1 being biclosed), Mel'nikova-Filinkov [2] (for the case of $n = 2$ and the coefficient operators A_0, A_1 commuting with each other), Sandefur [1] (for iterated equations), Neubrander [2] (for the case of $\mathcal{D}(A_{n-1}) \subset \mathcal{D}(A_i)$ $(0 \leq i \leq n-2))$, and references cited therein.

Theorems 1.4, 1.5 and 1.6 are based on Fattorini [3]. Lemma 2.1 is due to Liang-Xiao [10]. Remark 2.6 and Example 2.7 are in Xiao-Liang [4]. Ljubic [1] established a significant criterion for uniqueness of solutions of (ACP_1). Lemma 3.1, coming from Liang-Xiao [10], is an extension of this uniqueness theorem. Theorem 3.2 is from Xiao-Liang [12], as well as most of Section 2.4. Corollary 4.8, which is due to Xiao-Liang [5, 12] and appeared firstly in Xiao-Liang [2] for $n = 2$, generalizes Theorem 11 in Neubrander [2]. Theorem 5.1 is taken from Xiao [1]. Theorem 5.2 comes from Xiao-Liang [3]; see also Neubrander [5, Theorem 4.1] for a related result.

The proofs of the main results in this chapter, as well as in the most part of the book, base on the idea to deal with (ACP_n) straightforwardly for getting our conclusions. As mentioned in the preface, direct treatment enables one to get to the essence of higher order problems and obtain sharper results than the single reduction method or semigroup method. Here we make this point more explicitly in connection with the subject of this chapter.

Naturally, one can convert (ACP_n) to the first order problem

$$y'(t) = \mathbf{N}_n y(t), \quad y(0) = (u_0, \cdots, u_{n-1}), \tag{6.1}$$

on E^n, where

$$\mathbf{N}_n := \begin{pmatrix} 0 & I & 0 & \cdots & 0 \\ 0 & 0 & I & \cdots & 0 \\ \vdots & \vdots & \vdots & \ddots & \vdots \\ 0 & 0 & 0 & \cdots & I \\ -A_0 & -A_1 & -A_2 & \cdots & -A_{n-1} \end{pmatrix},$$

$$\mathcal{D}(\mathbf{N}_n) := \mathcal{D}(A_0) \times \cdots \times \mathcal{D}(A_{n-1}).$$

However, the operator matrix $\mathbf{N}_n$ is in general not closed. Thus one may be puzzled about how to transform the solutions of $y'(t) = \overline{\mathbf{N}}_n y(t)$ into those of (ACP_n). Even if $\mathbf{N}_n$ is closed, it does not generate a strongly continuous semigroup unless under very restrictive conditions. In fact, we have

Theorem 6.1 (Xiao-Liang [4]). *Suppose that E is a Banach space and N_n generates a strongly continuous semigroup on E^n. Then*
(i) $A_0, \cdots, A_{n-2} \in L(E)$;
(ii) $-A_{n-1}$ *is the generator of a strongly continuous semigroup.*

Proof. By virtue of the Phillips perturbation theorem, we obtain that

$$
\tilde{N}_n := \begin{pmatrix} 0 & 0 & \cdots & 0 \\ 0 & 0 & \cdots & 0 \\ \vdots & \vdots & \ddots & \vdots \\ -A_0 & -A_1 & \cdots & -A_{n-1} \end{pmatrix},
$$

$$
\mathcal{D}\left(\tilde{N}_n\right) := \mathcal{D}(A_0) \times \cdots \times \mathcal{D}(A_{n-1})
$$

is the generator of a strongly continuous semigroup on E^n. So, for sufficiently large λ, the range of

$$
\lambda - \tilde{N}_n = \begin{pmatrix} \lambda & 0 & \cdots & 0 \\ 0 & \lambda & \cdots & 0 \\ \vdots & \vdots & \ddots & \vdots \\ \lambda + A_0 & \lambda + A_1 & \cdots & \lambda + A_{n-1} \end{pmatrix},
$$

is the whole space E^n, i.e., for any $(v_0, v_1, \cdots, v_{n-1}) \in E^n$, there exists $u_i \in \mathcal{D}(A_i)$ $(0 \le i \le n-1)$ such that

$$
\left(\lambda - \tilde{N}_n\right) \begin{pmatrix} u_0 \\ u_1 \\ \vdots \\ u_{n-1} \end{pmatrix} = \begin{pmatrix} v_0 \\ v_1 \\ \vdots \\ v_{n-1} \end{pmatrix}.
$$

Hence $\lambda u_i = v_i$ $(0 \le i \le n-2)$. By the arbitrariness of v_i $(0 \le i \le n-2)$, we deduce that $\mathcal{D}(A_i) = E$ $(0 \le i \le n-2)$, that is $A_i \in L(E)$ $(0 \le i \le n-2)$.
On the other hand,

$$
\left(\lambda - \tilde{N}_n\right)^{-1} = \begin{pmatrix} \lambda^{-1} & 0 & \cdots & 0 \\ & \lambda^{-1} & \cdots & 0 \\ & & \ddots & \vdots \\ \# & & & (\lambda + A_{n-1})^{-1} \end{pmatrix},
$$

and it follows that $-A_{n-1}$ is the generator of a strongly continuous semigroup.

Another typical first order system, with a closed operator matrix, is

$$
y'(t) = M_n y(t), \quad y(0) = y_0, \tag{6.2}
$$

on E^n, where

$$
\mathbf{M}_n := \begin{pmatrix} -A_{n-1} & I & 0 & \cdots & 0 \\ -A_{n-2} & 0 & I & \cdots & 0 \\ \vdots & \vdots & \vdots & \ddots & \vdots \\ -A_1 & 0 & 0 & \cdots & I \\ -A_0 & 0 & 0 & \cdots & 0 \end{pmatrix}, \quad \mathcal{D}(\mathbf{M}_n) := \left(\bigcap_{i=0}^{n-1} \mathcal{D}(A_i) \right) \times E^{n-1}.
$$

There arose an impressive result by Neubrander [2], with the aid of this system.

Theorem 6.2 (Neubrander [2]). *Let $-A_{n-1}$ be a densely defined linear operator in a Banach space E and let A_i $(0 \le i \le n-2)$ be closed linear operators in E with $\mathcal{D}(A_{n-1}) \subset \mathcal{D}(A_i)$ for $0 \le i \le n-2$. Then the following two statements are equivalent:*

(i) *$-A_{n-1}$ is the generator of a strongly continuous semigroup on E.*

(ii) *$\rho(A_{n-1}) \ne \emptyset$ and for every initial value $(u_0, \cdots, u_{n-1}) \in (\mathcal{D}(A_{n-1}))^n$, (ACP_n) has a unique solution in*

$$
C^n(R^+, E) \bigcap C^{n-1}(R^+, [\mathcal{D}(A_{n-1})]).
$$

The idea of the proof is to show by the Phillips perturbation theorem that each of (i) and (ii) is equivalent to

(iii) $\mathbf{M}_n$ *is a generator of a strongly continuous semigroup on E^n.*

It seems that Theorem 6.2 is the best possible result what one could obtain for (ACP_n) by virtue of system (6.2) with property (iii), since we have

Theorem 6.3 (Xiao-Liang [4]). *The property (iii) implies that the restriction $-\mathcal{A}_{n-1}$ of $-A_{n-1}$ on $\bigcap_{i=0}^{n-1} \mathcal{D}(A_i)$, generates a strongly continuous semigroup on E.*

Proof. It can be verified (see also Neubrander [2, Proposition 7]) that

$$
R(\lambda; \mathbf{M}_n) =
$$

$$
\begin{pmatrix} \lambda^{n-1} R_\lambda & \lambda^{n-2} R_\lambda & \cdots & \lambda R_\lambda & R_\lambda \\ Y_1 R_\lambda & \lambda^{n-2} X_1 R_\lambda & \cdots & \lambda X_1 R_\lambda & X_1 R_\lambda \\ \lambda Y_2 R_\lambda & Y_2 R_\lambda & \cdots & \lambda X_2 R_\lambda & X_2 R_\lambda \\ \vdots & \vdots & \ddots & \vdots & \vdots \\ \lambda^{n-2} Y_{n-1} R_\lambda & \lambda^{n-3} Y_{n-1} R_\lambda & \cdots & Y_{n-1} R_\lambda & X_{n-1} R_\lambda \end{pmatrix}, \tag{6.3}
$$

where

$$
X_k = \lambda^{k-1}(\lambda + A_{n-1}) + \sum_{i=2}^{k} \lambda^{k-i} A_{n-i}, \quad 1 \le k \le n-1,
$$

$$Y_k = -\sum_{i=0}^{n-k-1} \lambda^i A_i, \quad 1 \le k \le n-1.$$

By hypothesis,

$$\lambda \longmapsto R(\lambda; \mathbf{M}_n) \in LT - \mathbf{L}(E^n).$$

Hence,

$$Y_i R_\lambda \in LT - \mathbf{L}(E), \quad 1 \le i \le n-1.$$

Consequently,

$$A_0 R_\lambda \in LT - \mathbf{L}(E),$$

$$\lambda^j A_j R_\lambda = (Y_{n-j-1} - Y_{n-j}) R_\lambda \in LT - \mathbf{L}(E), \quad 1 \le j \le n-2.$$

Now, an application of Theorem 1.1.11 yields that

$$(\lambda + \mathring{A}_{n-1})^{-1} = \lambda^{n-1} R_\lambda \left(I - \sum_{i=0}^{n-2} \lambda^i A_i R_\lambda \right)^{-1} \in LT - \mathbf{L}(E),$$

and so $-\mathring{A}_{n-1}$ generates a strongly continuous semigroup.

However, the continuous dependence of solutions on initial data obtained in Corollary 4.8 can not be directly derived with this system unless the coefficient operators are assumed to commute mutually (see Neubrander [2, Theorem 11]).

In view of Theorems 6.1 — 6.3, to make use of the theory of strongly continuous semigroups in the case when $-A_{n-1}$ or $-\mathring{A}_{n-1}$ does not generate a strongly continuous semigroup on E, one has to look for some phase space $F_0 \subset E^n$ such that $\mathbf{M}_n$ (or $\mathbf{N}_n$) generates a strongly continuous semigroup on F_0. But such phase spaces are generally hard to construct. Let us go into some details about it. We examine in the following, for instance, the possibility of proving the sufficiency of Theorem 2.3 for $n = 2$ via the system (6.2). Firstly, one gets that for $\lambda > a$,

$$(\lambda - \mathbf{M}_2)^{-1} = \begin{pmatrix} \lambda R_\lambda & R_\lambda \\ -A_0 R_\lambda & \lambda^{-1}(I - A_0 R_\lambda) \end{pmatrix}. \tag{6.4}$$

By hypothesis,

$$\left\| [\lambda R_\lambda]^{(m)} \right\|, \quad \left\| [\lambda^{-1} A_0 R_\lambda]^{(m)} \right\| \le Cm!(\lambda - a)^{-m-1}, \quad \lambda > a, \; m \in N_0. \tag{6.5}$$

Yet the application of the classical Hille-Yosida-Feller-Miyadera-Phillips theorem to $\mathbf{M}_2$ in the setting of the product space $E \times E$, in view of (6.4), requires that the term $-A_0 R_\lambda$ should satisfy the same estimate as that for λR_λ and $\lambda^{-1} A_0 R_\lambda$ in (6.5). This forces us to seek a suitable phase space to change the awkward situation. A natural consideration is the space $E_0 \times E$, where E_0 is the closure of $\mathcal{D}(A) \bigcap \mathcal{D}(B)$ with respect to the norm $||| \cdot |||$, defined by

$$|||u||| := \|u\| + \sup \left\{ \frac{(\mu - a)^{l+1}}{l!} \left\| (A_0 R_\mu u)^{(l)} \right\| ; \ l \in N_0, \ \mu > a \right\}.$$

Thus the term $-A_0 R_\lambda$, as an operator from E_0 to E, satisfies the required estimate. But in the meantime we are faced with the new problem of justifying the two estimates as follows

$$\left\| \left| [\lambda R_\lambda u]^{(m)} \right\| \right\| \leq Cm!(\lambda - a)^{-m-1} |||u|||,$$

$$\left\| \left| [R_\lambda u]^{(m)} \right\| \right\| \leq Cm!(\lambda - a)^{-m-1} \|u\|.$$

(6.6)

For the special case of $A_1 = 0$, this was realized in Watanabe [2]. However, when two coefficient operators are involved, the complications stemming from the lack of commutativity of A_0 and A_1 (among others) make the verification of (6.6) impractical. In order to avoid this inconvenience, one may resort to the theory of integrated semigroups. A successful example appears in Kellermann-Hieber [1, Theorem 3.5] and Neubrander [5, Theorem 4.1] (see also Neubrander [4], Sova [2]), where it is shown that A generates a strongly continuous cosine function on E if and only if

$$\begin{pmatrix} 0 & I \\ A & 0 \end{pmatrix}$$

generates a once integrated semigroup on $E \times E$. But for general (ACP_n), the situation is much more complicated. For example, the equivalent condition in Theorem 2.3 (for $n = 2$) do imply that $\mathbf{M}_2$ generates a once integrated semigroup, but this yields only a very restrictive set of initial data.

Chapter 3

Generalized wellposedness

Summary

Motivated by the ideas of integrated or regularized operator families, we in this chapter select to treat several types of higher order abstract Cauchy problems which are no longer wellposed or strongly wellposed in the sense of Section 2.1, but wellposed in some generalized sense. Our main purpose is to establish some concise and useful criteria for these (ACP_n) to be wellposed in some sense.

In Section 3.1, fixing $r \geq 0$ we give a set of conditions on $\lambda^{k-r-1} A_k R_\lambda$ $(1 \leq k \leq n)$ to lead to existence and uniqueness, as well as continuous dependence on initial data (in some sense) of solutions for (ACP_n). One of the conditions is that each of $\lambda^{k-r-1} A_k R_\lambda$ is a Laplace transform. When $n = 1$, $\lambda^{k-r-1} A_k R_\lambda$ reduces to $\lambda^{-r} R(\lambda; -A_0)$, and the condition reduces to the characterization for $-A_0$ to generate an r-times integrated semigroup.

Recall that Corollary 2.4.8 gave an interesting result about the equation whose 'principal' coefficient operator is the generator of a strongly continuous semigroup. There arises the problem of what if the generators of integrated semigroups in place of strongly continuous semigroups act as the 'principal' coefficient operators. Sections 3.2 and 3.3 are devoted to such problems. Section 3.4 deals with the case when the 'principal'coefficient operators are the generators of integrated cosine functions.

Section 3.5 defines C-wellposedness of (ACP_n), which is a reflection of the idea of C-regularized semigroups in higher order Cauchy problems. In particular, we consider complete (ACP_2) with differential operators as coefficient operators and explore the conditions under which such Cauchy problems are C-wellposed.

Finally in Section 3.6, we consider the Cauchy problem for $u^{(n)}(t) = Au(t)$. A classical result indicates that for $n \geq 3$, its wellposedness implies the boundedness of A. We will give an extension of this result, Theorem 6.5. In fact, the main aim of this section is to show conditions on A, which are valid for many unbounded operators, such that the underlying Cauchy problem is C-wellposed (in some sense).

Throughout this chapter, E denotes a Banach space.

3.1   Criteria for general (ACP_n)

Notation 1.1. For $f \in C(R^+, E)$, write

$$
J^m f(t) := \begin{cases} f(t), & \text{if } m = 0, \\[2mm] \displaystyle\int_0^t \frac{(t-s)^{m-1}}{(m-1)!} f(s)ds, & \text{if } m = 1, 2, 3, \cdots. \end{cases}
$$

Letting $A_0, \cdots, A_{n-1}$ be linear operators in E, we define $\mathbf{M}_n$ as in (2.6.2).

Theorem 1.2. *Let $r \in N_0$ and let $A_0, \cdots, A_{n-1}$ be closed linear operators in E. Suppose that $R_\lambda \in \mathbf{L}(E)$ for $\lambda > \omega_0 > 0$. If either*
(i) *for each $0 \le i \le n-1, \lambda^{i-r-1} A_i R_\lambda \in LT - \mathbf{L}(E)$,   or*
(ii) *for each $0 \le i \le n-1$,*

$$
\left\| (\lambda - \omega_0)^{m+1} \left[\lambda^{i-r} A_i R_\lambda \right]^{(m)} \frac{1}{m!} \right\| \le \text{const} \quad (\lambda > \omega_0, \ m \in N_0),
$$

or
(iii) *for each $0 \le i \le n-1$, $\left\| \lambda^{i-r+1} A_i R_\lambda \right\|$ is bounded on a right half plane,*
or
(iv) *$\bigcap_{i=0}^{n-1} \mathcal{D}(A_i)$ is dense in E, and for each $0 \le i \le n-1$,*

$$
\left\| (\lambda - \omega_0)^{m+1} \left[\lambda^{i-r-1} A_i R_\lambda \right]^{(m)} \frac{1}{m!} \right\| \le \text{const} \quad (\lambda > \omega_0, \ m \in N_0),
$$

then for each initial value $(u_0, u_1, \cdots, u_{n-1})$ with

$$
\left(\sum_{k=0}^{n-1} A_k u_k, \ \sum_{k=1}^{n-1} A_{k-1} u_k, \ \cdots, \ A_0 u_{n-2} + A_1 u_{n-1}, \ A_0 u_{n-1} \right) \in \mathcal{D}\left(\mathbf{M}_n^r \right), \quad (1.1)
$$

(ACP_n) has a unique solution $u(\cdot)$ satisfying that for $t \ge 0$.

$$
\|u(t)\| \le Ce^{\omega t} \begin{cases} \displaystyle\sum_{k=0}^{n-1} \|u_k\| + \sum_{k=0}^{n-2} \sum_{i=0}^{k} \|A_i u_k\|, & \text{if } r \le n-1, \\[5mm] \displaystyle\sum_{k=0}^{n-1} \left(\|u_k\| + \sum_{i=0}^{k} \|A_i u_k\| \right), & \text{if } r = n, \\[5mm] \displaystyle\sum_{k=0}^{n-1} \left(\|u_k\| + \sum_{m=0}^{r-n} \left\| x_{k,m}^{(1)} \right\| + \sum_{i=2}^{n} \left\| x_{k,r-n}^{(i)} \right\| \right), & \text{if } r \ge n+1, \end{cases}
$$
$$(1.2)$$

for some constants $C, \omega > 0$, *where for each* $1 \leq i \leq n, 0 \leq m \leq r, x_{k,m}^{(i)}$ *denotes the ith component of* $\mathbf{M}_n^m x_k$,

$$x_k := (A_k u_k, \ A_{k-1} u_k, \ \cdots, \ A_0 u_k, \ 0, \ \cdots, \ 0) \quad (0 \leq k \leq n-1). \tag{1.3}$$

Proof. Firstly, we show that either (ii) or (iii) or (iv) implies (i).

This is true for (ii) and (iii), by a direct application of Theorem 1.2.1 and Theorem 1.1.12 respectively. For (iv), observe that for $u \in \bigcap_{i=0}^{n-1} \mathcal{D}(A_i)$,

$$\lambda^{i-r-1} A_i R_\lambda u = \lambda^{-n+i-r-1} A_i u - \lambda^{-1} \sum_{j=0}^{n-1} \lambda^{-n+i+j-r} A_i R_\lambda (A_j u),$$

$$0 \leq i \leq n-1, \ \lambda > \omega_0.$$

Note that

$$-n+i-r-1 \leq -1, \quad -n+i+j-r \leq i-r-1$$

for $0 \leq i, \ j \leq n-1$. Making use of Theorem 1.2.1 yields that

$$\lambda^{i-r-1} A_i R_\lambda u \in LT - E, \quad u \in \bigcap_{i=0}^{n-1} \mathcal{D}(A_i); \tag{1.4}$$

also there exists an operator family $\{G(t)\}_{t \geq 0}$ with $G(0) = 0$, satisfying

$$\frac{1}{h} \|G(t+h) - G(t)\| \leq \text{const } e^{\omega_0(t+h)}, \quad t, \ h \geq 0 \tag{1.5}$$

such that

$$\lambda^{i-r-2} A_i R_\lambda u = \mathcal{L}[G(t)u](\lambda), \quad u \in E.$$

The last expression combined with (1.4) establishes by Theorem 1.1.9 that $G(\cdot)u$ is continuously differentiable for $u \in \bigcap_{i=0}^{n-1} \mathcal{D}(A_i)$, which is dense in E by hypothesis. It follows from (1.5) that $G(\cdot)u$ is continuously differentiable for $u \in E$ and

$$\|G'(t)u\| \leq \text{ const } e^{\omega_0 t} \|u\| \quad (t \geq 0, \ u \in E).$$

This verifies our claim.

Next, let hypothesis (i) hold, $(u_0, \ u_1, \ \cdots, \ u_{n-1})$ satisfy (1.1), and x_k be as in (1.3).

It can be verified easily that $\lambda \in \rho(\mathbf{M}_n)$ for λ sufficiently large and

$$(\lambda - \mathbf{M}_n)^{-1} = \begin{pmatrix} \lambda^{n-1} R_\lambda & \lambda^{n-2} R_\lambda & \cdots & \lambda R_\lambda & R_\lambda \\ & & * & & \end{pmatrix}$$

Then we see that for $0 \leq k, j \leq n-1, 1 \leq m \leq r$,

$$\lambda^{j-n} (\lambda - \mathbf{M}_n)^{-1} x_k$$

$$= \lambda^{j-n-1} x_k + \lambda^{j-n-2} \mathbf{M}_n x_k + \cdots + \lambda^{j-n-m} \mathbf{M}_n^{m-1} x_k$$

$$+ \lambda^{j-n-m} (\lambda - \mathbf{M}_n)^{-1} \mathbf{M}_n^m x_k.$$

Taking the first components on the two sides of the above equality yields that for $0 \leq k,\ j \leq n-1,\ 1 \leq m \leq r$,

$$\sum_{i=0}^{k} \lambda^{i-k+j-1} R_\lambda A_i u_k = \lambda^{j-n-1} A_k u_k + \cdots + \lambda^{j-n-m} x^{(1)}_{k,m-1}$$

$$+ \sum_{i=1}^{n} \lambda^{1-i} \lambda^{j-m-1} R_\lambda x^{(i)}_{k,m}.$$

Letting $m = r$ in the above equality, we see by hypothesis (i) that

$$A_j \sum_{i=0}^{k} \lambda^{i-k+j-1} R_\lambda A_i u_k \in LT - E \quad (0 \leq k,\ j \leq n-1),$$

which implies

$$\lambda^j A_j R_\lambda u_{n-1} = \lambda^{j-n} A_j u_{n-1} - A_j \sum_{i=0}^{n-1} \lambda^{i-n+j} R_\lambda A_i u_{n-1}$$

$$\in LT - E, \quad 0 \leq j \leq n-1.$$

Consequently, Lemma 2.2.1 applies and gives the existence result.

From the identity

$$\lambda^{n-1} R_\lambda = \lambda^{-1} - \sum_{i=0}^{n-1} \lambda^{i-1} A_i R_\lambda, \tag{1.6}$$

we see that

$$\lambda^{n-r-1} R_\lambda \in LT - L(E), \tag{1.7}$$

since by hypothesis $\lambda^{i-r-1} A_i R_\lambda \in LT - L(E)$ $(0 \leq i \leq n-1)$.

Let $F(t)$ be the determining function of $\lambda^{n-r-1} R_\lambda$, i.e.,

$$\lambda^{n-r-1} R_\lambda u = \mathcal{L}[F(t)u](\lambda) \quad (u \in E).$$

From the proof of Lemma 2.2.1, we know

$$\mathcal{L}[u(t)](\lambda) = \sum_{k=0}^{n-1} \left(\lambda^{-k-1} u_k - \sum_{i=0}^{k} \lambda^{i-k-1} R_\lambda A_i u_k \right).$$

Therefore, if $r = n$

$$\mathcal{L}[u(t)](\lambda) = \mathcal{L}\left[\sum_{k=0}^{n-1} \left(\frac{t^k}{k!} u_k - \sum_{i=0}^{k} J^{n-r-i+k} F(t) A_i u_k \right) \right](\lambda);$$

if $r \leq n - 1$, using (1.6) again, one obtains

$$\mathcal{L}[u(t)](\lambda)$$

$$= \sum_{k=0}^{n-2} \left(\lambda^{-k-1} u_k - \sum_{i=0}^{k} \lambda^{i-k-1} R_\lambda A_i u_k \right) + R_\lambda u_{n-1}$$

$$= \mathcal{L} \left[\sum_{k=0}^{n-2} \left(\frac{t^k}{k!} u_k - \sum_{i=0}^{k} J^{n-r-i+k} F(t) A_i u_k \right) + J^{n-r-1} F(t) u_{n-1} \right] (\lambda);$$

if $r \geq n + 1$, we have by virtue of (1.3) with $m = r - n$ that

$$\mathcal{L}[u(t)](\lambda) = \mathcal{L} \left[\sum_{k=0}^{n-1} \left(\frac{t^k}{k!} u_k - \frac{t^n}{n!} A_k u_k - \cdots - \frac{t^{r-1}}{(r-1)!} x_{k,r-n-1}^{(1)} \right. \right.$$

$$\left. \left. - \sum_{i=1}^{n} J^{i-1} F(t) x_{k,r-n}^{(i)} \right) \right] (\lambda).$$

From these observations, (1.2) follows immediately.

Finally, in view of Lemma 2.3.1, the uniqueness of solutions is derived from (1.7). The proof is complete.

Corollary 1.3. *Let the hypothesis in Theorem 1.2 hold. Then* (ACP_n) *has a unique solution for each* $(u_0, u_1, \cdots, u_{n-1})$ *satisfying*

$$u_k \in \mathcal{D}_k := \begin{cases} \displaystyle\bigcap_{i=0}^{k} \Omega_{k,i}, & \text{if } 0 \leq k \leq r-1, \\[4mm] \displaystyle\bigcap_{i=0}^{k-r} \mathcal{D}(A_i) \bigcap_{i=k-r+1}^{k} \Omega_{k,i}, & \text{if } k \geq r, \end{cases}$$

where

$$\Omega_{k,i} := \bigcap_{(i_1, \cdots, i_p) \in \Lambda_{k,i}} \mathcal{D} \left(A_{n-i_p} \cdots A_{n-i_1} A_i \right),$$

$$\Lambda_{k,i} := \left\{ (i_1, \cdots, i_p); \ \sum_{m=1}^{p} i_m \geq i - k + r > \sum_{m=1}^{p-1} i_m, \right.$$

$$\left. 1 \leq i_1, \cdots, i_p \leq n, \ p \in N \right\}.$$

Example 1.4. Consider the Cauchy problem

$$\begin{cases} u''(t) - Au'(t) - \left(aA^2 + bA + cI\right)u(t) = 0, \quad t \geq 0, \\ \\ u(0) = u_0, \ u'(0) = u_1, \end{cases} \tag{1.8}$$

where A is a linear operator in E, $a, b, c \in \mathbf{C}$, $a \neq 0$.

Set

$$f_{1,2}(\lambda) = \frac{1}{2a}\left(-\lambda - b \pm \left[(1 + 4a)\lambda^2 + 2b\lambda + b^2 - 4ac\right]^{1/2}\right),$$

$$g(\lambda) = \left[(1 + 4a)\lambda^2 + 2b\lambda + b^2 - 4ac\right]^{-1/2},$$

$$H_\omega = \{f_{1,2}(\lambda); \ \lambda \in \mathbf{C}, \ \text{Re}\lambda > \omega\} \quad (\omega > 0).$$

Suppose that there exist constants C, $\omega > 0$, $q \in \{-1, 0, 1, 2, \cdots\}$ such that for each $\lambda \in H_\omega$, $\lambda \in \rho(A)$ and

$$\|R(\lambda; \ A)\| \leq C|\lambda|^q.$$

Then there is a constant C_0 such that for $\text{Re}\lambda > \omega$,

$$\left\|\lambda^{-q+1} R_\lambda\right\| \leq |\lambda|^{-q+1}\|g(\lambda)[R(f_1(\lambda); \ A) - R(f_2(\lambda); \ A)]\|$$

$$\leq C_0,$$

$$\left\|\lambda^{-q} A R_\lambda\right\| \leq |\lambda|^{-q}\|g(\lambda)[(f_1(\lambda)R(f_1(\lambda); \ A) - f_2(\lambda)R(f_2(\lambda); \ A)]\|$$

$$\leq C_0,$$

where

$$R_\lambda = \left(\lambda^2 - \lambda A - aA^2 - bA - cI\right)^{-1}.$$

Further, making use of the identity

$$\left(aA^2 + bA + cI\right)R_\lambda = -I - \lambda A R_\lambda + \lambda^2 R_\lambda$$

gives that there is $C_1 > 0$ such that

$$\left\|\lambda^{-q-1}\left(aA^2 + bA + cI\right)R_\lambda\right\| \leq C_1 \quad (\text{Re}\lambda > \omega).$$

Now applying Corollary 1.3 with

$$n = 2, \ r = q + 2, \ A_1 = -A, \ A_0 = -aA^2 - bA - cI,$$

we conclude that (1.8) has a unique solution for each $(u_0, u_1) \in \mathcal{D}\left(A^{q+5}\right) \times \mathcal{D}\left(A^{q+4}\right)$. Here

$$\mathcal{D}_0 = \Omega_{0,0} = \mathcal{D}\left(A^{r+3}\right) = \mathcal{D}\left(A^{q+5}\right);$$

if $r = 1$, i.e., $q = -1$,

$$\mathcal{D}_1 = \mathcal{D}(A_0) \bigcap \Omega_{1,1} = \mathcal{D}(A^2) \bigcap \mathcal{D}(A^{r+2}) = \mathcal{D}(A^{r+2}) = \mathcal{D}(A^{q+4}),$$

if $r \geq 2$, i.e., $q \geq 0$,

$$\mathcal{D}_1 = \Omega_{1,0} \cap \Omega_{1,1} = \mathcal{D}(A^{r+2}) \bigcap \mathcal{D}(A^{r+2}) = \mathcal{D}(A^{q+4}).$$

3.2 The special case (I): coefficient operators relating to integrated semigroups

Let $r \in N_0$, and $B_0, \cdots, B_{n-2}$ be closable linear operators in E. Let $-A$ be the generator of an r-times integrated semigroup on E. Of concern is the following Cauchy problem

$$\begin{cases} u^{(n)}(t) + Au^{(n-1)}(t) + \sum_{i=0}^{n-2} B_i u^{(i)}(t) = 0, \quad t \geq 0, \\ \\ u^{(k)}(0) = u_k, \quad 0 \leq k \leq n-1. \end{cases} \tag{2.1}$$

Theorem 2.1. *Assume that* $B_0, \cdots, B_{n-2}$ *satisfy*
  (i) $\mathcal{D}(B_i) \supset \mathcal{D}(A), \ 0 \leq i \leq n-2,$
  (ii) $B_i Au = AB_i u, \ for \ each \ u \in \mathcal{D}(A^2), \ 0 \leq i \leq n-2.$
Then for every

$$u_k \in \bigcap_{i=0}^{k} \mathcal{D}\left(A^{(r-i)} B_{k-i}\right) \ (0 \leq k \leq n-2), \quad u_{n-1} \in \mathcal{D}(A^{r+1}), \tag{2.2}$$

(2.1) *has a unique solution* $u(\cdot)$ *and there exist constants* $M, \omega \geq 0$ *such that*

$$\|u(t)\| \ \leq \ Me^{\omega t}\left[\sum_{i=0}^{n-2}\left(\|u_i\| + \sum_{j=0}^{i}\sum_{l=0}^{(r-n-j)}\|A^l B_{i-j} u_i\|\right)\right.$$

$$\left. + \sum_{l=0}^{(r-n+1)}\|A^l u_{n-1}\|\right] \quad (t \geq 0), \tag{2.3}$$

where and in the sequel, for any real number b,

$$\langle b \rangle := \begin{cases} [b], & if \ b \geq 0, \\ \\ 0, & if \ b < 0. \end{cases}$$

Proof. Since $-A$ is the generator of an r-times integrated semigroup on E, there is a strongly continuous, exponentially bounded family $\{S(t)\}_{t\geq 0}$ of bounded operators on E such that for $u \in E$, λ sufficiently large,

$$\lambda^{-r}(\lambda + A)^{-1}u = \mathcal{L}[S(t)u](\lambda). \tag{2.4}$$

Take $\mu_0 \in \rho(A)$. For $t \geq 0$, $u \in E$, define

$$S(t; 0, r)u = S(t)u,$$

$$S(t; -1, r+1)u = \mu_0 \int_0^t S(s)u\,ds + S(t)u - \frac{t^r}{r!}u,$$

$$S(t; 1, r-1)u = \frac{t^{r+1}}{(r+1)!}u - A(\mu_0 - A)^{-1}S(t)u, \quad r \geq 1,$$

and for $r_0 \in N - \{1\}$, $r_1 \in N_0$, $r_0 + r_1 = r$,

$$
\begin{aligned}
S(t; r_0, r_1) &= (\mu_0 - A)^{-r_0}\frac{t^{r_1}}{r_1!} - \frac{t^{r_1+1}}{(r_1+1)!}A(\mu_0 - A)^{-r_0} \\[2mm]
&\quad + \cdots + (-1)^{r_0-1}\frac{t^{r-1}}{(r-1)!}A^{r_0-1}(\mu_0 - A)^{-r_0} \\[2mm]
&\quad + (-1)^{r_0}A^{r_0}(\mu_0 - A)^{-r_0}S(t).
\end{aligned}
$$

Clearly, there exist constants $C, \omega > 0$ such that for $r_0 \in N_0 \bigcup\{-1\}$, $r_1 \in N_0$, $r_0 + r_1 = r$,

$$\|S(t; r_0, r_1)\| \leq Ce^{\omega t}, \quad t \geq 0. \tag{2.5}$$

Since

$$\lambda^{-r-1}(\mu_0 - A)(\lambda + A)^{-1} = (\mu_0\lambda^{-r-1} + \lambda^{-r})(\lambda + A)^{-1} - \lambda^{-r-1},$$

$$\lambda^{-r+1}(\mu_0 - A)^{-1}(\lambda + A)^{-1} = \lambda^{-r}(\mu_0 - A)^{-1} - \lambda^{-r}A(\mu_0 - A)^{-1}(\lambda + A)^{-1},$$

$$\lambda^{-r_1}(\mu_0 - A)^{-r_0}(\lambda + A)^{-1}$$

$$= \lambda^{-r_1-1}(\mu_0 - A)^{-r_0} - \lambda^{-r_1-1}(\lambda + A)^{-1}A(\mu_0 - A)^{-r_0}$$

$$= \lambda^{-r_1-1}(\mu_0 - A)^{-r_0} - \lambda^{-r_1-2}A(\mu_0 - A)^{-r_0}$$

$$+ \lambda^{-r_1-2}(\lambda + A)^{-1}A^2(\mu_0 - A)^{-r_0}$$

$$= \cdots$$

$$= \quad \lambda^{-r_1-1}(\mu_0 - A)^{-r_0} - \lambda^{-r_1-2}A(\mu_0 - A)^{-r_0} + \cdots$$

$$+ (-1)^{r_0-1}\lambda^{-r_1-r_0}A^{r_0-1}(\mu_0 - A)^{-r_0}$$

$$+ (-1)^{r_0}\lambda^{-r_1-r_0}(\lambda + A)^{-1}A^{r_0}(\mu_0 - A)^{-r_0},$$

we have noting (2.4) that for $u \in E$, $\lambda > \omega$,

$$\lambda^{-r_1}(\mu_0 - A)^{-r_0}(\lambda + A)^{-1}u = \mathcal{L}[S(t; r_0, r_1)u](\lambda), \qquad (2.6)$$

where $r_0 \in N_0 \bigcup \{-1\}$, $r_1 \in N_0$, $r_0 + r_1 = r$.

By the identities

$$(-1)^j \left[(\lambda + A)^{-1} \right]^{j+1} = \frac{1}{j!} \left[(\lambda + A)^{-1} \right]^{(j)}, \quad j \in N_0, \ -\lambda \in \rho(A),$$

and (2.6), we obtain that for $k \in N_0$, $0 \le j \le k$, $0 \le m \le k(n-2) + j$ and $\lambda > \omega$,

$$\frac{(-1)^j}{\lambda^{k-j+m}} \left[(\lambda + A)^{-1} \right]^{j+1} u$$

$$= \quad \frac{1}{\lambda^{k-j+m}j!} \left[\lambda^{r_1}(\mu_0 - A)^{r_0}\mathcal{L}[S(t; r_0, r_1)u](\lambda) \right]^{(j)}$$

$$= \quad \frac{(\mu_0 - A)^{r_0}}{\lambda^{k-j+m}j!} \sum_{l=0}^{r_1+(j-r_1)} \left\{ \binom{j}{l} \lambda^{r_1-l}r_1(r_1 - 1)\cdots(r_1 - l + 1) \right.$$

$$\qquad\qquad (2.7)$$

$$\left. \cdot \mathcal{L}\left[(-t)^{j-l}S(t; r_0, r_1)u \right](\lambda) \right\}$$

$$= \quad \frac{\lambda^{r_1}(\mu_0 - A)^{r_0}}{\lambda^{k-j+m}} \sum_{l=0}^{r_1+(j-r_1)} \binom{r_1}{l} \frac{\lambda^{-l}}{(j-l)!}\mathcal{L}\left[(-t)^{j-l}S(t; r_0, r_1)u \right](\lambda),$$

$$u \in E,$$

with r_0, r_1 as in (2.6).

For k, j, m, r_0, r_1, u as above and for $0 \le l \le r_1 + \langle j - r_1 \rangle$, $t \ge 0$, define
$H_{kjml}(t; r_0, r_1)u$

$$= \begin{cases} \dfrac{1}{k!}(-t)^k S(t; r_0, r_1)u, & \text{if } j = k, m = l = 0, \\[2em] \dfrac{(-1^{j-l}}{(k-j+m+l-1)!(j-l)!} \displaystyle\int_0^t (t-s)^{k-j+m+l-1}s^{j-l}S(s; r_0, r_1)u\,ds, \\[1em] \qquad\qquad\qquad\qquad \text{otherwise,} \end{cases}$$

$$H_{kjm}(t; r_0, r_1) = \sum_{l=0}^{r_1+(j-r_1)} \binom{r_1}{l} H_{kjml}(t; r_0, r_1). \qquad (2.8)$$

Noting

$$\int_0^t (t-s)^l s^k ds = l!k! \frac{t^{k+l+1}}{(k+l+1)!}, \quad l, k \in N_0,$$

and using (2.5), we have that for some constant $M > 0$,

$$\|H_{kjml}(t; r_0, r_1)\| \leq M e^{\omega t} \frac{t^{k+m}}{(k+m)!}$$

$$\leq M e^{(\omega+1)t} \frac{t^k}{k!}, \quad t \geq 0,$$

and therefore

$$\|H_{kjm}(t; r_0, r_1)\| \leq \sum_{l=0}^{r_1} \binom{r_1}{l} M e^{(\omega+1)t} \frac{t^k}{k!}$$

$$= 2^{r_1} M e^{(\omega+1)t} \frac{t^k}{k!}.$$

Thus, it follows from (2.7) and (2.8) that for k, j, m, r_0, r_1, u as in (2.7), and for λ large enough,

$$\frac{(-1)^j}{\lambda^{k-j+m}} \left[(\lambda + A)^{-1} \right]^{j+1} u = \lambda^{r_1} (\mu_0 - A)^{r_0} \mathcal{L} \left[H_{kjm}(t; r_0, r_1) u \right] (\lambda).$$

Now, set

$$\widetilde{B}_i := B_i (\mu_0 - A)^{-1}, \quad 0 \leq i \leq n-2,$$

and let A_{kjm} (k, j, m as in (2.7)) be linear operators on E such that

$$\left[\sum_{i=0}^{n-2} \widetilde{B}_i \lambda^{i-n+2} \right]^k (1 + \mu_0 \lambda^{-1})^j = \sum_{m=0}^{k(n-2)+j} A_{kjm} \lambda^{-m}.$$

We define, for $t \geq 0$, r_0, r_1 as in (2.6),

$$H(t; r_0, r_1) = \sum_{k=0}^{\infty} \sum_{j=0}^{k} \binom{k}{j} \sum_{m=0}^{k(n-2)+j} A_{kjm} H_{kjm}(t; r_0, r_1),$$

then

$$\|H(t; r_0, r_1)\| \leq \sum_{k=0}^{\infty} \sum_{j=0}^{k} \binom{k}{j} \sum_{m=0}^{k(n-2)+j} \|A_{kjm}\| 2^{r_1} M e^{(\omega+1)t} \frac{t^k}{k!}$$

$$\leq \quad 2^{r+1} M e^{(\omega+1)t} \sum_{k=0}^{\infty} \frac{t^k}{k!} \sum_{j=0}^{k} \binom{k}{j} (1+|\mu_0|)^k \left(\sum_{i=0}^{n-2} \left\| \widetilde{B}_i \right\| \right)^k$$

$$= \quad 2^{r+1} M e^{(\omega+1)t} \sum_{k=0}^{\infty} \frac{t^k}{k!} (1+|\mu_0|)^k \left(\sum_{i=0}^{n-2} \left\| \widetilde{B}_i \right\| \right)^k 2^k$$

$$= \quad 2^{r+1} M \exp\left(2(1+|\mu_0|) \left(\sum_{i=0}^{n-2} \left\| \widetilde{B}_i \right\| \right) t + (\omega+1)t \right).$$

Consequently, for r_0, r_1 as in (2.6), $u \in E$, and λ large enough,

$$\lambda^{r_1} (\mu_0 - A)^{r_0} \mathcal{L}[H(t; r_0, r_1)u](\lambda)$$

$$= \quad \sum_{k=0}^{\infty} \sum_{j=0}^{k} \binom{k}{j} \sum_{m=0}^{k(n-2)+j} \lambda^{-m} A_{kjm} \frac{(-1)^j}{\lambda^{k-j}} \left[(\lambda + A)^{-1} \right]^{j+1}$$

$$= \quad \sum_{k=0}^{\infty} \left[\sum_{i=0}^{n-2} \widetilde{B}_i \lambda^{i-n+2} \right]^k \sum_{j=0}^{k} (1 + \mu_0 \lambda^{-1})^j \binom{k}{j} \frac{(-1)^j}{\lambda^{k-j}} \left[(\lambda + A)^{-1} \right]^{j+1}$$

$$= \quad \sum_{k=0}^{\infty} \left[\sum_{i=0}^{n-2} \widetilde{B}_i \lambda^{i-n+2} \right]^k \left[\lambda^{-1} - (1 + \mu_0 \lambda^{-1}) (\lambda + A)^{-1} \right]^k (\lambda + A)^{-1}$$

$$= \quad \sum_{k=0}^{\infty} \left[\sum_{i=0}^{n-2} \widetilde{B}_i \lambda^{i-n+1} (A - \mu_0)(\lambda + A)^{-1} \right]^k (\lambda + A)^{-1}$$

$$= \quad \sum_{k=0}^{\infty} \left[- \sum_{i=0}^{n-2} B_i \lambda^{i-n+1} (\lambda + A)^{-1} \right]^k (\lambda + A)^{-1}.$$

From this, we verify easily that for r_0, r_1, u, λ as above,

$$P_\lambda \lambda^{r_1} (\mu_0 - A)^{r_0} \mathcal{L}[H(t; r_0, r_1)u](\lambda) = \lambda^{n-1} u,$$

where

$$P_\lambda := \lambda^n + \lambda^{n-1} A + \sum_{i=0}^{n-2} \lambda^i B_i.$$

Since all the operators in the equality commute, we justify that for λ large enough, $R_\lambda := P_\lambda^{-1}$ exists and

$$\lambda^{n-1} R_\lambda u = \lambda^{r_1} (\mu_0 - A)^{r_0} \mathcal{L}[H(t; r_0, r_1)u](\lambda),$$

$$u \in E, r_0 \in N_0 \bigcup \{-1\}, \ r_1 \in N_0 \text{ with } r_0 + r_1 = r. \tag{2.9}$$

Taking $r_0 = 0$, (2.9) implies that there is $\delta > 0$ such that for each $u \in E$,

$$\lim_{\lambda \to \infty} e^{-\delta\lambda}\|R_\lambda u\| = 0.$$

By virtue of Lemma 2.4.3, we obtain that the (2.1) has at most one solution. Now let

$$u_k \in \bigcap_{i=0}^{k} \mathcal{D}\left(A^{(r-i)}B_{k-i}\right) \quad (0 \le k \le n-2), \quad u_{n-1} \in \mathcal{D}\left(A^{r+1}\right).$$

Using (2.9) we get that for λ sufficiently large,

$$\lambda^{n-1}(\mu_0 - A)R_\lambda u_{n-1} \;=\; (\mu_0 - A)^{-r}\lambda^{n-1}R_\lambda(\mu_0 - A)^{r+1}u_{n-1}$$

$$=\; \mathcal{L}\left[H(t; r, 0)(\mu_0 - A)^{r+1}u_{n-1}\right](\lambda),$$

$$\lambda^{n-1}(\mu_0 - A)\left(\lambda^{-(i+1)}R_\lambda B_{k-i}u_k\right)$$

$$=\; \lambda^{-(i+1)}(\mu_0 - A)^{-(r-i)+1}\lambda^{n-1}R_\lambda(\mu_0 - A)^{(r-i)}B_{k-i}u_k$$

$$=\; \mathcal{L}\left[H(t; (r-i)-1, i+1)(\mu_0 - A)^{(r-i)}B_{k-i}u_k\right](\lambda),$$

$$0 \le i \le k, \; 0 \le k \le n-2,$$

which implies by condition (i) that for $0 \le j \le n-2$, $0 \le i \le k$, $0 \le k \le n-2$,

$$\lambda \longmapsto \lambda^j B_j R_\lambda u_{n-1}, \quad \lambda \longmapsto \lambda^j B_j\left(\lambda^{-(i+1)}R_\lambda B_{k-i}u_k\right) \in LT - E.$$

Accordingly, applying Lemma 2.2.1 yields that (2.1) has a solution $u(\cdot)$ for every initial data $(u_0, \, u_1, \, \cdots, \, u_{n-1})$ as in (2.2). The proof of Lemma 2.2.1 shows that

$$\mathcal{L}[u(\cdot)](\lambda) = R_\lambda u_{n-1} + \sum_{k=0}^{n-2}\lambda^{-k-1}u_k + \sum_{k=0}^{n-2}\sum_{i=0}^{k}\lambda^{-(i+1)}R_\lambda B_{k-i}u_k. \qquad (2.10)$$

For $r_1 \ge r + 2$, $u \in E$, setting

$$H(t; -1, r_1)u = \int_0^t \frac{(t-s)^{r_1-r-2}}{(r_1-r-2)!}H(s; -1, r+1)u\,ds,$$

and for $r_1 \ge r + 1$, $u \in E$,

$$H(t; 0, r_1)u = \int_0^t \frac{(t-s)^{r_1-r-1}}{(r_1-r-1)!}H(s; 0, r)u\,ds,$$

we obtain by (2.8) and (2.9) that

$$u(t) = \sum_{k=0}^{n-2} \left[\frac{t^k}{k!} u_k + \sum_{i=0}^{k} H(t; \langle r-n-i \rangle, n+i)(\mu_0 - A)^{(r-n-i)} B_{k-i} u_k \right]$$
$$+ H(t; \langle r-n+1 \rangle, n-1)(\mu_0 - A)^{(r-n+1)} u_{n-1}.$$

Immediately, (2.3) follows. This ends the proof of the theorem.

Corollary 2.2. *Assume* $n = 2$ *and* $\mathcal{D}(B_0) \supset \mathcal{D}(A)$, $B_0 A u = A B_0 u$ $(u \in \mathcal{D}(A^2))$. *Then for every*
$$u_0 \in \mathcal{D}(A^r B_0), \quad u_1 \in \mathcal{D}(A^{r+1}),$$
(2.1) *has a unique solution* $u(t)$ *and there are constants* $M, \omega \geq 0$ *such that*
  (i) *if* $r = 1$,
$$\|u(t)\| \leq M e^{\omega t} (\|u_0\| + \|B_0 u_0\| + \|u_1\|),$$

  (ii) *if* $r \geq 2$,

$$\|u(t)\| \leq M e^{\omega t} \left(\|u_0\| + \sum_{l=0}^{r-2} \|A^l B_0 u_0\| + \sum_{l=0}^{r-1} \|A^l u_1\| \right).$$

Corollary 2.3. *Assume* B_i $(0 \leq i \leq n-2)$ *are as in Theorem 2.1 and* $r = n$. *Then for every*

$$u_i \in \bigcap_{j=0}^{i} \mathcal{D}\left(A^{n-j} B_{i-j}\right) \ (0 \leq i \leq n-2), \quad u_{n-1} \in \mathcal{D}\left(A^{n+1}\right),$$

(2.1) *has a unique solution* $u(t)$ *and there are constants* $M, \omega \geq 0$ *such that*

$$\|u(t)\| \leq M e^{\omega t} \left[\|u_{n-1}\| + \|A u_{n-1}\| + \sum_{i=0}^{n-2} \left(\|u_i\| + \sum_{j=0}^{i} \|B_j u_i\| \right) \right].$$

Corollary 2.4. *Assume* B_i $(0 \leq i \leq n-2)$ *are as in Theorem 2.1 and* $r = 1$. *Then for every*

$$u_i \in \mathcal{D}(A B_i) \bigcap_{j=0}^{i-1} \mathcal{D}(B_j) \ (0 \leq i \leq n-2), \quad u_{n-1} \in \mathcal{D}(A^2),$$

(2.1) *has a unique solution* $u(t)$ *and there are constants* $M, \omega \geq 0$ *such that*

$$\|u(t)\| \leq M e^{\omega t} \left[\|u_{n-1}\| + \sum_{i=0}^{n-2} \left(\|u_i\| + \sum_{j=0}^{i} \|B_j u_i\| \right) \right].$$

Example 2.5. Let $1 \leq p \leq \infty$, $\rho_1 \in R$, $\rho_2 > 0$, $c \in C$. Consider the following initial value problem in $L^p(R)$:

$$
\begin{cases}
\dfrac{\partial^2 u(t,x)}{\partial t^2} + \left(\rho_1 \dfrac{\partial^3}{\partial x^3} - \rho_2 \dfrac{\partial^2}{\partial x^2} \right) \dfrac{\partial u(t,x)}{\partial t} + c \dfrac{\partial^2 u(t,x)}{\partial x^2} = 0, \\
\\
\hspace{6cm} (t,x) \in R^+ \times R, \\
\\
u(0,x) = \varphi(x), \quad u_t(0,x) = \psi(x), \quad x \in R.
\end{cases}
\tag{2.11}
$$

Let

$$
B_0 = c \left(\frac{d}{dx} \right)^2, \qquad A = \rho_1 \left(\frac{d}{dx} \right)^3 - \rho_2 \left(\frac{d}{dx} \right)^2
$$

with maximal distributional domain. Then $-A$ generates a once integrated semigroup by Theorem 1.5.9. Therefore, making use of Corollary 2.2, we get that for every

$$
\varphi \in W^{5,p}(R), \quad \psi \in W^{6,p}(R),
$$

(2.11) has a unique solution

$$
u \in C^2(R^+, L^p(R)) \bigcap C^1 \left(R^+, W^{3,p}(R) \right)
$$

and there exist constants M, $\omega > 0$ such that

$$
\|u\|_{L^p(R)} \leq M e^{\omega t} (\|\varphi\|_{W^{2,p}(R)} + \|\psi\|_{L^p(R)}).
$$

Example 2.6. Consider the following initial value problem in $L^p(R^3)$ $\left(\frac{6}{5} < p < 6 \right)$:

$$
\begin{cases}
u_{ttt}(t,x) + i\rho \Delta u_{tt}(t,x) + \displaystyle\sum_{|\beta| \leq 2} b_\beta D^\beta u_t(t,x) \\
\\
\quad + \displaystyle\sum_{|\alpha| \leq 2} a_\alpha D^\alpha u(t,x) = 0, \quad (t,x) \in R^+ \times R^3, \\
\\
u(0,x) = \varphi(x), \quad u_t(0,x) = \psi(x), \quad u_{tt}(0,x) = \phi(x), \quad x \in R^3,
\end{cases}
\tag{2.12}
$$

where Δ denotes the Laplacian, $\rho \in R \setminus \{0\}$, a_α, $b_\beta \in C$ ($|\alpha|$, $|\beta| \leq 2$).
    Let

$$
A = i\rho \Delta, \quad B_0 = \sum_{|\alpha| \leq 2} a_\alpha D^\alpha, \quad B_1 = \sum_{|\beta| \leq 2} b_\beta D^\beta
$$

with distributional domains. Then $-A$ generates a once integrated semigroup by virtue of Theorem 1.5.9, since $3 \left| \frac{1}{2} - \frac{1}{p} \right| < 1$. Thus Corollary 2.4 (for $n = 3$) implies that (2.12) has a unique solution $u(t)$ for every

$$
\varphi(x), \quad \psi(x), \quad \phi(x) \in W^{4,p}(R^3)
$$

with

$$u \in C^3 \left(R^+, L^p(R^3)\right) \bigcap C^2 \left(R^+, W^{2,p}(R^3)\right),$$

$$\|u\|_{L^p(R^3)} \leq M e^{\omega t} \left(\|\varphi\|_{W^{2,p}(R^3)} + \|\psi\|_{W^{2,p}(R^3)} + \|\phi\|_{L^p(R^3)}\right),$$

for some constants $M, \omega > 0$.

3.3 The special case (II): coefficient operators relating to integrated semigroups (continuation)

Notation 3.1. If $m \in N$, B is a closed linear operator in the Banach space E, then E_m^B denotes the Banach space $\left(\mathcal{D}(B^m), \|u\|_m^B = \sum\limits_{i=0}^{m} \|B^i u\|\right)$.

Lemma 3.2. *Let $F(\cdot) \in LT - L(E)$ and let B be a closed linear operator on E. Suppose that there exists $b > 0$ such that for $\lambda > b$, $u \in \mathcal{D}(B)$, $BF(\lambda)u = F(\lambda)Bu$. Then $F(\cdot) \in LT - L\left(E_m^B\right)$ for any $m \in N$.*

Proof. By hypothesis, there is a strongly continuous $H(\cdot) : [0, \infty) \to L(E)$ with $\|H(t)\| \leq C e^{at}$ $(t \geq 0)$ for some $C, a > 0$ such that

$$F(\lambda)u = \int_0^\infty e^{-\lambda t} H(t) u \, dt \quad (\lambda > a, \ u \in E).$$

Hence for $\lambda > a$, $u \in \mathcal{D}(B)$,

$$BF(\lambda)u = F(\lambda)Bu = \int_0^\infty e^{-\lambda t} H(t) Bu \, dt.$$

By virtue of Theorem 1.1.10, we deduce that for any $t \geq 0$, $u \in \mathcal{D}(B)$,

$$H(t)u \in \mathcal{D}(B) \quad \text{and} \quad BH(t)u = H(t)Bu.$$

This implies that for each $m \in N$,

$$B^m H(t)u = H(t)B^m u \quad (t \geq 0, \ u \in \mathcal{D}(B^m)),$$

and $\{H(t)\}_{t \geq 0}$ is a strongly continuous, exponentially bounded operators on E_m^B.

Lemma 3.3. *Suppose that B is a closed linear operator in E, $p \in N$. Let $B_0 \in L\left(E_p^B, E\right)$ and $B^m F(\cdot) \in LT - L(E)$ for each $m \in \{0, \cdots, p\}$. Then $F(\cdot)B_0 \in LT - L\left(E_p^B\right)$.*

Proof. By hypothesis, for each $m \in \{0, \cdots, p\}$, there is a strongly continuous $H_m(\cdot) : [0, \infty) \to \mathbf{L}(E)$ with $\|H_m(t)\| \leq Ce^{at}$ $(t \geq 0)$ for some $C, a > 0$ such that

$$B^m F(\lambda)u = \int_0^\infty e^{-\lambda t} H_m(t)u\,dt \quad (\lambda > a, \ u \in E).$$

It follows from Theorem 1.1.10 that for $u \in E$, $t \geq 0$,

$$H_0(t)u \in \mathcal{D}(B^p) \quad \text{and} \quad B^m H_0(t)u = H_m(t)u \quad (0 \leq m \leq p).$$

Consequently, $\{H_0(t)B_0\}_{t \geq 0}$ is a strongly continuous, exponentially bounded family of bounded operators on E_p^B.

Throughout this section, p_0 will be a fixed positive integer, B a closed linear operator in E, and $\{a_j\}_{j=1}^{p_0}$ a sequence of mutually unequal and nonzero complex numbers such that each $-a_j B$ $(1 \leq j \leq p_0)$ is the generator of an r-times integrated semigroup, i.e.,

$$\lambda^{-r}(\lambda + a_j B)^{-1} \in LT - \mathbf{L}(E) \quad (1 \leq j \leq p_0). \tag{3.1}$$

Letting the characteristic polynomial of (ACP_n) take the form

$$P_\lambda := \sum_{i=0}^n \lambda^i A_i = \lambda^{n-p_0} \prod_{j=1}^{p_0} (\lambda + a_j B), \tag{3.2}$$

we consider the perturbed (ACP_n):

$$\begin{cases} u^{(n)}(t) + \sum_{i=0}^{n-1}(A_i + B_i)u^{(i)}(t) = 0, \quad t \geq 0, \\[2mm] u^{(k)}(0) = u_k, \quad 0 \leq k \leq n-1, \end{cases} \qquad (ACP_n)_{[B_{n-1}, \cdots, B_0]}$$

whose solutions are defined as in Section 2.4.

Write

$$\widetilde{P}_\lambda := P_\lambda + \sum_{i=0}^{n-1} \lambda^i B_i, \quad \lambda \in \mathbf{C}, \tag{3.3}$$

and $\widetilde{R}_\lambda := \widetilde{P}_\lambda^{-1}$ if the inverse exists.

Proposition 3.4. *Fix* $p \in \{1, \cdots, p_0\}$, $\lambda_0 \in \rho(B)$. *Let* $n \geq p_0 + 1$ *and let* B_0, $B_1, \cdots, B_{n-1}$ *be closable linear operators in* E. *Suppose*
  (i) $\mathcal{D}(B_i) \supset \mathcal{D}(B^p), 0 \leq i \leq n-1,$
  (ii) $B_i \mathcal{D}(B^p) \subset \mathcal{D}(B^{i-n+r+p+1}), i \geq n-r-p.$
Then $(ACP_n)_{[B_{n-1}, \cdots, B_0]}$ *has at most one solution for each initial data* $(u_0, \cdots, u_{n-1})$, *and for* $0 \leq i \leq k$, $0 \leq k \leq n-2$, $0 \leq j \leq n-1$,

$$\begin{cases} \mathcal{D}(B^{r+p}) \subset \mathcal{D}_{n-1}^j, \\[3mm] \bigcap_{i=0}^k \mathcal{D}(B^{(i+r-k+p-1)}(A_i + B_i)) \subset \mathcal{D}_k^j, \end{cases} \qquad \text{if} \ \ j \geq n-p+1; \tag{3.4}$$

$$\begin{cases} \mathcal{D}\left(B^{r+1}\right) \subset \mathcal{D}_{n-1}^j, \\ \bigcap\limits_{i=0}^{k} \mathcal{D}\left(B^{(i+r-k)}(A_i + B_i)\right) \subset \mathcal{D}_k^j, \end{cases} \quad \text{if} \quad 0 \le j \le n - p_0 - 1 \quad \text{or} \quad j = n - p.$$

$$(3.5)$$

Here for $0 \le j \le n - 1$,

$$\mathcal{D}_{n-1}^j := \left\{ u \in E; \ \lambda^j (\lambda_0 - B)^p \tilde{R}_\lambda u \in LT - E \right\},$$

$$\mathcal{D}_k^j := \left\{ u \in E; \ \lambda^{j+i-k-1} (\lambda_0 - B)^p \tilde{R}_\lambda (A_i + B_i) u \in LT - E \right.$$

$$\text{for each } 0 \le i \le k \right\}, \quad 0 \le k \le n - 2.$$

Moreover, there exist $C, \ \omega > 0$ *such that*

$$\sum_{j=0}^{p} \left\| B^j u(t) \right\| \le C e^{\omega t} \left[\sum_{j=0}^{(r-n+p+1)} \left\| B^j u_{n-1} \right\| \right.$$

$$+ \sum_{k=0}^{n-2} \sum_{i=0}^{k} \sum_{j=0}^{(i-k+r-n+p)} \left\| B^j (A_i + B_i) u_k \right\| \qquad (3.6)$$

$$\left. + \sum_{k=0}^{n-2} \sum_{j=0}^{p} \left\| B^j u_k \right\| \right] \quad (t \ge 0),$$

for any solution $u(\cdot)$ *of* $(ACP_n)_{[B_{n-1}, \cdots, B_0]}$ *with*

$$u_k \in \bigcap_{i=0}^{k} \mathcal{D}\left(B^{(i+r-k)}(A_i + B_i)\right) \bigcap \mathcal{D}(B^p) \quad (0 \le k \le n - 2),$$

$$u_{n-1} \in \mathcal{D}\left(B^{r+1}\right),$$

whenever the condition in Lemma 2.4.4 is satisfied.

Proof. Observe that for λ sufficiently large, $i, j \in \{1, \cdots, p_0\}$, $i \ne j$,

$$(\lambda + a_i B)^{-1} (\lambda + a_j B)^{-1}$$

$$= a_i (a_i - a_j)^{-1} \lambda^{-1} (\lambda + a_i B)^{-1} - a_j (a_i - a_j)^{-1} \lambda^{-1} (\lambda + a_j B)^{-1},$$

$$a_i B (\lambda + a_i B)^{-1} = I - \lambda (\lambda + a_i B)^{-1}. \qquad (3.7)$$

Accordingly, we see by (3.2) that given $m \in \{0, 1, \cdots, p_0\}$, there exist constants $C_0(m), C_1(m), \cdots, C_{p_0}(m)$ such that for λ sufficiently large,

$$B^m R_\lambda = \sum_{j=0}^{p_0} C_j(m) \lambda^{m-p_0+1} (\lambda + a_j B)^{-1}, \qquad (3.8)$$

where $R_\lambda := P_\lambda^{-1}$ (see (3.2)), $a_0 := 0$.

We shall show that for $0 \le i \le n - 1, 0 \le m \le p$,

$$\lambda^{i-n+p_0} B^m R_\lambda (\lambda_0 - B)^{-(i-n+r+p+1)} \in LT - \mathbf{L}(E). \tag{3.9}$$

In fact, for k any integer≥ 1, λ sufficiently large,

$$(\lambda + a_j B)^{-1}(\lambda_0 - B)^{-k}$$

$$= \lambda^{-1}(\lambda_0 - B)^{-k} - \lambda^{-1}(\lambda + a_j B)^{-1} a_j B(\lambda_0 - B)^{-k}$$

$$= \lambda^{-1}(\lambda_0 - B)^{-k} - \lambda^{-2} a_j B(\lambda_0 - B)^{-k}$$

$$+\lambda^{-2}(\lambda + a_j B)^{-1} a_j^2 B^2 (\lambda_0 - B)^{-k}$$

$$= \cdots \tag{3.10}$$

$$= \lambda^{-1}(\lambda_0 - B)^{-k} - \lambda^{-2} a_j B(\lambda_0 - B)^{-k} + \cdots$$

$$+(-1)^{k-1}\lambda^{-k} a_j^{k-1} B^{k-1}(\lambda_0 - B)^{-k}$$

$$+(-1)^k \lambda^{-k}(\lambda + a_j B)^{-1} a_j^k B^k (\lambda_0 - B)^{-k}, \quad 1 \le j \le p_0.$$

Since $B^l(\lambda_0 - B)^{-k} \in \mathbf{L}(E)$, $0 \le l \le k$, it follows by (3.1) that for k_1, k_2 any integers with $k_1, k_2 \le 0$, $k_1 + k_2 \le -r$

$$\lambda^{k_1}(\lambda + a_j B)^{-1}(\lambda_0 - B)^{k_2} \in LT - \mathbf{L}(E), \quad 1 \le j \le p_0. \tag{3.11}$$

Taking $m = 0$ in (3.8) yields that for λ sufficiently large,

$$\lambda^{p_0-1} R_\lambda = \sum_{j=0}^{p_0} C_j(0)(\lambda + a_j B)^{-1}, \tag{3.12}$$

which implies by (3.11) that

$$\lambda^{k_1+p_0-1} R_\lambda (\lambda_0 - B)^{k_2} \in LT - \mathbf{L}(E)$$

$$(k_1, \ k_2 \le 0, \ k_1 + k_2 \le -r). \tag{3.13}$$

On the other hand, (3.8) together with (3.1) shows that for k_1, k_2 any integers with $0 \le k_2 \le p_0$, $k_1 + k_2 \le -r$,

$$\lambda^{k_1+p_0-1} B^{k_2} R_\lambda \in LT - \mathbf{L}(E), \tag{3.14}$$

and therefore

$$\lambda^{k_1+p_0-1}(\lambda_0 - B)^{k_2} R_\lambda \in LT - \mathbf{L}(E). \tag{3.15}$$

Now, (3.9) follows from (3.13) and (3.14), noting that

$$B^m R_\lambda = \left[B^m (\lambda_0 - B)^{-p} \right] R_\lambda (\lambda_0 - B)^p.$$

By hypothesis,

$$B_i \in \mathbf{L}\left(E_p^B, E\right), \quad 0 \le i \le n-1;$$

moreover, for $i \ge n - r - p$,

$$(\lambda_0 - B)^{i-n+r+p+1} B_i \in \mathbf{L}\left(E_p^B, E\right),$$

since $B_i \mathcal{D}\left(B^p\right) \subset \mathcal{D}\left(B^{i-n+r+p+1}\right)$. Hence making use of Lemma 3.3, we obtain by (3.9) that

$$\sum_{i=0}^{n-1} \lambda^{i-n+p_0} R_\lambda B_i \in LT - \mathbf{L}\left(E_p^B\right). \tag{3.16}$$

This implies that, for λ sufficiently large, $\widetilde{P}_\lambda^{-1}$ exists and

$$\widetilde{R}_\lambda = \widetilde{P}_\lambda^{-1} = U_\lambda \cdot \lambda^{p_0-n} R_\lambda, \tag{3.17}$$

where

$$U_\lambda := \left[I + \sum_{i=0}^{n-1} \lambda^{i-n+p_0} R_\lambda B_i\right]^{-1}.$$

From (3.14), we know that there exists a positive integer q such that for each $u \in E$,

$$\lambda^{-q} \lambda^{p_0-n} R_\lambda u \in LT - E_p^B.$$

This combined with (3.16), (3.17) shows (using Theorem 1.1.11) that

$$\lambda^{-q} \widetilde{R}_\lambda u \in LT - E_p^B \quad \text{for each } u \in E,$$

which implies

$$\lambda^{-q} \widetilde{R}_\lambda u \in LT - E \quad (u \in E).$$

In conclusion, $(ACP_n)_{[B_{n-1}, \ldots, B_0]}$ has at most one solution for each initial value by virtue of Lemma 2.4.3.

From (3.13) and (3.15), we see (noting Lemma 3.2) that for q_1 any integer $\le p_0 - 1$,

$$\lambda^{q_1}(\lambda_0 - B)^{q_2} R_\lambda \in LT - \mathbf{L}\left(E_p^B\right), \tag{3.18}$$

where

$$q_2 = p - (q_1 - p_0 + r + p + 1).$$

Applying Theorem 1.1.11 again, it follows from (3.16) and (3.18) that for q_1, q_2 as above,

$$\lambda^{q_1} U_\lambda(\lambda_0 - B)^{q_2} R_\lambda \in LT - \mathbf{L}\left(E_p^B\right). \tag{3.19}$$

Now, for λ sufficiently large, $0 \le i \le k$, $0 \le k \le n-2$, $n-p+1 \le j \le n-1$, and

$$u \in \mathcal{D}\left(B^{(i+r-k+p-1)}(A_i + B_i)\right),$$

we write $\lambda^{j+i-k-1}(\lambda_0 - B)^p \widetilde{R}_\lambda(A_i + B_i)u$ as (see (3.17))

$$\lambda^{j-n+1}(\lambda_0 - B)^p \left[\lambda^{i-k+p_0-2} U_\lambda R_\lambda\right] (A_i + B_i)u.$$

Observe by (3.19) that for $0 \leq i \leq k$, $0 \leq k \leq n - 2$, and

$$u \in \mathcal{D}\left(B^{(i+r-k+p-1)}(A_i + B_i)\right),$$

we have

$$\left[\lambda^{i-k+p_0-2} U_\lambda(\lambda_0 - B)^{p-(i+r-k+p-1)} R_\lambda\right]$$

$$\cdot(\lambda_0 - B)^{(i+r-k+p-1)-p}(A_i + B_i)u$$

$$\in \quad LT - E_p^B,$$

noting

$$(\lambda_0 - B)^{(i+r-k+p-1)-p}(A_i + B_i)u \in \mathcal{D}\left(B^p\right).$$

Moreover, note that $j - n + 1 \leq 0$. We see that

$$\lambda^{j+i-k-1}(\lambda_0 - B)^p \widetilde{R}_\lambda(A_i + B_i)u \in LT - E$$

so that the second inclusion relation in (3.4) holds. Similarly, we have that, for $0 \leq i \leq k$, $0 \leq k \leq n - 2$, $0 \leq j \leq n - p_0 - 1$ or $j = n - p$, and $u \in \bigcap_{i=0}^{k} \mathcal{D}\left(B^{(i+r-k)}(A_i + B_i)\right),$

$$\lambda^{j+i-k-1}(\lambda_0 - B)^p \widetilde{R}_\lambda(A_i + B_i)u$$

$$= \quad \lambda^{j+p-n}(\lambda_0 - B)^p \left[\lambda^{i-k+p_0-p-1} U_\lambda(\lambda_0 - B)^{p-(i+r-k)} R_\lambda\right]$$

$$\cdot(\lambda_0 - B)^{(i+r-k)-p}(A_i + B_i)u$$

$$\in \quad LT - E.$$

Therefore the second inclusion relation in (3.5) holds.

Next, for $n - p + 1 \leq j \leq n - 1$, $u \in \mathcal{D}\left(B^{r+p}\right)$,

$$\lambda^j(\lambda_0 - B)^p \widetilde{R}_\lambda u$$

$$= \quad \lambda^{j-n+1}(\lambda_0 - B)^p \left[\lambda^{p_0-1} U_\lambda(\lambda_0 - B)^{-r} R_\lambda\right] (\lambda_0 - B)^r u$$

$$\in \quad LT - E,$$

which implies that

$$\mathcal{D}\left(B^{r+p}\right) \subset \mathcal{D}_{n-1}^j \quad (n - p + 1 \leq j \leq n - 1);$$

for $0 \leq j \leq n - p_0 - 1$ or $j = n - p$, $u \in \mathcal{D}\left(B^{r+1}\right)$,

$$\lambda^j (\lambda_0 - B)^p \widetilde{R}_\lambda u$$

$$= \lambda^{j+p-n} (\lambda_0 - B)^p \left[\lambda^{p_0-p} U_\lambda (\lambda_0 - B)^{p-r-1} R_\lambda\right] (\lambda_0 - B)^{r-p+1} u$$

$$\in LT - E,$$

which implies that

$$\mathcal{D}\left(B^{r+1}\right) \subset \mathcal{D}_{n-1}^j \quad (0 \leq j \leq n - p_0 - 1 \ \ or \ \ j = n - p).$$

Finally, let $u(\cdot)$ be a solution of $(ACP_n)_{[B_{n-1}, \cdots, B_0]}$ with

$$u_k \in \bigcap_{i=0}^{k} \mathcal{D}\left(B^{(i+r-k)}(A_i + B_i)\right) \quad (0 \leq k \leq n - 2), \quad u_{n-1} \in \mathcal{D}\left(B^{r+1}\right).$$

Observe

$$\widetilde{R}_\lambda u_{n-1} = \left[\lambda^{p_0-n} U_\lambda (\lambda_0 - B)^{p-(r-n+p+1)} R_\lambda\right] (\lambda_0 - B)^{(r-n+p+1)-p} u_{n-1},$$

$$\lambda^{i-k-1} \widetilde{R}_\lambda (A_i + B_i) u_k = \left[\lambda^{i-k+p_0-n-1} U_\lambda (\lambda_0 - B)^{p-(i-k+r-n+p)} R_\lambda\right]$$

$$\cdot (\lambda_0 - B)^{(i-k+r-n+p)-p} (A_i + B_i) u_k$$

$$(0 \leq i \leq k, \ 0 \leq k \leq n - 2),$$

and note (in view of (3.19))

$$\lambda^{p_0-n} U_\lambda (\lambda_0 - B)^{p-(r-n+p+1)} R_\lambda \in LT - \mathbf{L}\left(E_p^B\right),$$

$$\lambda^{i-k+p_0-n-1} U_\lambda (\lambda_0 - B)^{p-(i-k+r-n+p)} R_\lambda \in LT - \mathbf{L}\left(E_p^B\right).$$

We see from the proofs of Lemmas 2.2.1 and 2.4.4 that, for λ sufficiently large,

$$\int_0^\infty e^{-\lambda t} u(t) dt = \widetilde{R}_\lambda u_{n-1} + \sum_{k=0}^{n-2} \lambda^{-k-1} u_k - \sum_{k=0}^{n-2} \sum_{i=0}^{k} \lambda^{i-k-1} \widetilde{R}_\lambda (A_i + B_i) u_k,$$

whenever the condition in Lemma 2.4.4 is satisfied. (3.6) follows immediately.

Proposition 3.4, combined with Lemma 2.4.4, leads us to the following theorems.

Theorem 3.5. *Let* $n \geq 2$ *and* $-B$ *generate an* r-*times integrated semigroup on* E. *Suppose that* $B_0, \cdots, B_{n-2}$ *are closable linear operators in* E *such that*

$$\mathcal{D}(B_i) \supset \mathcal{D}(B) \quad (0 \leq i \leq n - 2),$$

$$B_i \mathcal{D}(B) \subset \mathcal{D}\left(B^{i-n+r+2}\right) \quad (i \geq n - r - 1).$$

Then the Cauchy problem

$$
\begin{cases}
u^{(n)}(t) + Bu^{(n-1)}(t) + \displaystyle\sum_{i=0}^{n-2} B_i u^{(i)}(t) = 0, \quad t \geq 0 \\[4mm]
u^{(k)}(0) = u_k, \quad 0 \leq k \leq n-1
\end{cases}
\tag{3.20}
$$

has a unique solution $u(\cdot)$ *for*

$$
u_k \in \bigcap_{i=0}^{k} \mathcal{D}\left(B^{(i+r-k)} B_i\right) \quad (0 \leq k \leq n-2), \qquad u_{n-1} \in \mathcal{D}\left(B^{r+1}\right),
$$

and (3.6) *holds for* $p = 1$.

Proof. Take $p_0 = p = 1, B_{n-1} = 0$ in Proposition 3.4. Noting $n - p_0 - 1 = n - 2, n - p = n - 1$, we have by (3.5) that

$$
\mathcal{D}\left(B^{r+1}\right) \subset \mathcal{D}_{n-1}^{j}, \quad \text{for any } 0 \leq j \leq n-1;
$$

$$
\bigcap_{i=0}^{k} \mathcal{D}\left(B^{(i+r-k)} B_i\right) \subset \mathcal{D}_k^j, \quad \text{for any } 0 \leq j \leq n-1, \ 0 \leq k \leq n-2.
$$

Corollary 3.6. *Let* $n \geq 2$ *and* $-B$ *generate a strongly continuous semigroup on* E. *Suppose that* $B_0, \cdots, B_{n-2}$ *are closable linear operators in* E *such that*

$$
\mathcal{D}(B_i) \supset \mathcal{D}(B) \quad (0 \leq i \leq n-2).
$$

Then the Cauchy problem (3.20) *has a unique solution* $u(\cdot)$ *for any*

$$
u_k \in \bigcap_{i=0}^{k} \mathcal{D}(B_i) \ (0 \leq k \leq n-2), \qquad u_{n-1} \in \mathcal{D}(B),
$$

and there are constants $C, \omega > 0$ *such that for* $t \geq 0$,

$$
\|u(t)\| + \|Bu(t)\| \leq Ce^{\omega t}\left(\|u_{n-1}\| + \sum_{k=0}^{n-2}(\|u_k\| + \|Bu_k\|)\right),
$$

if $u_0, \cdots, u_{n-2} \in \mathcal{D}(B)$.

(See also Corollary 2.4.8 for a related result).

Proof. Take $r = 0$ and apply Theorem 3.5, noting that there exists a constant C' such that

$$
\|B_i u\| \leq C'(\|u\| + \|Bu\|) \quad (0 \leq i \leq n-2),
$$

since $\mathcal{D}(B_i) \supset \mathcal{D}(B)$.

Corollary 3.7. *Let $n \geq 2$ and $-B$ generate an n-times integrated semigroup on E. Suppose that $B_0, \cdots, B_{n-2}$ are closable linear operators in E such that for any $0 \leq i \leq n-2$,*

$$\mathcal{D}(B_i) \supset \mathcal{D}(B) \quad and \quad B_i\mathcal{D}(B) \subset \mathcal{D}\left(B^{i+2}\right).$$

Then the Cauchy problem (3.20) has a unique solution $u(\cdot)$ for any

$$u_k \in \bigcap_{i=0}^{k} \mathcal{D}\left(B^{i+n-k}B_i\right) \quad (0 \leq k \leq n-2), \quad u_{n-1} \in \mathcal{D}\left(B^{n+1}\right),$$

and there are constants $C, \omega > 0$ such that for $t \geq 0$,

$$\|u(t)\| + \|Bu(t)\| \leq Ce^{\omega t}\left(\|u_{n-1}\| + \|Bu_{n-1}\| + \left\|B^2 u_{n-1}\right\|\right.$$

$$\left. + \sum_{k=0}^{n-2}(\|BB_k u_k\| + \|u_k\| + \|Bu_k\|)\right),$$

if $u_0, \cdots, u_{n-2} \in \mathcal{D}(B)$.

Proof. Take $r = n$ and apply Theorem 3.5.

Corollary 3.8. *Let $n \geq 2$ and $-B$ generate a once integrated semigroup on E. Suppose that $B_0, \cdots, B_{n-2}$ are closable linear operators in E such that for $0 \leq i \leq n-2$,*

$$\mathcal{D}(B_i) \supset \mathcal{D}(B) \quad and \quad B_{n-2}\mathcal{D}(B) \subset \mathcal{D}(B).$$

Then the Cauchy problem (3.20) has a unique solution $u(\cdot)$ for any

$$u_0 \in \mathcal{D}(BB_0),$$

$$u_k \in \bigcap_{i=0}^{k-1} \mathcal{D}(B_i)\bigcap\mathcal{D}(BB_k) \quad (1 \leq k \leq n-2),$$

$$u_{n-1} \in \mathcal{D}\left(B^2\right),$$

and there are constants $C, \omega > 0$ such that for $t \geq 0$,

$$\|u(t)\| + \|Bu(t)\| \leq Ce^{\omega t}(\|u_1\| + \|Bu_1\| + \|u_0\| + \|Bu_0\|), \quad if \quad n = 2,$$

$$\|u(t)\| + \|Bu(t)\| \leq Ce^{\omega t}\left(\|u_{n-1}\| + \sum_{k=0}^{n-2}(\|u_k\| + \|Bu_k\|)\right), \quad if \quad n \geq 3,$$

with $u_0, \cdots, u_{n-2} \in \mathcal{D}(B)$.

Proof. Take $r = 1$ and apply Theorem 3.5.

Theorem 3.9. *Let* $n \geq 3$. *Suppose that both* B *and* $-B$ *are the generators of* r-*times integrated semigroups on* E, *and* $B_0, \cdots, B_{n-3}$ *are closable linear operators in* E *such that*

$$\mathcal{D}(B_i) \supset \mathcal{D}\left(B^2\right) \quad (0 \leq i \leq n-3),$$

$$B_i \mathcal{D}\left(B^2\right) \subset \mathcal{D}\left(B^{i-n+r+3}\right) \quad (i \geq n-r-2).$$

Then the Cauchy problem

$$\begin{cases} u^{(n)}(t) - B^2 u^{(n-2)}(t) + \displaystyle\sum_{i=0}^{n-3} B_i u^{(i)}(t) = 0, \quad t \geq 0 \\ u^{(k)}(0) = u_k, \quad 0 \leq k \leq n-1 \end{cases} \tag{3.21}$$

has a unique solution $u(\cdot)$ *for*

$$u_k \in \bigcap_{i=0}^{k} \mathcal{D}\left(B^{(i+r-k)} B_i\right) \quad (0 \leq k \leq n-3),$$

$$u_{n-2} \in \mathcal{D}\left(B^{r+2}\right), \quad u_{n-1} \in \mathcal{D}\left(B^{r+1}\right),$$

and

$$\sum_{l=0}^{2} \left\|B^l u(t)\right\| \leq Ce^{\omega t}\left[\sum_{l=0}^{(r-n+3)} \left\|B^l u_{n-1}\right\| + \sum_{l=0}^{(r-n+4)} \left\|B^l u_{n-2}\right\|\right.$$

$$+ \sum_{k=0}^{n-3}\sum_{i=0}^{k} \sum_{l=0}^{(i-k+r-n+2)} \left\|B^l B_i u_k\right\| \tag{3.22}$$

$$\left.+ \sum_{k=0}^{n-3}\sum_{l=0}^{2} \left\|B^l u_k\right\|\right] \quad (t \geq 0),$$

for some $C, \omega > 0$, *if* $u_k \in \mathcal{D}\left(B^2\right)$, $0 \leq k \leq n-3$.

Proof. Take $p_0 = p = 2, B_{n-1} = B_{n-2} = 0$ in Proposition 3.4. Noting $n - p_0 - 1 = n - 3, n - p = n - 2$, and $A_{n-1} = 0$, we have by (3.5) and Lemma 2.4.4 that for

$$u_{n-1} \in \mathcal{D}\left(B^{r+1}\right), \quad u_k \in \bigcap_{i=0}^{k} \mathcal{D}\left(B^{(i+r-k)} B_i\right) \quad (0 \leq k \leq n-2),$$

(3.21) has a solution.

Moreover, we claim that u_{n-2} can take values arbitrarily in $\mathcal{D}\left(B^{r+2}\right)$. In fact, for $0 \leq j \leq n-2, u \in \mathcal{D}\left(B^{r+2}\right)$,

$$\lambda^j(\lambda_0 - B)^2 \sum_{i=0}^{n-2} \lambda^{i-n+1} \widetilde{R}_\lambda (A_i + B_i) u$$

$$= \lambda^{j-n+1}(\lambda_0 - B)^2 u - \lambda^{j+1}(\lambda_0 - B)^2 \widetilde{R}_\lambda u$$

$$= \lambda^{j-n+1}(\lambda_0 - B)^2 u - \lambda^{j-n+2}(\lambda_0 - B)^2 \left[\lambda U_\lambda (\lambda_0 - B)^{-r} R_\lambda\right](\lambda_0 - B)^r u$$

$$\in LT - E,$$

by (3.19).

(3.22) follows from (3.6) and the following observations. If $u_{n-2}(\cdot)$ is the solution of (3.21) with

$$u_0 = u_1 = \cdots = u_{n-3} = u_{n-1} = 0, \quad u_{n-2} \in \mathcal{D}\left(B^{r+2}\right),$$

then from the proof of Lemma 2.2.1, we see that for λ sufficiently large,

$$\int_0^\infty e^{-\lambda t} u_{n-2}(t) dt$$

$$= \lambda^{-n+1} u_{n-2} - \sum_{i=0}^{n-2} \lambda^{i-n+1} \widetilde{R}_\lambda (A_i + B_i) u_{n-2}$$

$$= \lambda \widetilde{R}_\lambda u_{n-2}$$

$$= \left[\lambda^{3-n} U_\lambda (\lambda_0 - B)^{2-(r-n+4)} R_\lambda\right](\lambda_0 - B)^{(r-n+4)-2} u_{n-2};$$

(3.19) implies that

$$\lambda^{3-n} U_\lambda (\lambda_0 - B)^{2-(r-n+4)} R_\lambda \in LT - \mathbf{L}\left(E_2^B\right).$$

Corollary 3.10. *Suppose that both B and $-B$ generate r-times integrated semigroups on E. Suppose that B_0 is a closable linear operator in E such that*

$$\mathcal{D}(B_0) \supset \mathcal{D}\left(B^2\right) \quad and \quad B_0 \mathcal{D}\left(B^2\right) \subset \mathcal{D}(B).$$

Then the Cauchy problem

$$\begin{cases} u'''(t) - B^2 u'(t) + B_0 u(t) = 0 \quad (t \geq 0) \\ u^{(i)}(0) = u_i \quad (i = 0, 1, 2) \end{cases}$$

has a unique solution $u(\cdot)$ *for any*

$$u_0 \in \mathcal{D}(BB_0), \quad u_1 \in \mathcal{D}\left(B^3\right), \quad u_2 \in \mathcal{D}\left(B^2\right),$$

and there are constants $C, \omega > 0$ *such that for* $t \geq 0$,

$$\|u(t)\| + \|Bu(t)\| + \left\|B^2 u(t)\right\|$$

$$\leq \quad Ce^{\omega t}\left(\|u_2\| + \|Bu_2\| + \|u_1\| + \|Bu_1\| + \left\|B^2 u_1\right\|\right.$$

$$\left. + \|u_0\| + \|Bu_0\| + \left\|B^2 u_0\right\|\right),$$

if $u_0 \in \mathcal{D}\left(B^2\right)$.

Proof. Take $n = 3, r = 1$ and apply Theorem 3.9.

For arbitrary p_0, p, we have

Theorem 3.11. *Suppose that the hypotheses in Proposition 3.4 hold, except that* (ii) *is replaced by*
(ii)' $B_i\mathcal{D}\left(B^m\right) \subset \mathcal{D}\left(B^{i-n+r+m+1}\right)$ $(i \geq n - r - m, \ p \leq m \leq p_0)$.
Then $(ACP_n)_{[B_{n-1}, \cdots, B_0]}$ *has a unique solution* $u(\cdot)$ *for any*

$$u_k \in \bigcap_{i=0}^{k} \mathcal{D}\left(B^{(i+r-k+p-1)}(A_i + B_i)\right) \quad (0 \leq k \leq n - 2),$$

$$u_{n-1} \in \mathcal{D}\left(B^{r+p}\right),$$

and (3.6) *holds if, in addition,* $u_k \in \mathcal{D}\left(B^p\right)$ $(0 \leq k \leq n - 2)$.

Proof. The condition (ii)' implies that (3.5) holds also for $j \in \{n - p_0, \cdots, n - p\}$. This combined with (3.4) justifies our conclusion.

Example 3.12. Let B be the generator of a once integrated semigroup on E. Consider the Cauchy problem

$$\begin{cases} u^{(4)}(t) - Bu^{(3)}(t) + \delta B^2 u^{(2)}(t) + C_0 u^{(2)}(t) + C_1 Bu^{(1)}(t) \\ \\ \qquad + C_2 u^{(1)}(t) + C_3 Bu(t) + C_4 u(t) = 0 \quad (t \geq 0), \qquad (3.23) \\ \\ u^{(k)}(0) = u_k \quad (k = 0, 1, 2, 3), \end{cases}$$

where $C_0, \cdots, C_4$ are any complex numbers, $\delta \in [0, \frac{1}{4})$. Then (3.23) has a unique solution whenever $u_0, u_1, u_3 \in \mathcal{D}\left(B^2\right)$, $u_2 \in \mathcal{D}\left(B^3\right)$, and

$$\|u(t)\| \leq Ce^{\omega t}\left(\|u_0\| + \|Bu_0\| + \|u_1\| + \|Bu_1\|\right.$$

$$\left. + \|u_2\| + \|Bu_2\| + \left\|B^2 u_2\right\| + \|u_3\|\right) \quad (t \geq 0),$$

for some C, $\omega > 0$.

The assertion above can be checked by taking

$$r = 1, \quad p_0 = 2, \quad p = 1, \quad a_{1,2} = -\frac{1}{2}\left(1 \pm \sqrt{1 - 4\delta}\right),$$

$$B_0 = C_3 B + C_4 I, \quad B_1 = C_1 B + C_2 I, \quad B_2 = C_0 I, \quad B_3 = 0,$$

and then applying Theorem 3.11.

3.4 The special case (III): coefficient operators relating to integrated cosine functions

Notation 4.1. Let A be a closed linear operator in E. By $[\mathcal{D}(A)]$, we denote the Banach space $\mathcal{D}(A)$ endowed with the graph norm

$$\|u\|_{[\mathcal{D}(A)]} := \|u\| + \|Au\|.$$

This section deals with the following Cauchy problem:

$$\begin{cases} u^{(n+2)}(t) = Au^{(n)}(t) + \displaystyle\sum_{i=0}^{n-1} B_i u^{(i)}(t) & (t \geq 0), \\[2mm] u^{(j)}(0) = u_j & (0 \leq j \leq n+1), \end{cases} \tag{4.1}$$

where $B_0, \cdots, B_{n-1}$ are closable linear operators in E and A is the generator of an r-times integrated cosine function ($r \in N_0$) on E.

Theorem 4.2. *Let* $A, B_0, \cdots, B_{n-1}$ *be as above. Assume that for each* $0 \leq i \leq n-1, \mathcal{D}(A) \subset \mathcal{D}(B_i)$ *such that*

$$B_i \mathcal{D}(A) \subset \mathcal{D}\left(A^{\langle \frac{1}{2}(i-n+r+2)\rangle}\right) \quad (n-r \leq i \leq n-1).$$

Then (4.1) *has a unique solution* $u(\cdot)$ *for any*

$$u_j \in \bigcap_{i=0}^{j} \mathcal{D}\left(A^{\langle \frac{1}{2}(r-i+1)\rangle} B_{j-i}\right) \quad (0 \leq j \leq n-1),$$

$$u_n \in \mathcal{D}\left(A^{\langle \frac{1}{2}(r+3)\rangle}\right), \quad u_{n+1} \in \mathcal{D}\left(A^{\langle \frac{1}{2}(r+2)\rangle}\right).$$

Moreover, there exist constants $C, \omega > 0$ such that for $t \geq 0$,

$$\|u(t)\| + \|Au(t)\| \ \leq \ Ce^{\omega t}\Bigg\{ \sum_{l=0}^{\langle\frac{1}{2}(r+3)\rangle} \|A^l u_n\| + \sum_{l=0}^{\langle\frac{1}{2}(r+2)\rangle} \|A^l u_{n+1}\|$$

$$+ \sum_{j=0}^{n-1}\sum_{i=0}^{j}\sum_{l=0}^{\langle\frac{1}{2}(r-i+1)\rangle} \|A^l B_{j-i} u_j\| \tag{4.2}$$

$$+ \sum_{i=0}^{n-1}(\|u_i\| + \|Au_i\|)\Bigg\}.$$

Proof. First, we have

$$\lambda^{1-r}\left(\lambda^2 - A\right)^{-1} \in LT - L(E), \tag{4.3}$$

since A is the generator of an r-times integrated cosine function. Take $a \in \rho(A)$. Observe that for k any positive integer, $u \in \mathcal{D}\left(A^k\right)$,

$$\left(\lambda^2 - A\right)^{-1} u \ = \ \left\{\lambda^{-2} + \lambda^{-2}A\left(\lambda^2 - A\right)^{-1}\right\} u$$

$$= \ \cdots$$

$$= \ \left\{\lambda^{-2} + \cdots + \lambda^{-2k}A^{k-1} + \lambda^{-2k}A^k\left(\lambda^2 - A\right)^{-1}\right\} u.$$

It follows from (4.3) that for k any integer with $0 \leq k \leq \frac{1}{2}r$,

$$\lambda^{1-r}\lambda^{2k}\left(\lambda^2 - A\right)^{-1}(a - A)^{-k} \in LT - L(E), \tag{4.4}$$

noting $A^l(a-A)^{-k} \in L(E)$, $(l = 1, \cdots, k)$. Therefore for k any integer with $0 \leq k \leq \frac{1}{2}(r+2)$,

$$\lambda^{1-r}\lambda^{2k-2}A\left(\lambda^2 - A\right)^{-1}(a - A)^{-k} \in LT - L(E), \tag{4.5}$$

because of the following equalities: for $1 \leq k \leq \frac{1}{2}(r+2)$,

$$\lambda^{1-r}\lambda^{2k-2}A\left(\lambda^2 - A\right)^{-1}(a - A)^{-k}$$

$$= \ \lambda^{1-r}\lambda^{2(k-1)}\left(\lambda^2 - A\right)^{-1}(a - A)^{-(k-1)}A(a-A)^{-1},$$

$$\lambda^{1-r}\lambda^{-2}A\left(\lambda^2 - A\right)^{-1} = \lambda^{1-r}\left(\lambda^2 - A\right)^{-1} - \lambda^{1-r}\lambda^{-2}.$$

From (4.4) and (4.5), it follows that for $0 \leq i \leq n-1, m = 0, 1$,

$$\lambda^{i-n}A^m\left(\lambda^2 - A\right)^{-1}(a - A)^{-\langle\frac{1}{2}(i-n+r+2)\rangle} \in LT - L(E),$$

by noting

$$2 \cdot \left[\frac{1}{2}(i - n + r + 2) \right] \geq i - n + r + 1.$$

Consequently, for $0 \leq i \leq n - 1$,

$$\lambda^{i-n} \left(\lambda^2 - A \right)^{-1} (a - A)^{-\langle \frac{1}{2}(i-n+r+2) \rangle} \in LT - L(E, [\mathcal{D}(A)]), \qquad (4.6)$$

by an application of Theorem 1.1.10. By hypothesis,

$$(a - A)^{\langle \frac{1}{2}(i-n+r+2) \rangle} B_i \in L([\mathcal{D}(A)], E) \quad (0 \leq i \leq n - 1),$$

which implies, by (4.6), that

$$\lambda^{i-n} \left(\lambda^2 - A \right)^{-1} B_i \in LT - L([\mathcal{D}(A)]) \quad (0 \leq i \leq n - 1).$$

Now let

$$\sum_{i=0}^{n-1} \lambda^{i-n} \left(\lambda^2 - A \right)^{-1} B_i u = \mathcal{L}[x(t)u](\lambda), \qquad (4.7)$$

for $u \in \mathcal{D}(A)$, λ sufficiently large, where $\{x(t)\}_{t \geq 0}$ is a strongly continuous, exponentially bounded family of bounded linear operators on $[\mathcal{D}(A)]$. Set

$$V_\lambda = \left\{ I - \sum_{i=0}^{n-1} \lambda^{i-n} \left(\lambda^2 - A \right)^{-1} B_i \right\}^{-1}.$$

Then for $u \in \mathcal{D}(A)$, λ sufficiently large,

$$(V_\lambda - I)u = \mathcal{L} \left[\sum_{i=1}^{\infty} (x(t))^{*i} u \right] (\lambda). \qquad (4.8)$$

Using (4.4) and (4.5) again, we obtain that for $p = -2, -1, 0, \cdots, n - 1$,

$$\lambda^{-p-1}(a - A)^{q(p)} \left(\lambda^2 - A \right)^{-1} \in LT - L(E), \qquad (4.9)$$

where

$$q(p) := \begin{cases} -\left\langle \frac{1}{2}(r - p - 1) \right\rangle, & \text{if } p \leq r - 1, \\ \\ 1, & \text{if } p \geq r. \end{cases} \qquad (4.10)$$

Now for $u \in E$, λ sufficiently large, p, $q(p)$ as above, let $y_p(t)$ be such that

$$\lambda^{-p-1}(a - A)^{q(p)} \left(\lambda^2 - A \right)^{-1} u = \mathcal{L}[y_p(t)u](\lambda). \qquad (4.11)$$

We see easily that

$$y_p(t)(a - A)^{-1} = (a - A)^{-1} y_p(t) \quad (p = -2, -1, 0, 1, \cdots, n - 1),$$

which implies

$$Ay_p(t)u = y_p(t)Au \quad (u \in \mathcal{D}(A),\ t \geq 0,\ -2 \leq p \leq n-1).$$

So $\{y_p(t)\}_{t \geq 0}$ is also a strongly continuous, exponentially bounded family of bounded linear operators on $[\mathcal{D}(A)]$. It follows from (4.8) and (4.11) that for $u \in \mathcal{D}(A)$, λ sufficiently large, $-2 \leq p \leq n-1$, $q(p)$ as in (4.10),

$$\lambda^{-p-1} V_\lambda (a-A)^{q(p)} \left(\lambda^2 - A\right)^{-1} u = \mathcal{L}\left[z_p(t)u\right](\lambda), \tag{4.12}$$

where

$$z_p(t) = y_p(t) + y_p(t) * \sum_{m=1}^{\infty} (x(t))^{*m}.$$

Setting, for λ sufficiently large

$$R_\lambda := \left(\lambda^{n+2} - \lambda^n A - \sum_{i=0}^{n-1} \lambda^i B_i\right)^{-1},$$

then $\lambda^n R_\lambda = V_\lambda \left(\lambda^2 - A\right)^{-1}$. Consequently, we have by (4.12) that

$$\lambda^n R_\lambda \left(u_{n+1} + \lambda u_n + \sum_{j=0}^{n-1} \sum_{p=0}^{j} \lambda^{-p-1} B_{j-p} u_j\right) = \mathcal{L}[v(t)](\lambda) \tag{4.13}$$

where

$$
\begin{aligned}
v(t) \ := \ & z_{-1}(t)(a-A)^{\langle \frac{1}{2} r \rangle} u_{n+1} + z_{-2}(t)(a-A)^{\langle \frac{1}{2}(r+1) \rangle} u_n \\
& + \sum_{j=0}^{n-1} \sum_{p=0}^{j} z_p(t)(a-A)^{q(p)} B_{j-p} u_j \tag{4.14}
\end{aligned}
$$

$$\in \ \ C(R^+, [\mathcal{D}(A)]),$$

noting $(a-A)^{\langle \frac{1}{2} r \rangle} u_{n+1},\ (a-A)^{\langle \frac{1}{2}(r+1) \rangle} u_n \in \mathcal{D}(A)$ and

$$(a-A)^{q(p)} B_{j-p} u_j \in \mathcal{D}(A) \quad (0 \leq p \leq j,\ 0 \leq j \leq n-1),$$

by hypothesis. From (4.13) and the identity

$$\lambda^n R_\lambda = \lambda^{-2} \lambda^n A R_\lambda + \lambda^{-2} \sum_{i=0}^{n-1} \lambda^i B_i R_\lambda + \lambda^{-2},$$

we deduce by the uniqueness theorem for Laplace transforms that for $t \geq 0$,

$$
\begin{aligned}
v(t) \ = \ & \int_0^t (t-s) A v(s) ds + \int_0^t (t-s) \sum_{i=0}^{n-1} B_i J^{n-i} v(s) ds \\
& + t u_{n+1} + u_n + \int_0^t (t-s) I_0(s) ds,
\end{aligned}
$$

where for $s \geq 0$,

$$I_0(s) := \sum_{j=0}^{n-1} \sum_{p=0}^{j} \frac{s^p}{p!} B_{j-p} u_j.$$

Hence

$$\begin{cases} v''(t) = Av(t) + \sum_{i=0}^{n-1} B_i J^{n-i} v(t) + I_0(t) & (t \geq 0), \\ \\ v(0) = u_n, \quad v'(0) = u_{n+1}. \end{cases}$$

Accordingly,

$$u(t) := \sum_{j=0}^{n-1} \frac{t^j}{j!} u_j + J^n v(t) \quad (t \geq 0)$$

is a solution of (4.1), and (4.2) follows from (4.14).

For uniqueness, we observe from (4.4) and (4.5) that

$$\lambda^{-r-1} \left(\lambda^2 - A \right)^{-1} \in LT - \mathbf{L}(E, [\mathcal{D}(A)]).$$

So

$$\lambda^{-r-1} \lambda^n R_\lambda = \lambda^{-r-1} V_\lambda \left(\lambda^2 - A \right)^{-1} \in LT - \mathbf{L}(E, [\mathcal{D}(A)])$$

due to (4.8). Thus R_λ is polynomially bounded, which completes the proof by virtue of Lemma 2.4.3.

3.5 *C*-wellposedness

Throughout this section, $A_0, \cdots, A_{n-1}$ be linear operators in E and C is a bounded, injective operator on E such that $A_i = C^{-1} A_i C$, $0 \leq i \leq n-1$.

We write

$$\rho_C(A_0, \cdots, A_{n-1})$$

$$:= \{\lambda \in \mathbf{C}; \ P_\lambda^{-1} \text{ exists}, \ \mathcal{D}(R_\lambda) \supset \mathcal{R}(C), \ R_\lambda C \in \mathbf{L}(E)\},$$

$$\rho(A_0, \cdots, A_{n-1}) := \rho_I(A_0, \cdots, A_{n-1}).$$

Definition 5.1. The tuple $\{S_0(t), \cdots, S_{n-1}(t)\}_{t \geq 0}$ of strongly continuous families of bounded linear operators on E is called a strong C-propagation family for (ACP_n) if
  (i) C commutes with $S_k(t)$ for each $t \geq 0$, $0 \leq k \leq n-1$;
  (ii) for each $u \in E$, $1 \leq k \leq n-1$, $S_k(\cdot)u \in C^k(R^+, E)$, $S_{n-1}^{(k-1)}(t)u \in \mathcal{D}(A_k)$ $(t \geq 0)$ and $A_k S_{n-1}^{(k-1)}(\cdot)u \in C(R^+, E)$;

(iii) for each $u \in E$ and $t \geq 0$, $\int_0^t S_{n-1}(s)u\,ds \in \mathcal{D}(A_0)$ such that

$$A_0 \int_0^t S_{n-1}(s)u\,ds = Cu - S_{n-1}^{(n-1)}(t)u - \sum_{k=1}^{n-1} A_k S_{n-1}^{(k-1)}(t)u,$$

$$S_{n-1}^{(j)}(0) = 0 \quad (0 \leq j \leq n-2);$$

(iv) there exist constants $M, \omega > 0$ such that

$$\|S_0(t)\|, \quad \left\|S_k^{(k)}(t)\right\|, \quad \left\|A_k S_{n-1}^{(k-1)}(t)\right\| \leq M e^{\omega t}, \quad t \geq 0, \ 1 \leq k \leq n-1;$$

(v) any solution $u(\cdot)$ of (ACP_n) with initial values $u_0, \cdots, u_{n-1} \in \mathcal{R}(C)$ can be expressed as

$$u(t) = \sum_{k=0}^{n-1} S_k(t)C^{-1}u_k, \quad t \geq 0. \tag{5.1}$$

Definition 5.2. The Cauchy problem (ACP_n) is called strongly C-wellposed if there exists a strong C-propagation family for (ACP_n).

Immediately, we know that any solution $u(\cdot)$ of (ACP_n) is unique and

$$u(t) = C^{-1} \sum_{k=0}^{n-1} S_k(t)u_k, \quad t \geq 0, \tag{5.2}$$

whenever (ACP_n) is strongly C-wellposed. Indeed, $Cu(\cdot)$ is also a solution of (ACP_n) with initial values $u^{(k)}(0) = Cu_k \in \mathcal{R}(C)$ $(0 \leq k \leq n-1)$, since C commutes with each of A_k. Hence

$$Cu(t) = \sum_{k=0}^{n-1} S_k(t)u_k,, \quad t \geq 0,$$

by (5.1). Then (5.2) follows.

Remark. When $\bigcap_{i=0}^{n-1} \mathcal{D}(A_i)$ is dense in E and $C = I$, the definition here of strong C-wellposedness coincides with that of strong wellposedness in Section 2.2. This can be seen from equality (2.1.8) and the following result.

Proposition 5.3. *Let (ACP_n) be strongly C-wellposed. Then*
(i) *(ACP_n) has a solution for every*

$$u_k \in C\left(\bigcap_{i=0}^{k} \mathcal{D}(A_i)\right), \quad 0 \leq k \leq n-1;$$

(ii) *for $t \geq 0$,*

$$S_0(t)u = Cu - \int_0^t S_{n-1}(s)A_0 u \, ds, \quad u \in \mathcal{D}(A_0),$$

and for $1 \leq k \leq n-1$,

$$S_k'(t)u = S_{k-1}(t)u - S_{n-1}(t)A_k u, \quad u \in \bigcap_{i=0}^{k} \mathcal{D}(A_i);$$

(iii) $(\omega, \infty) \subset \rho_C(A_0, \cdots, A_{n-1})$ *and for $\lambda > \omega$*

$$\lambda^{n-1} R_\lambda C u = \mathcal{L}\left[S_{n-1}^{(n-1)}(t)u\right](\lambda), \quad u \in E,$$

$$\lambda^{k-1} A_k R_\lambda C u = \mathcal{L}\left[A_k S_{n-1}^{(k-1)}(t)u\right](\lambda), \quad 1 \leq k \leq n-1, \ u \in E,$$

$$\lambda^{k-1} R_\lambda C A_k u = \mathcal{L}\left[S_{k-1}^{(k-1)}(t)u - S_k^{(k)}(t)u\right](\lambda),$$

$$1 \leq k \leq n-1, \ u \in \bigcap_{i=0}^{k} \mathcal{D}(A_i),$$

$$\lambda^{-1} A_0 R_\lambda C u = \mathcal{L}\left[A_0 \int_0^t S_{n-1}(s)u \, ds\right](\lambda), \quad u \in E,$$

$$\lambda^{-1} R_\lambda C A_0 u = \mathcal{L}\left[Cu - S_0(t)u\right](\lambda), \quad u \in \mathcal{D}(A_0).$$

Proof. It is easy to verify, by Definition 5.1 (iii), that for each $u \in \mathcal{D}(A_0)$,

$$v_0(t; \, u) := Cu - \int_0^t S_{n-1}(s)A_0 u \, ds$$

is a solution of (ACP_n) with initial values $u_0 = Cu$, $u_j = 0$ $(1 \leq j \leq n-1)$; for each $u \in \bigcap_{i=0}^{k} \mathcal{D}(A_i)$, $1 \leq k \leq n-1$,

$$v_k(t; \, u) := \int_0^t (v_{k-1}(s; \, u) - S_{n-1}(s)A_k u) ds$$

is a solution of (ACP_n) with initial values $u_k = Cu$, $u_j = 0$ $(0 \leq j \leq n-1$, $j \neq k)$, by an inductive argument. Then part (i) follows immediately and part (ii) is derived from (5.1).

Take the Laplace transform to the two sides of the first equality in (iii) of Definition 5.1. We obtain, noting Definition 5.1 (iv), that for $u \in E$, $\lambda > \omega$,

$$\mathcal{L}\left[A_0 \int_0^t S_{n-1}(s)u \, ds\right](\lambda)$$

$$= \lambda^{-1} C u - \mathcal{L}\left[S_{n-1}^{(n-1)}(t)u\right](\lambda) - \sum_{k=1}^{n-1} \mathcal{L}\left[A_k S_{n-1}^{(k-1)}(t)u\right](\lambda).$$

Integrating by parts and using the closedness of A_k $(0 \leq k \leq n-1)$, we get

$$P_\lambda \mathcal{L}\left[S_{n-1}(t)u\right](\lambda) = Cu, \qquad u \in E, \; \lambda > \omega.$$

Thus, proceeding similarly as in the proof of the implication (i) $\implies$ (ii) in Theorem 2.2.2 (with (5.2) in place of (2.1.2)), we obtain part (iii). This finishes the proof.

Remark. We now pay attention to the first equality in (iii) of Proposition 5.3. When $n = 1$, it reduces to

$$(\lambda + A_0)^{-1} Cu = \mathcal{L}\left[S_0(t)u\right](\lambda), \qquad u \in E, \; \lambda > \omega,$$

and so one is getting a C-regularized semigroup $S_0(t)$. When $n = 2$ and $A_1 = 0$, it reduces to

$$\lambda(\lambda^2 + A_0)^{-1} Cu = \mathcal{L}\left[S_1'(t)u\right](\lambda), \qquad u \in E, \; \lambda > \omega,$$

and so one is getting a C-regularized cosine function $S_1'(t)$.

Theorem 5.4. *The Cauchy problem* (ACP_n) *is strongly C-wellposed if and only if the following statements hold:*
  (i) $(\omega, \infty) \subset \rho_C(A_0, \cdots, A_{n-1})$ *for some* $\omega > 0$;
  (ii) *for* $0 \leq k \leq n - 2$,

$$\lambda^{n-1} R_\lambda C, \; \lambda^{k-1} A_k R_\lambda C \in LT - \mathbf{L}(E), \tag{5.3}$$

and there exists a strongly continuous function $T_k(\cdot) : R^+ \to \mathbf{L}(E)$ *satisfying that* $\|T_k(t)\| \leq M e^{\omega t}$ $(t \geq 0)$ *for some* $M > 0$, $T_k(t)$ *commutes with C for each* $t \geq 0$, *and for* $\lambda > \omega$,

$$\lambda^{k-1} R_\lambda C A_k u = \mathcal{L}[T_k(t)u](\lambda), \qquad u \in \bigcap_{i=0}^{k} \mathcal{D}(A_i).$$

Proof. The "only if" part is given by Proposition 5.3.
    The "if" part. Denote by $T_n(\cdot)$ the determining function of $\lambda \mapsto \lambda^{n-1} R_\lambda C$. It is clear that (5.3) holds also for $k = n - 1$ and

$$\lambda^{n-2} R_\lambda C A_{n-1} u = \mathcal{L}\left[u - T_n(t)u - \sum_{k=0}^{n-2} T_k(t)u\right](\lambda), \qquad u \in \bigcap_{i=0}^{n-1} \mathcal{D}(A_i).$$

For $t \geq 0$, $u \in E$, define

$$S_k(t)u = \frac{t^k}{k!}Cu - \sum_{i=0}^{k} \int_0^t \frac{(t-s)^{k-i}}{(k-i)!}T_i(s)u\,ds, \quad 0 \leq k \leq n-2,$$

$$S_{n-1}(t)u = \int_0^t \frac{(t-s)^{n-2}}{(n-2)!}T_n(s)u\,ds.$$

Arguing similarly as in the proof of Theorem 2.2.2, we can verify that for every

$$u_k \in C\left(\bigcap_{i=0}^{k} \mathcal{D}(A_i)\right), \quad 0 \leq k \leq n-1,$$

(ACP_n) has a solution and the tuple $\{S_0(t), \cdots, S_{n-1}(t)\}_{t \geq 0}$ is a strong C-propagation family for (ACP_n). The proof is complete.

In the sequel, E is one of the Banach spaces $L^p(R^n)$ $(1 \leq p \leq \infty)$, $C_0(R^n)$, $C_b(R^n)$ or $UC_b(R^n)$ (the space of uniformly continuous and bounded functions). Given a complex polynomial $p(x) = \sum_{|\beta| \leq m} a_\beta(ix)^\beta$ on R^n, we define

$$p(D) = \sum_{|\beta| \leq m} a_\beta D^\beta = \sum_{|\beta| \leq m} a_\beta \left(\frac{\partial}{\partial x_1}\right)^{\beta_1} \cdots \left(\frac{\partial}{\partial x_n}\right)^{\beta_n}$$

with

$$\mathcal{D}(p(D)) = \left\{ f \in E; \ \sum_{|\beta| \leq m} a_\beta D^\beta f \in E \right\}.$$

It is easy to see that $p(D)$ is a closed operator in E and $p(D)f = \mathcal{F}^{-1}\left(p\hat{f}\right)$ for all $f \in \mathcal{D}(p(D))$.

Define

$$n_E := \begin{cases} n\left|\dfrac{1}{2} - \dfrac{1}{p}\right|, & \text{if } E = L^p(R^n) \ (1 < p < \infty), \\[2mm] \dfrac{n}{2}, & \text{otherwise.} \end{cases}$$

With a given $G(x) \in \mathcal{F}L^1$, we associate a bounded linear operator $\mathbf{T}\langle G(x)\rangle$ on E as follows

$$\mathbf{T}\langle G(x)\rangle f := \mathcal{F}^{-1}G * f = \mathcal{F}^{-1}\left(G\hat{f}\right), \quad \text{for all } f \in E.$$

Assuming $H(x) \in \mathcal{M}_p$ $(1 < p < \infty)$, we define as in Section 1.5

$$\mathbf{T}\langle H(x)\rangle : f \mapsto \mathcal{F}^{-1}\left(H\hat{f}\right), \quad \text{for all } f \in \mathcal{S}(R^n),$$

which extends to a bounded linear operator on $L^p(R^n)$ $(1 < p < \infty)$.

By Δ, we will denote the Laplacian $\sum_{i=1}^{n} \frac{\partial^2}{\partial x_i^2}$. For each $z \in \mathbf{C}$, we will write

$$\sqrt{z} := |z|^{\frac{1}{2}} e^{\frac{1}{2} i \arg z}, \quad -\pi < \arg z \leq \pi,$$

so that $\operatorname{Re}\sqrt{z} \geq 0$.

Theorem 5.5. *Let $p(x)$, $q(x)$ be complex polynomials of degrees l, m respectively on R^n. Write $h = \max\{2l, m\}$. Assume*

$$\sup_{x \in R^n} \operatorname{Re}\left(-p(x) + \sqrt{p^2(x) - 4q(x)}\right) < \infty.$$

Let $A_0 = q(D)$, $A_1 = p(D)$. Then (ACP_2) is strongly $(I - \Delta)^{-\kappa}$-wellposed for

$$\kappa \begin{cases} \geq \dfrac{1}{4}(n_E + 1)h, & \text{if } E = L^p(R^n) \ (1 < p < \infty), \\[2mm] > \dfrac{1}{4}(n_E + 1)h, & \text{otherwise.} \end{cases} \tag{5.4}$$

If in addition, there exists $r \in (0, h]$ such that

$$|p^2(x) - 4q(x)| \geq C_0 |x|^r, \quad |x| \geq L_0 \tag{5.5}$$

for some C_0, $L_0 > 0$, then the κ can be improved as

$$\kappa \begin{cases} \geq \dfrac{1}{4}(n_E h + h - r), & \text{if } E = L^p(R^n) \ (1 < p < \infty), \\[2mm] > \dfrac{1}{4}(n_E h + h - r), & \text{otherwise.} \end{cases} \tag{5.6}$$

Proof. For $\lambda \in R$, define

$$\mathcal{D}\left(\tilde{P}_\lambda\right) = \left\{ f \in E; \ \mathcal{F}^{-1}\left[\left(\lambda^2 + p(x)\lambda + q(x)\right)\hat{f}\right] \in E \right\},$$

$$\tilde{P}_\lambda f = \mathcal{F}^{-1}\left[\left(\lambda^2 + p(x)\lambda + q(x)\right)\hat{f}\right] \quad \text{for all } f \in \mathcal{D}\left(\tilde{P}_\lambda\right).$$

Clearly, $\tilde{P}_\lambda$ is a closed operator in E and

$$P_\lambda \subset \tilde{P}_\lambda, \quad \mathcal{D}(A_1) \bigcap \mathcal{D}\left(\tilde{P}_\lambda\right) \subset \mathcal{D}(A_0), \quad \mathcal{D}(A_0) \bigcap \mathcal{D}\left(\tilde{P}_\lambda\right) \subset \mathcal{D}(A_1). \tag{5.7}$$

Write

$$\omega := \frac{1}{2} \sup_{x \in R^n} \operatorname{Re}\left(-p(x) + \sqrt{p^2(x) - 4q(x)}\right).$$

Then $\omega < \infty$ by hypothesis. We note that for each $\lambda > \omega$,

$$\left(\lambda^2 + p(x)\lambda + q(x)\right)^{-1} \in C^\infty(R^n).$$

For each $\lambda > \omega$, put

$$\mathcal{D}\left(\tilde{R}_\lambda\right) = \left\{f \in E; \ \ \mathcal{F}^{-1}\left[\left(\lambda^2 + p(x)\lambda + q(x)\right)^{-1}\hat{f}\right] \in E\right\},$$

$$\tilde{R}_\lambda f = \mathcal{F}^{-1}\left[\left(\lambda^2 + p(x)\lambda + q(x)\right)^{-1}\hat{f}\right] \ \ \text{for all} \ f \in \mathcal{D}\left(\tilde{R}_\lambda\right).$$

It is easy to see that

$$\tilde{P}_\lambda \tilde{R}_\lambda f = f \ \ \left(f \in \mathcal{D}\left(\tilde{R}_\lambda\right)\right), \ \ \ \tilde{R}_\lambda \tilde{P}_\lambda f = f \ \ \left(f \in \mathcal{D}\left(\tilde{P}_\lambda\right)\right).$$

Whence, $\tilde{P}_\lambda$ is injective and $\tilde{P}_\lambda^{-1} = \tilde{R}_\lambda$ for each $\lambda > \omega$. As a consequence, $\tilde{R}_\lambda$ is a closed operator in E for $\lambda > \omega$.

Set

$$c_\kappa(x) = \left(1 + |x|^2\right)^{-\kappa}, \ \ x \in R^n,$$

$$\mu_\pm(x) = \frac{1}{2}\left(-p(x) \pm \sqrt{p^2(x) - 4q(x)}\right), \ \ x \in R^n. \tag{5.8}$$

We have that for each multiindex β,

$$\left|D^\beta c_\kappa(x)\right| \leq \text{ const } (1 + |x|)^{-2\kappa - |\beta|}, \ \ x \in R^n. \tag{5.9}$$

This shows by Lemma 1.5.3 that $c_\kappa(x) \in \mathcal{F}L^1$ when $\kappa \neq 0$. Let

$$C_\kappa := \begin{cases} I, & \text{if } E = L^2(R^n), \\ \\ T\langle c_\kappa(x)\rangle, & \text{otherwise.} \end{cases}$$

Then $C_\kappa = (I - \Delta)^{-\kappa}$.

Next, set

$$\mathcal{P}(x) = \begin{pmatrix} 0 & \left(1 + |x|^2\right)^{\frac{m}{4}} \\ \\ -\left(1 + |x|^2\right)^{-\frac{m}{4}}q(x) & -p(x) \end{pmatrix}, \ \ \ x \in R^n.$$

We recall Theorem 2 of Friedman [1, p. 169], which states:

Letting A be an $m \times m$ matrix and $H(A)$ the convex hull (in the complex plane) of all the eigenvalues of A, then for any entire function $f(z)$,

$$\|f(A)\| \leq \sum_{j=0}^{m-1} z^j \|A\|^j \max_{z \in H(A)} \left|f^{(j)}(z)\right|.$$

It follows that

$$\left\|e^{t\mathcal{P}(x)}\right\| \leq \text{const } \left(1 + t + t|x|^{\frac{k}{2}}\right) e^{\omega t}, \quad t \geq 0, \ x \in R^n. \tag{5.10}$$

In order to get a better estimate on $\left\|e^{t\mathcal{P}(x)}\right\|$ for $|x| \geq L_0$ in the case of (5.5) holding, we put

$$g_t(x) := \frac{1}{\sqrt{p^2(x) - 4q(x)}} \left(e^{t\mu_+(x)} - e^{t\mu_-(x)}\right), \quad t \geq 0, \ |x| \geq L_0,$$

$$w_t(x) := e^{t\mu_+(x)} + e^{t\mu_-(x)}, \quad t \geq 0, \ |x| \geq L_0.$$

Clearly, for $\lambda > \omega$,

$$\int_0^\infty e^{-\lambda t} g_t(x)dt = \left(\lambda^2 + p(x)\lambda + q(x)\right)^{-1},$$

$$\int_0^\infty e^{-\lambda t} w_t(x)dt = \left(2\lambda + p(x)\right)\left(\lambda^2 + p(x)\lambda + q(x)\right)^{-1}.$$

From this and the easily verified equality

$$(\lambda - \mathcal{P}(x))^{-1} =$$

$$\begin{pmatrix} (\lambda + p(x))\left(\lambda^2 + p(x)\lambda + q(x)\right)^{-1} & \\ - (1 + |x|^2)^{-\frac{m}{4}} q(x)\left(\lambda^2 + p(x)\lambda + q(x)\right)^{-1} & \\ & (1 + |x|^2)^{\frac{m}{4}}\left(\lambda^2 + p(x)\lambda + q(x)\right)^{-1} \\ & \lambda\left(\lambda^2 + p(x)\lambda + q(x)\right)^{-1} \end{pmatrix},$$

$$\lambda > \omega, \tag{5.11}$$

it follows, by the uniqueness theorem for Laplace transforms, that if we write

$$e^{t\mathcal{P}(x)} = \begin{pmatrix} v_{11}(x;\ t) & v_{12}(x;\ t) \\ v_{21}(x;\ t) & v_{22}(x;\ t) \end{pmatrix}, \quad t \geq 0, \ x \in R^n, \tag{5.12}$$

then for $t \geq 0$, $|x| \geq L_0$,

$$v_{11}(x;\ t) = \frac{1}{2}\left(w_t(x) + p(x)g_t(x)\right),$$

$$v_{22}(x;\ t) = \frac{1}{2}\left(w_t(x) - p(x)g_t(x)\right),$$

$$v_{12}(x;\ t) = (1 + |x|^2)^{\frac{m}{4}} g_t(x),$$

$$v_{21}(x;\ t) = -(1 + |x|^2)^{-\frac{m}{4}} q(x)g_t(x).$$

Obviously,

$$|w_t(x)| \leq 2e^{\omega t}, \quad t \geq 0, \ |x| \geq L_0,$$

and by (5.5),

$$|g_t(x)| \leq 2C_0^{-\frac{1}{2}}|x|^{-\frac{r}{2}}e^{\omega t}, \quad t \geq 0, \ |x| \geq L_0.$$

This combined with (5.10) yields that for all $t \geq 0$, $x \in R^n$,

$$|v_{11}(x; t)|, \ |v_{22}(x; t)| \leq \text{const } (1+t)(1+|x|)^{l-\frac{r}{2}}e^{\omega t},$$

$$|v_{12}(x; t)|, \ |v_{21}(x; t)| \leq \text{const } (1+t)(1+|x|)^{\frac{m}{2}-\frac{r}{2}}e^{\omega t},$$

and therefore by (5.12),

$$\left\| e^{t\mathcal{P}(x)} \right\| \leq \text{const } (1+t)(1+|x|)^{\frac{1}{2}(h-r)}e^{\omega t}, \quad t \geq 0, \ x \in R^n, \tag{5.13}$$

valid for the case of (5.5). In fact for the otherwise case, (5.13) also holds by (5.10), if we let $r = 0$ (here and in the sequel). Now, note that for each multiindex β

$$\left\| D^\beta \mathcal{P}(x) \right\| \leq \text{const } (1+|x|)^{\frac{h}{2}-|\beta|}, \quad x \in R^n. \tag{5.14}$$

Then using Leibniz's formula, we deduce by (5.13) and (5.14) that for each multiindex β,

$$\left\| D^\beta e^{t\mathcal{P}(x)} \right\|$$

$$\leq \quad \text{const } (1+t)^{|\beta|+1}(1+|x|)^{(\frac{h}{2}-1)|\beta|+\frac{1}{2}(h-r)}e^{\omega t}, \quad t \geq 0, \ x \in R^n.$$

This implies by (5.12) that for each multiindex β,

$$\left| D^\beta v_{11}(x; t) \right|, \ \left| D^\beta v_{22}(x; t) \right|, \ \left| D^\beta v_{12}(x; t) \right|$$

$$\leq \quad \text{const } (1+t)^{|\beta|+1}(1+|x|)^{(\frac{h}{2}-1)|\beta|+\frac{1}{2}(h-r)}e^{\omega t}, \quad t \geq 0, \ x \in R^n. \tag{5.15}$$

Set

$$v_0(x; t) = \left(1+|x|^2\right)^{-\frac{m}{4}} v_{12}(x; t), \quad t \geq 0, \ x \in R^n. \tag{5.16}$$

Then combining (5.15) with (5.9) shows, by Leibniz's formula, that for each multiindex β,

$$\left| D^\beta \left[v_0(x; t)c_\kappa(x) \right] \right|, \ \left| D^\beta \left[v_{11}(x; t)c_\kappa(x) \right] \right|, \ \left| D^\beta \left[v_{22}(x; t)c_\kappa(x) \right] \right|$$

$$\leq \quad \text{const } (1+t)^{|\beta|+1}(1+|x|)^{(\frac{h}{2}-1)|\beta|+\frac{1}{2}(h-r)-2\kappa}e^{\omega t}, \quad t \geq 0, \ x \in R^n.$$

Therefore, we deduce by virtue of Lemmas 1.5.2 and 1.5.4 that, if

$$\kappa \geq \frac{1}{4}\left(hn \left| \frac{1}{2} - \frac{1}{p} \right| + h - r \right),$$

$1 < p < \infty$, then

$$v_0(x;\ t)c_\kappa(x),\ v_{11}(x;\ t)c_\kappa(x),\ v_{22}(x;\ t)c_\kappa(x) \in \mathcal{M}_p$$

and

$$\|v_0(x;\ t)c_\kappa(x)\|_{\mathcal{M}_p},\ \|v_{11}(x;\ t)c_\kappa(x)\|_{\mathcal{M}_p},\ \|v_{22}(x;\ t)c_\kappa(x)\|_{\mathcal{M}_p}$$

$$\leq\ \mathrm{const}\ (1+t)^{1+n\left|\frac{1}{2}-\frac{1}{p}\right|}e^{\omega t},\quad t \geq 0;$$

if $\kappa > \frac{1}{4}\left(\frac{1}{2}hn + h - r\right)$, then

$$v_0(x;\ t)c_\kappa(x),\ v_{11}(x;\ t)c_\kappa(x),\ v_{22}(x;\ t)c_\kappa(x) \in \mathcal{F}L^1,$$

being continuous in $t \in [0,\ \infty)$ under the norm of $\mathcal{F}L^1$, and

$$\|v_0(x;\ t)c_\kappa(x)\|_{\mathcal{F}L^1},\ \|v_{11}(x;\ t)c_\kappa(x)\|_{\mathcal{F}L^1},\ \|v_{22}(x;\ t)c_\kappa(x)\|_{\mathcal{F}L^1}$$

$$\leq\ \mathrm{const}\ (1+t)^{1+\frac{n}{2}}e^{\omega t},\quad t \geq 0.$$

Accordingly, putting

$$V_0(t) = \mathbf{T}\langle v_0(x;\ t)c_\kappa(x)\rangle,\quad t \geq 0,$$

$$V_{11}(t) = \mathbf{T}\langle v_{11}(x;\ t)c_\kappa(x)\rangle,\quad t \geq 0,$$

$$V_{22}(t) = \mathbf{T}\langle v_{22}(x;\ t)c_\kappa(x)\rangle,\quad t \geq 0,$$

we have that

$$A_1 V_0(t) = V_{11}(t) - V_{22}(t),\quad t \geq 0, \tag{5.17}$$

and

$$\|V_0(t)\|,\ \|V_{11}(t)\|,\ \|V_{22}(t)\| \leq\ \mathrm{const}\ (1+t)^{1+n_E}e^{\omega t},\quad t \geq 0; \tag{5.18}$$

moreover, when

$$E = L^1(R^n),\ \ L^\infty(R^n),\ \ C_0(R^n),\ C_b(R^n),\ \text{or}\ UC_b(R^n),$$

$$t \mapsto V_0(t),\ \ t \mapsto V_{11}(t),\ \ t \mapsto V_{22}(t)\quad (\text{for } t \geq 0)$$

are continuous in the uniform operator topology. On the other hand, we observe that for each $t_0 \in R^+$, $\phi \in \mathcal{S}(R^n)$,

$$\lim_{t \to t_0} v_0(x;\ t)c_\kappa(x)\hat{\phi} = v_0(x;\ t_0)c_\kappa(x)\hat{\phi}$$

under the topology of $\mathcal{S}(R^n)$, and therefore

$$\lim_{t \to t_0} \mathcal{F}^{-1}\left(v_0(x;\ t)c_\kappa(x)\hat{\phi}\right) = \mathcal{F}^{-1}\left(v_0(x;\ t_0)c_\kappa(x)\hat{\phi}\right)$$

under the topology of $S(R^n)$. This indicates that

$$\lim_{t \to t_0} V_0(t)\phi = V_0(t_0)\phi$$

under the topology of $S(R^n)$, and so under the norm of $L^p(R^n)$ $(1 < p < \infty)$. Thus, (5.18) and the denseness of $S(R^n)$ in $L^p(R^n)$ $(1 < p < \infty)$ together yield that $V_0(\cdot)$ is strongly continuous when $E = L^p(R^n)$ $(1 < p < \infty)$. So do $V_{11}(\cdot)$ and $V_{22}(\cdot)$ by a similar argument.

Finally, define

$$J_\lambda f = \int_0^\infty e^{-\lambda t} V_0(t) f dt, \quad K_\lambda f = \int_0^\infty e^{-\lambda t} V_{22}(t) f dt, \quad \lambda > \omega + 1, \ f \in E.$$

We note by (5.11), (5.12) and (5.16) that for $\lambda > \omega + 1$, $x \in R^n$,

$$\int_0^\infty e^{-\lambda t} v_0(x; \ t) dt = \left(\lambda^2 + p(x)\lambda + q(x) \right)^{-1},$$

$$\int_0^\infty e^{-\lambda t} v_{11}(x; \ t) dt = \left(\lambda + p(x) \right) \left(\lambda^2 + p(x)\lambda + q(x) \right)^{-1},$$

$$\int_0^\infty e^{-\lambda t} v_{22}(x; \ t) dt = \lambda \left(\lambda^2 + p(x)\lambda + q(x) \right)^{-1}.$$

From this, we obtain using Fubini's theorem that for $\lambda > \omega + 1$, $\phi \in C_c^\infty(R^n)$, and

$$f \in \begin{cases} S(R^n), & \text{if } E = L^p(R^n) \ (1 < p < \infty), \\ E, & \text{otherwise}, \end{cases}$$

$$\left\langle J_\lambda f, \ \left(\lambda^2 + p(-D)\lambda + q(-D) \right) \phi \right\rangle$$

$$= \int_0^\infty e^{-\lambda t} \left\langle \mathcal{F}^{-1} \left(v_0(x; \ t) c_\kappa(x) \hat{f} \right), \ \left(\lambda^2 + p(-D)\lambda + q(-D) \right) \phi \right\rangle dt$$

$$= \int_0^\infty e^{-\lambda t} \left[\mathcal{F}^{-1} \left(v_0(x; \ t) c_\kappa(x) \hat{f} \right) * \left(\lambda^2 + p(D)\lambda + q(D) \right) \phi_- \right] (0) dt$$

$$= \int_0^\infty e^{-\lambda t} \mathcal{F}^{-1} \left(v_0(x; \ t) c_\kappa(x) \hat{f} \left(\lambda^2 + p(x)\lambda + q(x) \right) \hat{\phi}_- \right) (0) dt$$

$$= \mathcal{F}^{-1} \left(c_\kappa(x) \hat{f} \hat{\phi}_- \right) (0) = \left(\mathcal{F}^{-1} \left(c_\kappa(x) \hat{f} \right) * \phi_- \right) (0)$$

$$= \left\langle C_\kappa f, \ \phi \right\rangle, \quad \text{where } \phi_-(x) = \phi(-x).$$

Similarly, we get that for λ, f, ϕ as above,

$$\left\langle K_\lambda f, \ \left(\lambda^2 + p(-D)\lambda + q(-D) \right) \phi \right\rangle = \left\langle \lambda C_\kappa f, \ \phi \right\rangle.$$

In conclusion, for λ, f as above,

$$\tilde{P}_\lambda J_\lambda f = C_\kappa f, \quad \tilde{P}_\lambda K_\lambda f = \lambda C_\kappa f.$$

Using the closedness of $\tilde{P}_\lambda$ and the denseness of $\mathcal{S}(R^n)$ in $L^p(R^n)$, we infer that the above equalities hold for all $f \in E$ in any case. Consequently

$$J_\lambda f = \tilde{R}_\lambda C_\kappa f, \quad K_\lambda f = \lambda \tilde{R}_\lambda C_\kappa f, \quad \lambda > \omega + 1, \ f \in E. \qquad (5.19)$$

The first equality together with (5.17), (5.18) implies that for any $\lambda > \omega + 1$, $f \in E$,

$$\begin{cases} \mathcal{R}\left(\tilde{R}_\lambda C_\kappa\right) \subset \mathcal{D}(A_1), \\[2mm] A_1 \tilde{R}_\lambda C_\kappa f = \displaystyle\int_0^\infty e^{-\lambda t} \left(V_{11}(t) - V_{22}(t)\right) f \, dt. \end{cases}$$

Combining this with (5.7) establishes that $\mathcal{R}\left(\tilde{R}_\lambda C_\kappa\right) \subset \mathcal{D}(A_0)$. Therefore

$$\mathcal{R}\left(C_\kappa\right) \subset \mathcal{D}(R_\lambda), \quad R_\lambda C_\kappa = \tilde{R}_\lambda C_\kappa, \quad \lambda > \omega + 1. \qquad (5.20)$$

Accordingly, we obtain by (5.19) that for $\lambda > \omega + 1$, $f \in E$,

$$\begin{cases} A_1 R_\lambda C_\kappa f = \displaystyle\int_0^\infty e^{-\lambda t} \left(V_{11}(t) - V_{22}(t)\right) f \, dt, \\[3mm] \lambda R_\lambda C_\kappa f = \displaystyle\int_0^\infty e^{-\lambda t} V_{22}(t) f \, dt. \end{cases} \qquad (5.21)$$

Moreover, it is plain that

$$C_\kappa^{-1} A_0 C_\kappa f = \mathcal{F}^{-1}\left\{ \frac{1}{c_\kappa(x)} \mathcal{F}\mathcal{F}^{-1} \left[q(x)\mathcal{F}\mathcal{F}^{-1}\left(c_\kappa(x)\mathcal{F}f\right) \right] \right\}$$

$$= \mathcal{F}^{-1}\{q(x)\mathcal{F}f\} = A_0 f, \quad f \in \mathcal{D}\left(C_\kappa^{-1} A_0 C_\kappa\right) = \mathcal{D}(A_0),$$

$$C_\kappa^{-1} A_1 C_\kappa f = A_1 f, \quad f \in \mathcal{D}\left(C_\kappa^{-1} A_1 C_\kappa\right) = \mathcal{D}(A_1),$$

$$\tilde{P}_\lambda A_0 R_\lambda C_\kappa = A_0 C_\kappa \quad \text{on } \mathcal{D}(A_0),$$

$$\tilde{P}_\lambda A_1 R_\lambda C_\kappa = A_1 C_\kappa \quad \text{on } \mathcal{D}(A_1).$$

This shows by (5.20) that

$$\begin{cases} R_\lambda C_\kappa A_0 u = A_0 R_\lambda C_\kappa u, \quad u \in \mathcal{D}(A_0), \\[2mm] R_\lambda C_\kappa A_1 u = A_1 R_\lambda C_\kappa u, \quad u \in \mathcal{D}(A_1). \end{cases}$$

Thus, recalling (5.21), we can apply Theorem 5.4 to obtain the desired results.

Remark. Let $p(x) \equiv 0$ and recall the remark after Proposition 5.3. Then Theorem 5.5 gives a result for regularized cosine functions. See also Corollary 5.7.

A complex polynomial $p(x) = \sum_{|\beta| \le l} a_\beta x^\beta$ on R^n is called strongly elliptic if

$$\mathrm{Re} \sum_{|\beta|=l} a_\beta x^\beta > 0, \quad x \in R^n \setminus \{0\}.$$

Theorem 5.6. *Suppose that $p_1(x)$, $p_2(x)$, $q_1(x)$, $q_2(x)$ are real polynomials of degrees l_1, l_2, m_1, m_2 respectively on R^n. Let $A_0 = q_1(D) + iq_2(D)$, $A_1 = p_1(D) + ip_2(D)$. Then for any κ with*

$$\kappa \begin{cases} \ge n \left| \dfrac{1}{2} - \dfrac{1}{p} \right|, & \text{if } E = L^p(R^n) \ (1 < p < \infty), \\[2mm] > \dfrac{n}{2}, & \text{otherwise,} \end{cases} \tag{5.22}$$

(1) (ACP_2) *is strongly $(I - \Delta)^{-\frac{1}{2}l_2\kappa}$-wellposed, provided either*
(i) $l_1 = 0$, $m_1 < 2l_2$, $m_2 \le l_2$, *and $p_2(x)$ is elliptic,* or
(ii) $l_1 = 0$, $m_1 = 2l_2$, $m_2 \le l_2$, $p_2(x)$ *is elliptic, and $q_1(x) \ge 0$ ($|x| \ge L_0$) for* some $L_0 > 0$;
(2) (ACP_2) *is strongly $(I - \Delta)^{-\frac{1}{4}m_1\kappa}$-wellposed, provided either*
(iii) $l_1 = 0$, $l_2 \le \frac{1}{2}m_1$, $m_2 \le \frac{1}{2}m_1$, *and $q_1(x)$ is strongly elliptic,* or
(iv) $0 < l_1 < \frac{1}{2}m_1$, $l_2 < \frac{1}{2}m_1$, $m_2 < l_1 + \frac{1}{2}m_1$, *and $p_1(x)$, $q_1(x)$ are strongly* elliptic;
(3) (ACP_2) *is strongly $(I - \Delta)^{-\frac{1}{2}l_1\kappa}$-wellposed, provided*
(v) $l_2 \le \frac{1}{2}l_1$, $m_1 \le l_1$, $m_2 \le \frac{3}{2}l_1$, *and $p_1(x)$ is strongly elliptic.*

Proof. For $x \in R^n$, let

$$p(x) = p_1(x) + ip_2(x), \quad q(x) = q_1(x) + iq_2(x),$$

and let

$$r_1(x) = p_1^2(x) - p_2^2(x) - 4q_1(x), \quad r_2(x) = 2p_1(x)p_2(x) - 4q_2(x).$$

Then

$$p^2(x) - 4q(x) = r_1(x) + ir_2(x), \quad x \in R^n.$$

It can be verified that

$$\sqrt{p^2(x) - 4q(x)} = s_1(x) + is_2(x), \quad x \in R^n,$$

where

$$
s_1(x) = \begin{cases} \dfrac{\sqrt{2}}{2}\left(r_1(x) + \sqrt{r_1^2(x) + r_2^2(x)}\right)^{\frac{1}{2}}, & \text{if } r_1(x) \geq 0, \\[3mm] \dfrac{\sqrt{2}}{2}|r_2(x)|\left(\sqrt{r_1^2(x) + r_2^2(x)} - r_1(x)\right)^{-\frac{1}{2}}, & \text{if } r_1(x) < 0, \end{cases}
$$

$$
s_2(x) = \begin{cases} \dfrac{\sqrt{2}}{2}r_2(x)\left(r_1(x) + \sqrt{r_1^2(x) + r_2^2(x)}\right)^{-\frac{1}{2}}, & \text{if } r_1(x) > 0, \\[3mm] \dfrac{\sqrt{2}}{2}\text{sign}(r_2(x))\left(\sqrt{r_1^2(x) + r_2^2(x)} - r_1(x)\right)^{\frac{1}{2}}, & \text{if } r_1(x) \leq 0. \end{cases}
$$

Keeping this in mind and recalling the definition of strong ellipticity, we begin the following discussion. When condition (i) or (ii) holds, we have that

$$
\max\{2\deg p,\ \deg q\} = 2l_2
$$

and

$$
|p^2(x) - 4q(x)| \geq |r_1(x)| \geq C_0|x|^{2l_2}, \quad |x| \geq L_0,
$$

for some C_0, $L_0 > 0$; $r_1(x) < 0$ and

$$
\left(\sqrt{r_1^2(x) + r_2^2(x)} - r_1(x)\right)^{-\frac{1}{2}} \leq \text{const } |x|^{-l_2}
$$

for $|x|$ sufficiently large; $\deg r_2 \leq l_2$, so that

$$
s_1(x) \leq \text{const }\left(1 + |x|^{l_2}|x|^{-l_2}\right)
$$

and therefore

$$
\sup_{x \in R^n} \text{Re}\left(-p(x) + \sqrt{p^2(x) - 4q(x)}\right) < \infty. \tag{5.23}
$$

When condition (iii) holds, we have that

$$
\max\{2\deg p,\ \deg q\} = m_1
$$

and

$$
|p^2(x) - 4q(x)| \geq |r_1(x)| \geq C_1|x|^{m_1}, \quad |x| \geq L_1, \tag{5.24}
$$

for some C_1, $L_1 > 0$; $r_1(x) < 0$ for $|x|$ sufficiently large,

$$
\deg r_2 \leq \max\{l_2,\ m_2\} \leq \frac{1}{2}m_1,
$$

so that

$$
s_1(x) \leq \text{const }\left(1 + |x|^{\frac{1}{2}m_1}|x|^{-\frac{1}{2}m_1}\right),
$$

and therefore (5.23) is satisfied.

When condition (iv) holds, we have that

$$\max\{2\deg p,\ \deg q\} = m_1$$

and (5.24) is satisfied; $r_1(x) < 0$ for $|x|$ sufficiently large,

$$\deg r_2 \leq \max\{l_1 + l_2,\ m_2\},$$

so that

$$s_1(x) \leq \text{const}\ \left(1 + |x|^{\max\{l_1+l_2,\ m_2\}}|x|^{-\frac{1}{2}m_1}\right),$$

and therefore (5.23) is satisfied noting

$$\max\{l_1 + l_2,\ m_2\} - \frac{1}{2}m_1 < l_1$$

as well as the strong ellipticity of $p_1(x)$.

Finally, let condition (v) hold. We have that

$$\max\{2\deg p,\ \deg q\} = 2l_1$$

and

$$\left|p^2(x) - 4q(x)\right| \geq |r_1(x)| \geq C_2|x|^{2l_1},\quad |x| \geq L_2,$$

for some $C_2,\ L_2 > 0$; $r_1(x) > 0$ for $|x|$ sufficiently large,

$$\deg r_2 \leq \max\{l_1 + l_2,\ m_2\}.$$

Observe

$$\text{Re}\left(-p(x) + \sqrt{p^2(x) - 4q(x)}\right)$$

$$= -p_1(x) + \frac{\sqrt{2}}{2}\left(r_1(x) + \sqrt{r_1^2(x) + r_2^2(x)}\right)^{\frac{1}{2}}$$

$$= \frac{1}{\left[2p_1(x) + \sqrt{2}\left(r_1(x) + \sqrt{r_1^2(x) + r_2^2(x)}\right)^{\frac{1}{2}}\right]}$$

$$\cdot \frac{r_2^2(x) - 4(p_1^2(x)p_2^2(x) + 4p_1^2(x)q_1(x))}{\left[\sqrt{r_1^2(x) + r_2^2(x)} + p_1^2(x) + p_2^2(x) + 4q_1(x)\right]}$$

$$\leq \text{const}\ \left(1 + |x|^{\max\{2(l_1+l_2),\ 2m_2,\ 2l_1+m_1\}}|x|^{-l_1-2l_1}\right),\quad x \in R^n.$$

We see that (5.23) is satisfied.

Consequently, using Theorem 5.5 leads to the results as required.

Corollary 5.7. *Let $q(x)$ be strongly elliptic, and*

$$\deg(\text{Im}[q(x)]) \leq \frac{1}{2} \deg(\text{Re}[q(x)]).$$

Then for any κ as in (5.22), the incomplete second order Cauchy problem

$$\begin{cases} u''(t) + q(D)u(t) = 0, & t \geq 0, \\ u(0) = u_0, \quad u'(0) = u_1, \end{cases}$$

is strongly $(I - \Delta)^{-\frac{1}{4}\kappa \deg(\text{Re}[q(x)])}$-wellposed, or equivalently, $-q(D)$ is the generator of a $(I - \Delta)^{-\frac{1}{4}\kappa \deg(\text{Re}[q(x)])}$-regularized cosine function.

Proof. Use Theorem 5.6 (ii) and note the remark after Proposition 5.3.

Example 5.8. We consider the damped Klein-Gordon equation in one dimension

$$\begin{cases} u_{tt} + au_{tx} - \rho u_{xx} + \gamma u = 0, & t \geq 0, \ x \in R \\ u(0, x) = \phi(x), \quad u_t(0, x) = \psi(x), & x \in R \end{cases} \tag{5.25}$$

in $E = L^p(R)$ $(1 \leq p \leq \infty)$, $C_0(R)$, $C_b(R)$ or $UC_b(R)$, where $a \in R$, $\rho, \gamma > 0$.
    Take

$$p_1(x) = 0, \quad p_2(x) = ax,$$

$$q_1(x) = \rho x^2 + \gamma, \quad q_2(x) = 0,$$

$$l_1 = 0, \quad l_2 = 1, \quad m_1 = 2, \quad m_2 = 0.$$

Then we can apply Theorem 5.6 (2) to conclude that the Cauchy problem (5.25) is strongly $(I - \Delta)^{-\frac{1}{2}\kappa}$-wellposed for

$$\kappa \begin{cases} \geq \left| \dfrac{1}{2} - \dfrac{1}{p} \right|, & \text{if } E = L^p(R^n) \ (1 < p < \infty), \\[2mm] > \dfrac{1}{2}, & \text{otherwise.} \end{cases}$$

3.6   The case $u^{(n)}(t) = Au(t)$

We consider the Cauchy problem for the following higher order abstract differential equation

$$\begin{cases} u^{(n)}(t) = Au(t), & t \geq 0, \\ u^{(i)}(0) = u_i, & 0 \leq i \leq n - 1, \end{cases} \tag{6.1}$$

where A is a linear closed operator in E.

It is well known that in the case of $n = 1$, (6.1) is wellposed if and only if A generates a strongly continuous semigroup; in the case of $n = 2$, the wellposedness of (6.1) is equivalent to A being the generator of a strongly continuous cosine function However, for $n \geq 3$, (6.1) is not wellposed unless A is bounded. This was proved by Chazarain [1, 2] and Fattorini [1] three decades ago.

In this section we show that it is possible that for $n \geq 3$, (6.1) is C-wellposed (in some sense) for some bounded injective linear operator C from E to E, even if A is unbounded. We give explicit conditions on A, which are valid for many unbounded operators, ensuring that (6.1) ($n \geq 2$) be C-wellposed (Theorem 6.2).

On the other hand, as shown in Section 1.6, if A generates an r-times ($r \in N_0$) integrated semigroup, then for $u_0 \in \mathcal{D}(A^{r+1})$, (6.1) with $n = 1$ has a unique solution, satisfying the estimate

$$\sup_{t \geq 0} \left\{ e^{-\omega t} \| u(t) \| \right\} < \infty, \quad \text{for some } \omega \geq 0;$$

and if A generates an r-times integrated cosine function, a similar conclusion holds as well for (6.1) with $n = 2$.

A problem arises naturally then: is it possible for us to find a class of unbounded operators, each serving as A, such that (6.1) ($n \geq 3$) has the property (weaker than wellposedness) that, for $u_0, \cdots, u_{n-1} \in \mathcal{D}(A^r)$ ($r \in N$), (6.1) ($n \geq 3$) has a unique solution with the estimate

$$\sup_{t \geq 0} \left\{ e^{-\omega t} \| u(t) \| \right\} < \infty, \quad \text{for some } \omega \geq 0, \qquad (6.2)$$

or the estimate (weaker than (6.2))

$$\sup_{t \geq 0} \left\{ e^{-\omega t} \left\| \int_0^t \frac{1}{q!} (t - s)^q u(s) ds \right\| \right\} < \infty, \quad \text{for some } \omega \geq 0, \ q \in N_0 \ ?$$

We shall show that its answer is generally negative (Theorem 6.5).

Theorem 6.5 can be also regarded as a development of the classial result of Chazarain and Fattorini mentioned above. In the result here, we do not only weaken the wellposedness assumption of the Chazarain and Fattorini theorem , but also delete the density of $\mathcal{D}(A)$ in E which is implied by the wellposedness.

Define

$$\Sigma_\theta = \{ z \in \mathbf{C}; \ z \neq 0, \ |\arg z| < \theta \}, \quad \theta \in (0, \pi].$$

Definition 6.1. A linear operator B in E is called nonnegative if $(-\infty, 0)$ is contained in $\rho(B)$ with $\{ \lambda \| (\lambda + B)^{-1} \|; \ \lambda > 0 \}$ bounded.

Let B be a nonnegative operator. Then, it can be shown using the power series expansion of $(\lambda + B)^{-1}$ that for some $\phi_0 \in (0, \frac{\pi}{2}]$, $\Sigma_{\phi_0} \subset \rho(-B)$ with

$\{\lambda\|(\lambda+B)^{-1}\|;\ \lambda\in\Sigma_{\phi_0}\}$ bounded; see Balakrishnan [1, Lemma 6.1] or Fattorini [6, p. 365] for the proof.

Theorem 6.2. *Let $n \geq 2$. Suppose that there exist $z_0 \in \mathbf{C}$, $z_0 \neq 0$, $0 < \theta < \frac{\pi}{2}$ such that*

$$e^{i\,\arg z_0}\,(|z_0|+\Sigma_\theta) \subset \rho(A).\qquad(6.3)$$

In addition, there exist $\alpha \geq -1$, $C > 0$ such that

$$\left\|\lambda^{-\alpha}(\lambda-A)^{-1}\right\| \leq C,\quad \text{for every } \lambda\in e^{i\,\arg z_0}\,(|z_0|+\Sigma_\theta).\qquad(6.4)$$

Then there exists a one parameter family $\{C_\epsilon\}_{\epsilon>0}$ of bounded injective operators on E such that

(i) *for each $\epsilon > 0$, the (6.1) has a unique solution $u(\cdot)$ with initial data u_0, $\cdots$, $u_{n-1} \in \mathcal{R}(C_\epsilon)$; moreover,*

$$\|u(t)\| \leq M(t)\sum_{i=0}^{n-1}\left\|C_\epsilon^{-1}u_i\right\|,\quad t \geq 0$$

for some nonnegative and locally bounded function $M(t)$.

(ii) *$\bigcup_{\epsilon>0} C_\epsilon\left(\mathcal{D}\left(A^{[\alpha]+2}\right)\right)$ is dense in $\mathcal{D}\left(A^{[\alpha]+2}\right)$.*

Proof. Let $\arg z_0 = \beta$, $A_0 = e^{-i\beta}A$. If $\mu \in |z_0| + \Sigma_\theta$, then $\lambda = e^{i\beta}\mu \in e^{i\beta}(|z_0|+\Sigma_\theta)$. Hence $\lambda \in \rho(A)$. By

$$(\mu-A_0)^{-1} = e^{i\beta}(\lambda-A)^{-1},$$

we have that $\mu\in\rho(A_0)$, i.e., $|z_0|+\Sigma_\theta\subset\rho(A_0)$, and (6.4) gives

$$\sup_{\mu\in|z_0|+\Sigma_\theta}\left\|\mu^{-\alpha}(\mu-A_0)^{-1}\right\|$$

$$\leq \sup_{\lambda\in e^{i\beta}(|z_0|+\Sigma_\theta)}\left\|\left(e^{-i\beta}\lambda\right)^{-\alpha}\left(e^{i\beta}(\lambda-A)^{-1}\right)\right\|$$

$$< \infty.$$

Take

$$b\in(|z_0|,\ \infty),\quad \sigma\in\left(\frac{1}{n},\ \frac{\pi}{2}(\pi-\theta)^{-1}\right)$$

(please note that $\frac{1}{n} < \frac{\pi}{2}(\pi-\theta)^{-1}$ due to $n \geq 2$). Let $\phi\in(0,\theta)$, and let Γ be a curve, contained in $\rho(A_0)$, coming from infinity along the ray $\arg(\lambda-|z_0|) = -\phi$, leaving point $P(|z_0|,\ 0)$ to its left side, point $Q(b,\ 0)$ to its right side, and passing from this ray to $\arg(\lambda-|z_0|) = \phi$, and then going off to infinity along the last ray.

For $z\in\mathbf{C}$, $k=0,\ \cdots,\ n-1$, $\epsilon>0$, set

$$S_k(z;\ \epsilon) = \frac{1}{2\pi i}\int_\Gamma \exp\left(-\epsilon(b-\lambda)^\sigma\right)\sum_{j=0}^\infty \frac{z^{nj+k}\lambda^j}{(nj+k)!}(\lambda-A_0)^{-1}d\lambda,\qquad(6.5)$$

where we take the branch of $(b - \lambda)^\sigma$ such that it is holomorphic off the half-line $[b, \infty)$ and positive for $\lambda < b$. Let Ω denote the region lying to the left of the path Γ. Then

$$\lambda - b \in \{z \in \mathbf{C}; \ \phi < \arg z < 2\pi - \phi\}, \ \text{if } \lambda \in \Omega \bigcup \Gamma,$$

and therefore

$$b - \lambda \in \{z \in \mathbf{C}; \ \phi - \pi < \arg z < \pi - \phi\}, \ \text{if } \lambda \in \Omega \bigcup \Gamma.$$

Thus for any $\lambda \in \Omega \bigcup \Gamma$,

$$|\exp\left(-\varepsilon(b - \lambda)^\sigma\right)| \le \exp\left(-\varepsilon \cos(\sigma(\pi - \theta))|b - \lambda|^\sigma\right). \tag{6.6}$$

On the other hand, for $z \in \mathbf{C}$, $k = 0, 1, \cdots, n - 1$,

$$\left\| \sum_{j=0}^{\infty} \frac{z^{nj+k} \lambda^j}{(nj+k)!} \right\| \le |z|^k e^{|z| \|\lambda\|^{\frac{1}{n}}}. \tag{6.7}$$

In view of

$$\frac{1}{n} < \sigma, \ \ \varepsilon \cos(\sigma(\pi - \theta)) > 0, \tag{6.8}$$

and (6.6), (6.7), we obtain that the integral in (6.5) is well defined.

Moreover, by (6.5) – (6.8), we have that for every k, z, ε, $S_k(z; \varepsilon)$ commutes with A_0, and for $0 \le l \le n - 1$,

$$\lim_{m \to \infty} S_k^{(l)}(z; \varepsilon) u_m = 0, \ \text{as } u_m \to 0, \tag{6.9}$$

uniformly on compact sets of $\mathbf{C}$.

Observe that for $z \in \mathbf{C}$, $0 \le k, l \le n - 1$,

$$\frac{d^n}{dz^n} \sum_{j=0}^{\infty} \frac{z^{nj+k} \lambda^j}{(nj+k)!} = \lambda \sum_{j=0}^{\infty} \frac{z^{nj+k} \lambda^j}{(nj+k)!}, \tag{6.10}$$

$$\frac{d^l}{dz^l} \sum_{j=0}^{\infty} \frac{z^{nj+k} \lambda^j}{(nj+k)!} \bigg|_{z=0} = \begin{cases} 1, & \text{if } l = k, \\ 0, & \text{if } l \ne k. \end{cases} \tag{6.11}$$

We have that for $z \in \mathbf{C}$, $0 \le k \le n - 1$, $\varepsilon > 0$,

$$\frac{d^n}{dz^n} S_k(z; \varepsilon) = A_0 S_k(z; \varepsilon) + \frac{1}{2\pi i} \int_\Gamma \exp\left(-\varepsilon(b - \lambda)^\sigma\right) \sum_{j=0}^{\infty} \frac{z^{nj+k} \lambda^j}{(nj+k)!} d\lambda. \tag{6.12}$$

We see that the integral on the right-hand side of (6.12) vanishes, by a deformation of contour. So for $z \in \mathbf{C}$, $0 \le k, l \le n - 1$, $\varepsilon > 0$,

$$\begin{cases} \dfrac{d^n}{dz^n} S_k(z; \varepsilon) = A_0 S_k(z; \varepsilon), \\[2mm] S_k^{(l)}(0; \varepsilon) = \begin{cases} C_\varepsilon, & \text{if } l = k, \\ 0, & \text{if } l \ne k, \end{cases} \end{cases} \tag{6.13}$$

where

$$C_\varepsilon := \frac{1}{2\pi i} \int_\Gamma \exp\left(-\varepsilon(b-\lambda)^\sigma\right)(\lambda - A_0)^{-1} d\lambda, \quad \varepsilon > 0.$$

Set

$$\widetilde{S}_k(z;\ \varepsilon) = e^{-i\frac{k\beta}{n}} S_k\left(e^{i\frac{\beta}{n}}z;\ \varepsilon\right), \quad z \in \mathbf{C},\ 0 \le k \le n-1,\ \varepsilon > 0.$$

Then by (6.13), for $z \in \mathbf{C}$, $0 \le k$, $l \le n-1$, $\varepsilon > 0$,

$$\begin{cases} \dfrac{d^n}{dz^n}\widetilde{S}_k\left(z;\ \varepsilon\right) = A\widetilde{S}_k\left(z;\ \varepsilon\right), \\[4mm] \widetilde{S}_k^{(l)}\left(0;\ \varepsilon\right) = \begin{cases} C_\varepsilon, & \text{if } l = k, \\[2mm] 0, & \text{if } l \ne k. \end{cases} \end{cases} \tag{6.14}$$

Thus for each $\varepsilon > 0$ and every $v_0,\ v_1,\ \cdots,\ v_{n-1} \in \mathcal{R}\left(C_\varepsilon\right)$,

$$u(t) := \sum_{k=0}^{n-1} \widetilde{S}_k\left(t;\ \varepsilon\right) u_k, \quad t \ge 0,$$

where $u_k \in E$ satisfying $C_\varepsilon u_k = v_k$ ($0 \le k \le n-1$), is a solution of (6.1) with $u^{(k)}(0) = v_k$ ($0 \le k \le n-1$). Combining this with (6.9), we obtain (i).

Observing that

$$(\lambda - A_0)^{-1}u \;=\; \sum_{j=1}^{[\alpha]+2} \frac{1}{\lambda^j} A_0^{j-1}u + \frac{1}{\lambda^{[\alpha]+2}}(\lambda - A_0)^{-1} A_0^{[\alpha]+2} u,$$

$$\text{for any } u \in \mathcal{D}\left(A_0^{[\alpha]+2}\right),$$

and

$$\left\|\frac{1}{\lambda^{[\alpha]+2}}(\lambda - A_0)^{-1}\right\| \le \frac{1}{|\lambda|^{[\alpha]-\alpha+2}}, \quad \lambda \in |z_0| + \Sigma_\theta,$$

we obtain by shifting the path of integration again that

$$\lim_{\varepsilon \to 0+} C_\varepsilon u = u, \quad \text{for each } u \in \mathcal{D}\left(A^{[\alpha]+2}\right) = \mathcal{D}\left(A_0^{[\alpha]+2}\right). \tag{6.15}$$

Moreover, it can be verified from the resolvent identity and Fubini's theorem that, $\{C_\varepsilon\}_{\varepsilon>0}$ satisfies the semigroup property

$$C_{\varepsilon_1+\varepsilon_2} = C_{\varepsilon_1} C_{\varepsilon_2},$$

and can be extended analytically to the sector $\Sigma_{\frac{\pi}{2}-\sigma(\pi-\theta)}$. Now let $C_{\varepsilon_0}u = 0$ for some $\varepsilon_0 > 0$, $u \in E$. Then the semigroup property implies $C_\varepsilon u = 0$ for all $\varepsilon \ge \varepsilon_0$, and so the analyticity implies $C_\varepsilon u = 0$ for all $\varepsilon > 0$. But C_ε commutes with $(\lambda - A_0)^{-1}$, $\lambda \in \rho(A_0)$. This indicates by (6.15),

$$\left[(\lambda - A_0)^{-1}\right]^{[\alpha]+2} u = \lim_{\varepsilon \to 0+} C_\varepsilon \left[(\lambda - A_0)^{-1}\right]^{[\alpha]+2} u = 0.$$

Hence $u = 0$. Thus we see that C_ϵ is injective for each $\epsilon > 0$.

Now we prove the uniqueness of the solution of (6.1). To this end, suppose that $u(t)$ is a solution of (6.1) with $u_0 = \cdots = u_{n-1} = 0$, take $t \geq 0$, $\epsilon > 0$ and consider the function

$$f(s) = \sum_{j=0}^{n-1} \frac{d^{n-j-1}}{dz^{n-j-1}} \tilde{S}_0(z;\ \epsilon) \Bigg|_{z=t-s} u^{(j)}(s), \quad 0 \leq s \leq t. \tag{6.16}$$

Immediately,

$$f(0) = 0, \quad f(t) = C_\epsilon u^{(n-1)}(t).$$

By (6.14) and the fact that $u(t)$ is a solution of (6.1), we obtain that for $0 \leq s \leq t$,

$$f'(s) = -\tilde{S}_0^{(n)}(t-s;\ \epsilon)\, u(s) + \tilde{S}_0(t-s;\ \epsilon)\, u^{(n)}(s) = 0,$$

It follows that for each $t \geq 0$, $f(t) \equiv 0$, i.e., for each $t \geq 0$, $\epsilon > 0$, $C_\epsilon u^{(n-1)}(t) = 0$. Since C_ϵ is injective, we have $u^{(n-1)}(t) = 0$, $t \geq 0$. So

$$u(t) \equiv 0, \quad \text{for all } t \geq 0.$$

This ends the proof.

Corollary 6.3. *For* $n \geq 2$, *if there exists* $a > 0$, $-\pi < \phi \leq \pi$ *such that* $a - e^{-i\phi}A$ *is nonnegative, then the conclusion of Theorem 6.2 holds.*

Proof. Since $a - e^{-i\phi}A$ is nonnegative, we obtain that for some $\theta \in (0,\ \frac{\pi}{2})$, $C_\theta > 0$, $a + \Sigma_\theta \subset \rho(e^{-i\phi}A)$, and

$$\left\| \lambda(\lambda - A)^{-1} \right\| \leq C_\theta, \quad \text{for every } \lambda \in a + \Sigma_\theta. \tag{6.17}$$

Therefore, $e^{i\phi}(a + \Sigma_\theta) \subset \rho(A)$, and

$$\sup \left\{ \left\| \lambda(\lambda - A)^{-1} \right\|;\ \lambda \in e^{i\phi}(a + \Sigma_\theta) \right\} < \infty. \tag{6.18}$$

Consequently, in view of Theorem 6.2 with $\alpha = -1$, we know that Corollary 6.3 is true.

Corollary 6.4. *Assume that* μA *is the generator of an* r-*times* $(r \in N_0, \mu \in \mathbf{C})$ *integrated semigroup. Then the conclusion of Theorem 6.2 holds for* $n \geq 2$.

Proof. From the definition of integrated semigroups (see Section 1.3), we deduce that the hypotheses in Theorem 6.2 hold. So Corollary 6.4 holds.

Theorem 6.5. *Let* $n \geq 3$, $r \in N$, $q \in N_0$, *and* $[a, \infty) \subset \rho(A)$ *for some* $a > 0$. *For each* $u_0, \cdots, u_{n-1} \in D(A^r)$, (6.1) *has a unique solution* $u(t)$ *satisfying the estimate*

$$\sup_{t \geq 0} \left\{ e^{-at} \left\| \int_0^t \frac{1}{q!}(t-s)^q u(s)\,ds \right\| \right\} < \infty \tag{6.19}$$

if and only if A is bounded.

Proof. "$\Longrightarrow$". Consider the following Cauchy problem

$$\begin{cases} x'(t) = Bx(t), \quad t \geq 0, \\ \\ x(0) = x_0, \end{cases} \tag{6.20}$$

in the product space E^n. Here

$$B := \begin{pmatrix} 0 & I & 0 & \cdots & 0 \\ 0 & 0 & I & \cdots & 0 \\ \vdots & \vdots & \vdots & \ddots & \vdots \\ A & 0 & 0 & \cdots & 0 \end{pmatrix}_{n \times n}, \quad \mathcal{D}(B) := \mathcal{D}(A) \times E^{n-1}.$$

It is not difficult to see that $\left[a^{\frac{1}{n}}, \infty \right) \subset \rho(B)$, and for each $\lambda \in \left[a^{\frac{1}{n}}, \infty \right)$,

$$(\lambda - B)^{-1} = \begin{pmatrix} \lambda^{n-1}(\lambda^n - A)^{-1} & \cdots & (\lambda^n - A)^{-1} \\ A(\lambda^n - A)^{-1} & \cdots & \lambda(\lambda^n - A)^{-1} \\ \vdots & \ddots & \vdots \\ \lambda^{n-2}A(\lambda^n - A)^{-1} & \cdots & \lambda^{n-1}(\lambda^n - A)^{-1} \end{pmatrix}.$$

Note that for $u_0, u_1, \cdots, u_{n-1} \in \mathcal{D}(A)$,

$$B^i \left(u_0, u_1, \cdots, u_{i-1}, u_i, u_{i+1}, \cdots, u_{n-1} \right)^T$$

$$= \left(u_i, u_{i+1}, \cdots, u_{n-1}, Au_0, Au_1, \cdots, Au_{i-1} \right)^T,$$

where T denotes the transpose. We then have

$$\mathcal{D}\left(B^{nr+1} \right) \subset \left(\mathcal{D}(A^r) \right)^n.$$

Accordingly, we obtain by (6.19) that for each initial value in $\mathcal{D}\left(B^{nr+1} \right)$, (6.20) has a unique solution

$$x(t) = \left(u(t), u'(t), \cdots, u^{(n-1)}(t) \right)^T, \quad t \geq 0$$

satisfying

$$\sup_{t \geq 0} \left\{ e^{-(a+1)t} \left\| \int_0^t \frac{1}{(n+q-1)!}(t-s)^{n+q-1} x(s)ds \right\| \right\}$$

$$\leq \sup_{t \geq 0,\ 0 \leq i \leq n-1} \left\{ e^{-(a+1)t} \left\| \frac{(q+i)!}{(n+q-1)!} t^{n-i-1} \right. \right.$$

$$\left. \left. \cdot \int_0^t \frac{1}{(q+i)!}(t-s)^{q+i} u^{(i)}(s)ds \right\| \right\}$$

$$< \infty.$$

By applying Corollary 1.6.4 to the Cauchy problem (6.20), we see that B is the generator of a $(n(r+1)+q+1)$-times integrated semigroup. Therefore, There exist constants $M,\ b > 1$ and a strongly continuous family $\{H(t)\}_{t \geq 0}$ of bounded linear operators on E, satisfying

$$\|H(t)\| \leq Me^{bt}, \quad \text{for all } t \geq 0$$

and

$$\lambda^{-n(r+1)-q-1}(\lambda^n - A)^{-1} u = \int_0^\infty e^{-\lambda t} H(t) u\, dt, \quad \text{Re}\lambda > b,\ u \in E.$$

This implies that for $\text{Re}\lambda > b + 1$,

$$\left\| (\lambda^n - A)^{-1} \right\| \leq M |\lambda|^{n(r+1)+q+1}. \tag{6.21}$$

Write

$$\mathcal{G} := \{ \lambda^n;\ \lambda \in \mathbf{C},\ \text{Re}\lambda > b+1 \}.$$

It follows from (6.21) that $\mathcal{G} \subset \rho(A)$ and for each $\eta \in \mathcal{G}$,

$$\left\| (\eta - A)^{-1} \right\| \leq M |\eta|^{r+2+[\frac{q+1}{n}]}. \tag{6.22}$$

For any

$$z \in \mathcal{G}_0 := \left\{ z \in \mathbf{C};\ |z| > \left((b+1)\sec\frac{\pi}{n} \right)^n \right\},$$

if we set $\lambda = |z|^{\frac{1}{n}} e^{i\frac{\arg z}{n}}$, then by $|\arg z| \leq \pi$, we have that

$$\text{Re}\lambda = |z|^{\frac{1}{n}} \cos\frac{\arg z}{n} > (b+1)\frac{\cos\frac{\arg z}{n}}{\cos\frac{\pi}{n}} \geq (b+1)$$

and $z = \lambda^n$. This shows that $\mathcal{G}_0 \subset \mathcal{G} \subset \rho(A)$, i.e.,

$$\sigma(A) \subset \left\{ z \in \mathbf{C};\ |z| \leq \left((b+1)\sec\frac{\pi}{n} \right)^n \right\}.$$

Moreover, according to (6.22), we get

$$\|(z - A)^{-1}\| \le M|z|^{r+2+[\frac{q+1}{n}]}, \quad z \in \mathcal{G}_0. \tag{6.23}$$

Denote $m = r + 2 + [\frac{q+1}{n}] + 2$. Clearly, for each $u \in \mathcal{D}(A^m)$, $z \in \rho(A)$,

$$(z - A)^{-1}u = \sum_{i=0}^{m-1} \frac{1}{z^{i+1}} A^i u + \frac{1}{z^m}(z - A)^{-1} A^m u.$$

Thus, taking L a circle centered at the origin and enclosing $\sigma(A)$, (6.23) implies that

$$\frac{1}{2\pi i} \int_L (z - A)^{-1} u dz = u, \quad u \in \mathcal{D}(A^m).$$

Consequently, for some $z_0 \in \rho(A)$,

$$
\begin{aligned}
(z_0 - A)^{-m} u &= \frac{1}{2\pi i} \int_L (z - A)^{-1} (z_0 - A)^{-m} u dz \\
&= (z_0 - A)^{-m} \frac{1}{2\pi i} \int_L (z - A)^{-1} u dz, \quad u \in E.
\end{aligned}
$$

Since $(z_0 - A)^{-m}$ is injective, we get that

$$\frac{1}{2\pi i} \int_L (z - A)^{-1} dz = I. \tag{6.24}$$

For $u \in E$, set

$$Bu = \frac{1}{2\pi i} \int_L A(z - A)^{-1} u dz.$$

It is clear that $B \in \mathbf{L}(E)$, and (6.24) shows that

$$Au = Bu \quad \text{for} \quad u \in \mathcal{D}(A).$$

We say that $\mathcal{D}(A)$ must be E. In fact, if this is false, then there exists a $u \in E$ such that $u \notin \mathcal{D}(A)$. Taking $\mu \in \rho(A) \bigcap \rho(B)$ (please note that $B \in \mathbf{L}(E)$ and $\sigma(A)$ is bounded, which implies that $\rho(A) \bigcap \rho(B) \ne \emptyset$), then both $\mu - B$ and $\mu - A$ are injective and $\mathcal{R}(\mu - A) = E$. Hence,

$$
\begin{aligned}
(\mu - B)u \notin (\mu - B)(\mathcal{D}(A)) &= \{(\mu - B)v; \ v \in \mathcal{D}(A)\} \\
&= \{(\mu - A)v; \ v \in \mathcal{D}(A)\} \\
&= (\mu - A)(\mathcal{D}(A)) = E.
\end{aligned}
$$

A contradiction! So $\mathcal{D}(A) = E$, which implies $A \in \mathbf{L}(E)$. This ends the proof.

"⟸". It is easy to verify that for $u_0, \cdots, u_{n-1} \in \mathbf{L}(E)$,

$$u(t) := \sum_{k=0}^{n-1} \sum_{j=0}^{\infty} t^{nj+k} \frac{A^j u_k}{(nj+k)!}$$

is a solution of (6.1) satisfying (6.19).

To show uniqueness, let $u(\cdot)$ be a solution of (6.1) with $u_i = 0$ $(0 \le i \le n-1)$. Then for each $t \ge 0$

$$u(t) = \frac{1}{(nj-1)!} \int_0^t (t-s)^{nj-1} A^j u(s) ds, \quad j \in N.$$

Hence for each $t > 0$ fixed,

$$\|u(t)\| \le t \frac{t^{nj-1}}{(nj-1)!} \|A\|^j \max_{0 \le s \le t} \|u(s)\|$$

$$\longrightarrow 0, \quad \text{as } j \longrightarrow \infty.$$

Namely $u(t) \equiv 0$ on R^+. Theorem 6.5 is then proved.

3.7 Notes

Using the theory of integrated semigroups, Neubrander [5] studied firstly the generalized (or nonstandard) wellposedness of (ACP_2) by the semigroup method and set up some interesting results. Section 3.1 comes from Xiao-Liang [15]. For $n = 2$, Theorem 1.2 improves the corresponding result in Neubrander [5, §5].

Sections 3.2 and 3.3 are taken from Liang-Xiao [12] and [10] respectively, which are stimulated by the works of deLaubenfels [3], Kellermann-Hieber [1] and Neubrander [2]. As in Neubrander [2], one might adopt the system (2.6.2) (with $-B_0, \cdots, -B_{n-1}, -A$ replacing $-A_0, \cdots, -A_{n-2}, -A_{n-1}$, respectively) to deal with the problem (2.1). Let

$$\mathcal{M}_0 = \begin{pmatrix} -A & 0 & 0 & \cdots & 0 \\ -B_{n-2} & 0 & 0 & \cdots & 0 \\ \vdots & \vdots & \vdots & \ddots & \vdots \\ -B_1 & 0 & 0 & \cdots & 0 \\ -B_0 & 0 & 0 & \cdots & 0 \end{pmatrix}, \quad \mathcal{P} = \begin{pmatrix} 0 & I & 0 & \cdots & 0 \\ 0 & 0 & I & \cdots & 0 \\ \vdots & \vdots & \vdots & \ddots & \vdots \\ 0 & 0 & 0 & \cdots & I \\ 0 & 0 & 0 & \cdots & 0 \end{pmatrix}.$$

Then $\mathcal{M}_n = \mathcal{M}_0 + \mathcal{P}$ and it can be verified under the condition of Theorem 2.1 that $\mathcal{M}_0$ generates an r-times integrated semigroup. Now recall that the addition of a bounded operator U to the generator U_0 of an r-times integrated semigroup may fail to preserve this property of U_0, unless additional conditions are imposed on U. As far as we know, there exist two typical conditions of such kind (cf. Kellermann-Hieber [1, Section 3], Neubrander [5, Corollary 2.5]):

(a) U commutes with the resolvent of U_0;

(b) $\mathcal{R}(U) \subset \mathcal{D}(U_0^r)$.

However, neither of them is satisfied by the perturbation operator $\mathcal{P}$. Therefore, a direct application of the perturbation theorems for integrated semigroups seems impractical in this situation.

Integrated cosine functions were introduced by Arendt-Kellermann [1]. The content of Section 3.4 is from Xiao-Liang [15].

The concept of C-wellposedness of (ACP_1) and (ACP_2) was introduced by deLaubenfels [4]. The definition of C-wellposedness of (ACP_n) given in Section 3.5 is somewhat different from that in deLaubenfels [4] when $n = 2$. Section 3.5 is from Xiao-Liang [23].

In deLaubenfels [9] (see Chapters XIII and XIV) and Hieber-Holderrieth-Neubrander [1], arbitrary systems of constant coefficient partial differential operators are dealt with by introducing a matrix of differential operators

$$\mathcal{A} := (p_{i,j}(D))_{k \times k};$$

with the usual matrix reduction of (ACP_2) to (ACP_1), with

$$\mathcal{A} := \begin{pmatrix} 0 & I \\ -q(D) & -p(D) \end{pmatrix},$$

the related theorems in deLaubenfels [9] and Hieber-Holderrieth-Neubrander [1] will produce a similar result as Theorem 5.5. By comparison, the α in Theorem 5.5 is sharper. Moreover, there is also another advantage of Theorem 5.5. In order to illustrate this, we write

$$p(x) = \sum_{|\beta| \leq l} a_\beta (ix)^\beta, \quad q(x) = \sum_{|\beta| \leq m} b_\beta (ix)^\beta.$$

From Theorem 5.5 one gets the information that the solution $u(\cdot)$ satisfies

$$t \longmapsto \sum_{|\beta| \leq l} a_\beta D^\beta u'(t), \quad t \longmapsto \sum_{|\beta| \leq m} b_\beta D^\beta u(t) \in C(R^+, E).$$

On the other hand, we note that $\mathcal{A}$ is not closed in general. Thus using the related theorems in deLaubenfels [9] and Hieber-Holderrieth-Neubrander [1] with the operator matrix $\mathcal{A}$ shows merely that

$$t \longmapsto \sum_{|\beta| \leq l} a_\beta D^\beta u'(t) + \sum_{|\beta| \leq m} b_\beta D^\beta u(t) \in C(R^+, E),$$

without giving the information whether

$$t \longmapsto \sum_{|\beta| \leq l} a_\beta D^\beta u'(t), \quad t \longmapsto \sum_{|\beta| \leq m} b_\beta D^\beta u(t) \in C(R^+, E), \quad t \geq 0.$$

Section 3.6 is taken from Xiao-Liang [17]. Related results can be found in deLaubenfels [6], Goldstein-deLaubenfels-Sandefur [1] (see also deLaubenfels [9, Examples 14.3 and 25.8]).

Chapter 4

Analyticity and parabolicity

Summary

In Section 4.1, we characterize, in terms of the estimates of $\lambda^{n-1}R_\lambda$, $\lambda^{k-1}A_kR_\lambda$, $\lambda^{k-1}\overline{R_\lambda A_k}$ $(1 \leq k \leq n-1)$, those (ACP_n) whose propagators can be extended analytically to the sector Σ_θ (for a fixed $\theta \in (0, \frac{\pi}{2}]$) satisfying appropriate conditions there; such behavior of (ACP_n) is called analytic wellposedness (in Σ_θ), which will be made precise in Definition 1.2. We also treat perturbation problems about analytic wellposedness in this section. A new type of perturbation operators is introduced, besides that given in Section 2.4.

Section 4.2 is devoted to the parabolicity of higher order equations, a property defined in terms of the estimates of $\lambda^{n-1}R_\lambda$, $\lambda^{k-1}A_kR_\lambda$ $(1 \leq k \leq n-1)$. One can see that (ACP_n) with this property is analytically solvable (Definition 2.2 , Theorem 2.3). We pay attention to the following type of equations

$$u^{(n)}(t) + \sum_{i=1}^{n-1} c_i A^{k_i} u^{(n-i)}(t) + Au(t) = 0, \qquad (*)$$

where $c_i \in \mathbf{C}$ $(1 \leq i \leq n-1)$ and A is a nonnegative operator in E. A sufficient and necessary condition for $(*)$ to be parabolic is obtained, provided $k_1 > k_2 - k_1 > \cdots > 1 - k_{n-1} > 0$. Also some perturbation theorems are presented, and among them Theorem 2.11 takes care of general cases. Finally in Theorem 2.17, a sharp criterion is given for the case of $n = 3$, A nonnegative in a strict sense.

Complete (ACP_2) with differential operators as coefficient operators, last seen in Section 3.5, reappear in Section 4.3. The objective is to explore conditions for analytic wellposedness or analytic solvability of such (ACP_2).

In Section 4.4, we are concerned with the entire solutions of (ACP_n), that is, the solutions which can be extended analytically to the whole complex plane. The purpose is to find some conditions ensuring that (ACP_n) have a unique entire solution for every initial value in a dense set. These conditions turn out to be particularly satisfied for parabolic higher order equations.

As in Chapter 3, E denotes a Banach space.

4.1   Analyticity

Theorem 1.1. *Let $A_0, \cdots, A_{n-1}$ be closed linear operators in E. For every fixed $\theta \in (0, \frac{\pi}{2}]$, $a \in R$, the following statements are equivalent:*

(i) (ACP_n) *is strongly wellposed. The propagators $S_k(\cdot)$ $(0 \le k \le n-1)$ can be extended analytically to Σ_θ, $S_{n-1}^{(k-1)}(z)E \subset \mathcal{D}(A_k)$ $(z \in \Sigma_\theta)$ and $A_k S_{n-1}^{(k-1)}(\cdot)$ are analytic in Σ_θ $(1 \le k \le n-1)$. For each $\theta' \in (0, \theta)$, there exists $C_{\theta'} > 0$ such that for each $z \in \Sigma_{\theta'}$,*

$$\left\| S_k^{(k)}(z) \right\| \le C_{\theta'} e^{a\mathrm{Re}z} \quad (0 \le k \le n-1), \tag{1.1}$$

$$\left\| A_k S_{n-1}^{(k-1)}(z) \right\| \le C_{\theta'} e^{a\mathrm{Re}z} \quad (1 \le k \le n-1). \tag{1.2}$$

(ii) $\bigcap_{i=0}^{n-1} \mathcal{D}(A_i)$ *is dense in E. For each $\theta' \in (0, \theta)$, there exists $M_{\theta'} > 0$ such that for each $\lambda \in a + \Sigma_{\frac{\pi}{2}+\theta'}$, $R_\lambda \in \mathbf{L}(E)$, $R_\lambda A_k$ $(0 \le k \le n-1)$ is closable and for each $1 \le k \le n-1$,*

$$\left\| \lambda^{n-1} R_\lambda \right\|, \quad \left\| \lambda^{k-1} A_k R_\lambda \right\|, \quad \left\| \lambda^{k-1} \overline{R_\lambda A_k} \right\| \le M_{\theta'} |\lambda - a|^{-1}. \tag{1.3}$$

Moreover, in this case, we have that for each $z \in \Sigma_\theta$,

$$S_{n-1}^{(n)}(z) + \sum_{i=0}^{n-1} A_i S_{n-1}^{(i)}(z) = 0, \tag{1.4}$$

$$S_k^{(n)}(z)u + \sum_{i=0}^{n-1} A_i S_k^{(i)}(z)u = 0 \quad \left(u \in \bigcap_{i=0}^{k} \mathcal{D}(A_i), \ 0 \le k \le n-2 \right), \tag{1.5}$$

and for each $\theta' \in (0, \theta)$, as $z \to 0$ $(z \in \Sigma_{\theta'})$,

$$S_k^{(k)}(z)u \longrightarrow u \quad (u \in E, \ 0 \le k \le n-1), \tag{1.6}$$

$$S_k^{(i)}(z)u \longrightarrow 0 \quad \left(u \in \bigcap_{i=0}^{k} \mathcal{D}(A_i), \ 0 \le k, \ i \le n-1, \ i \ne k \right), \tag{1.7}$$

$$S_{n-1}^{(i)}(z)u \longrightarrow 0 \quad (u \in E, \ 0 \le i \le n-2). \tag{1.8}$$

Proof. (i) $\Longrightarrow$ (ii). Define

$$\rho_0(A_0, \cdots, A_{n-1})$$

$$:= \Big\{ \lambda \in \mathbf{C}; \ R_\lambda \in \mathbf{L}(E), \ R_\lambda A_k \text{ are closable} \tag{1.9}$$

$$\text{and } \overline{R_\lambda A_k} \in \mathbf{L}(E) \text{ for } 0 \le k \le n-1 \Big\}.$$

Since (ACP_n) is strongly wellposed with (1.1), (1.2) holding, we have by Remark 2.2.5 that

$$\{\lambda \in \mathbf{C}; \quad \mathrm{Re}\lambda > a_0\} \subset \rho_0(A_0, \cdots, A_{n-1}),$$

for some $a_0 \geq a$. Let $\lambda_0 \in \rho_0(A_0, \cdots, A_{n-1})$, then $A_k R_{\lambda_0} \in L(E)$ due to $\mathcal{D}(P_\lambda) = \bigcap_{i=0}^{n-1} \mathcal{D}(A_i)$. Hence for $\mu \in \mathbf{C}$ with $|\mu|$ small enough,

$$
\begin{aligned}
R_{\lambda_0 + \mu} &= (P_{\lambda_0} + P_{\lambda_0 + \mu} - P_{\lambda_0})^{-1} \\
&= \left[I + \sum_{k=1}^{n} \left((\lambda_0 + \mu)^k - \lambda_0^k \right) \overline{R_{\lambda_0} A_k} \right]^{-1} R_{\lambda_0} \\
&= R_{\lambda_0} \left[I + \sum_{k=1}^{n} \left((\lambda_0 + \mu)^k - \lambda_0^k \right) A_k R_{\lambda_0} \right]^{-1}.
\end{aligned}
$$

Consequently, $\rho_0(A_0, \cdots, A_{n-1})$ is an open subset of $\mathbf{C}$, and R_λ, $\overline{R_\lambda A_k}$, $A_k R_\lambda$ $(0 \leq k \leq n-1)$ are analytic in $\rho_0(A_0, \cdots, A_{n-1})$. Thus we have by Remark 2.2.5 that for every $\mathrm{Re}\lambda > a_0$, $u \in E$,

$$\lambda^{k-1} A_k R_\lambda u = \int_0^\infty e^{-(\lambda-a)t} e^{-at} A_k S_{n-1}^{(k-1)}(t) u \, dt \quad (1 \leq k \leq n), \qquad (1.10)$$

$$
\begin{cases}
\overline{R_\lambda A_0} u = u - \lambda \int_0^\infty e^{-(\lambda-a)t} e^{-at} S_0(t) u \, dt, \\
\\
\lambda^{k-1} \overline{R_\lambda A_k} u \\
\quad = \int_0^\infty e^{-(\lambda-a)t} e^{-at} \left[S_{k-1}^{(k-1)}(t) - S_k^{(k)}(t) \right] u \, dt \\
\\
\qquad (1 \leq k \leq n-1).
\end{cases}
\qquad (1.11)
$$

For every $\theta' \in (0, \theta)$, choose $\theta_1 = \frac{1}{2}(\theta + \theta')$. Clearly, for every $z \in \Sigma_{\theta_1}$,

$$\left\| e^{-az} A_k S_{n-1}^{(k-1)}(z) \right\| \leq C_{\theta_1} \quad (1 \leq k \leq n),$$

$$\| e^{-az} S_0(z) \| \leq C_{\theta_1},$$

$$\left\| e^{-az} \left[S_{k-1}^{(k-1)}(z) - S_k^{(k)}(z) \right] \right\| \leq 2 C_{\theta_1} \quad (1 \leq k \leq n-1).$$

Therefore, if $\mathrm{Re}\lambda > a_0$ and $\mathrm{Re}(\lambda - a)e^{-i\theta_1} > 0$, we can shift the path of integration in (1.10) and (1.11) to the ray $re^{-i\theta_1}$, $0 \leq r < \infty$; namely, for every

$u \in E$,

$$\lambda^{k-1} A_k R_\lambda u = \int_0^\infty e^{-i\theta_1} \exp\left(-(\lambda - a)e^{-i\theta_1}r\right)$$

$$\cdot \exp\left(-are^{-i\theta_1}\right) A_k S_{n-1}^{(k-1)} \left(re^{-i\theta_1}\right) u dr \qquad (1.12)$$

$$(1 \le k \le n),$$

$$\begin{cases} \overline{R_\lambda A_0} u = u - \lambda \int_0^\infty e^{-i\theta_1} \exp\left(-(\lambda - a)e^{-i\theta_1}r\right) \\ \\ \qquad\qquad \cdot \exp\left(-are^{-i\theta_1}\right) S_0 \left(re^{-i\theta_1}\right) u dr, \\ \\ \lambda^{k-1} \overline{R_\lambda A_k} u \\ \\ = \int_0^\infty e^{-i\theta_1} \exp\left(-(\lambda - a)e^{-i\theta_1}r\right) \exp\left(-are^{-i\theta_1}\right) \\ \\ \qquad \cdot \left[S_{k-1}^{(k-1)}\left(re^{-i\theta_1}\right) - S_k^{(k)}\left(re^{-i\theta_1}\right)\right] u dr \qquad (1 \le k \le n-1). \end{cases}$$

$$(1.13)$$

We say

$$\left\{\lambda \in \mathbf{C}; \ \ \mathrm{Re}(\lambda - a)e^{-i\theta_1} > 0\right\} \subset \rho_0(A_0, \cdots, A_{n-1}). \qquad (1.14)$$

In fact, if this is false, then there is a $\lambda_0 = a + \sigma_0 + i\tau_0$ satisfying $\sigma_0 > -\tau_0 \tan\theta_1$ ($\tau_0 \in R$) such that $\lambda_0 \notin \rho_0(A_0, \cdots, A_{n-1})$ and

$$J_1 \ := \ \left\{\lambda \in \mathbf{C}; \ \ \mathrm{Re}(\lambda - a) > \sigma_0, \ \mathrm{Im}\lambda = \tau_0\right\}$$

$$\subset \ \rho_0(A_0, \cdots, A_{n-1}).$$

Thus (1.12), (1.13) hold for $\lambda \in J_1$. Hence, for every $1 \le k \le n-1$, $\lambda \in J_1$,

$$\left\|\lambda^{n-1} R_\lambda\right\|, \ \ \left\|\lambda^{k-1} A_k R_\lambda\right\|, \ \ \left\|\lambda^{k-1} \overline{R_\lambda A_k}\right\|$$

$$\le \ 2C_{\theta_1} \left[\mathrm{Re}(\lambda - a)e^{-i\theta_1}\right]^{-1}$$

$$\le \ 2C_{\theta_1} \left[\mathrm{Re}(\lambda_0 - a)e^{-i\theta_1}\right]^{-1}.$$

Let

$$K \ = \ \sup\left\{\left\|\sum_{i=0}^{n-1} \lambda^i \lambda_0^{n-i-1} R_\lambda\right\| + \left\|\sum_{k=1}^{n-1}\sum_{i=0}^{k-1} \lambda^i \lambda_0^{k-i-1} A_k R_\lambda\right\|\right.$$

$$\left. + \left\|\sum_{k=1}^{n-1}\sum_{i=0}^{k-1} \lambda^i \lambda_0^{k-i-1} \overline{R_\lambda A_k}\right\|; \ \lambda \in J_1\right\}.$$

Then $K < \infty$. Take a $\lambda_1 \in J_1$, such that $\mathrm{Re}\lambda_1 = a + \sigma_0 + \frac{1}{2K}$. Then

$$
\begin{aligned}
R_{\lambda_0} &= R_{\lambda_1} \left\{ I + (\lambda_0 - \lambda_1) \left[\sum_{i=0}^{n-1} \lambda_1^i \lambda_0^{n-i-1} R_{\lambda_1} \right.\right. \\
&\qquad\qquad\qquad \left.\left. + \sum_{k=1}^{n-1}\sum_{i=0}^{k-1} \lambda_1^i \lambda_0^{k-i-1} A_k R_{\lambda_1} \right] \right\}^{-1} \\
&= \left\{ I + (\lambda_0 - \lambda_1) \left[\sum_{i=0}^{n-1} \lambda_1^i \lambda_0^{n-i-1} R_{\lambda_1} \right.\right. \\
&\qquad\qquad\qquad \left.\left. + \sum_{k=1}^{n-1}\sum_{i=0}^{k-1} \lambda_1^i \lambda_0^{k-i-1} \overline{R_{\lambda_1} A_k} \right] \right\}^{-1} R_{\lambda_1}.
\end{aligned}
$$

The first equality implies that $\mathcal{D}(R_{\lambda_0}) = E$, which justifies the second equality. Thus we see that $\lambda_0 \in \rho_0(A_0, \cdots, A_{n-1})$. A contradiction! Whence, (1.14) is true.

Accordingly, (1.12) and (1.13) hold for all $\lambda \in \{\lambda \in \mathbf{C};\ \mathrm{Re}(\lambda - a)e^{i\theta_1} > 0\}$. Therefore, $R_\lambda A_k$ $(0 \le k \le n-1)$ are closable and for $1 \le k \le n-1$, $\lambda^{n-1} R_\lambda$, $\lambda^{k-1}\overline{R_\lambda A_k}$, $\lambda^{k-1} A_k R_\lambda$ are bounded by

$$
2C_{\theta_1} \left[\mathrm{Re}(\lambda - a)e^{-i\theta_1} \right]^{-1} \le 2C_{\theta_1} \left[\sin(\theta_1 - \theta') \right]^{-1} |\lambda - a|^{-1},
$$

whenever

$$
0 \le \arg(\lambda - a) < \theta' + \frac{\pi}{2}.
$$

Letting $\mathrm{Re}\lambda > a_0$ and $\mathrm{Re}(\lambda - a)e^{i\theta_1} > 0$, we shift the path of integration in (1.10) and (1.11) to the ray $re^{i\theta_1}$, $0 \le r < \infty$ and argue similarly as in the previous case. Thus we can show that, whenever

$$
-(\theta' + \frac{\pi}{2}) < \arg(\lambda - a) \le 0,
$$

$R_\lambda A_k$ $(0 \le k \le n-1)$ are closable and for $1 \le k \le n-1$, $\lambda^{n-1} R_\lambda$, $\lambda^{k-1}\overline{R_\lambda A_k}$, $\lambda^{k-1} A_k R_\lambda$ are bounded by $2C_{\theta_1} \left[\sin(\theta_1 - \theta') \right]^{-1} |\lambda - a|^{-1}$.

The implication of (i) $\implies$ (ii) is then proved.

(ii) $\implies$ (i). Let $\varepsilon > 0$, $\frac{\pi}{2} < \phi_1 < \theta + \frac{\pi}{2}$. For every $0 \le k \le n-1$, define

$$
S_k(t) = \frac{1}{2\pi i} \int_\gamma e^{\lambda t} \sum_{i=k+1}^n \lambda^{i-k-1} \overline{R_\lambda A_i} d\lambda, \quad t > 0, \tag{1.15}
$$

where $\gamma = \gamma_1 \cup \gamma_2 \cup \gamma_3$ with

$$
\gamma_1 = \{a + re^{-i\phi_1};\ \varepsilon \le r < \infty\},
$$

$$
\gamma_2 = \{a + \varepsilon e^{i\phi};\ |\phi| \le \phi_1\},
$$

$$
\gamma_3 = \{a + re^{i\phi_1};\ \varepsilon \le r < \infty, \},
$$

and is oriented such that $\text{Im}\lambda$ increases along γ. By our assumption, the convergence in (1.15) for $t > 0$ is in the uniform operator topology and $S_k(\cdot)$ $(0 \leq k \leq n-1)$ are independent of ε and ϕ_1.

For every $\theta' \in (0, \theta)$, $z \in \Sigma_{\theta'}^+ := \{z \in \mathbf{C}; \ z \neq 0, \ 0 \leq \arg z \leq \theta'\}$, take

$$\phi_1 = \frac{\pi}{2} + \frac{1}{2}(\theta + \theta'), \quad \varepsilon = \frac{1}{|z|}.$$

By (1.3), we have that for every $0 \leq k \leq n-1$,

$$\left\| \frac{1}{2\pi i} \int_{\gamma_1} e^{\lambda z} \sum_{i=k+1}^{n} \lambda^{i-1} \overline{R_\lambda A_i} d\lambda \right\|$$

$$\leq \frac{n-k}{2\pi} M_{\theta'} e^{a\text{Re}z} \int_{\frac{1}{|z|}}^{\infty} e^{-r|z|\sin\frac{1}{2}(\theta-\theta')} \frac{1}{r} dr$$

$$= \frac{n-k}{2\pi} M_{\theta'} e^{a\text{Re}z} \int_{\sin\frac{1}{2}(\theta-\theta')}^{\infty} \frac{e^{-r}}{r} dr,$$

$$\left\| \frac{1}{2\pi i} \int_{\gamma_2} e^{\lambda z} \sum_{i=k+1}^{n} \lambda^{i-1} \overline{R_\lambda A_i} d\lambda \right\| \leq \frac{n-k}{2\pi} M_{\theta'} e^{a\text{Re}z} \int_{-\phi_1}^{\phi_1} e^{\cos(\phi+\arg z)} d\phi,$$

$$\left\| \frac{1}{2\pi i} \int_{\gamma_3} e^{\lambda z} \sum_{i=k+1}^{n} \lambda^{i-1} \overline{R_\lambda A_i} d\lambda \right\| \leq \frac{n-k}{2\pi} M_{\theta'} e^{a\text{Re}z} \int_{\sin\frac{1}{2}(\theta+\theta')}^{\infty} \frac{e^{-r}}{r} dr.$$

Hence, $S_k(\cdot)$ $(0 \leq k \leq n-1)$ can be extended analytically to $\Sigma_{\theta'}^+$,

$$S_k(z) = \frac{1}{2\pi i} \int_{\gamma} e^{\lambda z} \sum_{i=k+1}^{n} \lambda^{i-k-1} \overline{R_\lambda A_i} d\lambda, \quad z \in \Sigma_{\theta'}^+, \qquad (1.16)$$

and the estimate (1.1) holds. Similarly, by (1.3) we can obtain the estimate (1.2).

Choose ε such that $\varepsilon > -a$. According to (1.16), for every $0 \leq k \leq n-1$,

$$S_k(z)u = \frac{1}{2\pi i} \int_{\gamma} e^{\lambda z} \lambda^{-k-1} \left(u - \sum_{i=0}^{k} \lambda^i R_\lambda A_i u \right) d\lambda$$

$$\left(u \in \bigcap_{i=0}^{k} \mathcal{D}(A_i), \ z \in \Sigma_{\theta'}^+ \right). \qquad (1.17)$$

Therefore, by a deformation of contour, we deduce that as $z \to 0$ $(z \in \Sigma_{\theta'}^+)$,

$$S_k^{(k)}(z)u \to u \quad \left(u \in \bigcap_{i=0}^{k} \mathcal{D}(A_i), \ 0 \leq k \leq n-1 \right), \qquad (1.18)$$

$$A_k S_{n-1}^{(k-1)}(z)u \to 0 \quad \left(u \in \bigcap_{i=0}^{n-1} \mathcal{D}(A_i), \ 1 \le k \le n-1 \right), \tag{1.19}$$

since $\lambda^n R_\lambda$, $\lambda^i A_i R_\lambda$ $(0 \le i \le n-1)$ are bounded on the region lying to the right of the path γ. Making use of (1.1), (1.2) which have been proved, we see that (1.18) and (1.19) are also valid for all $u \in E$.

For $z \in \Sigma_{\theta'}^- = \{ z \in \mathbb{C}; \ z \ne 0, \ -\theta' < \arg z \le 0 \}$, a similar argument can be taken.

We next show that (ACP_n) is strongly wellposed.

For $\mu > a$, choose ε such that $\varepsilon < \mu - a$. By (1.15), for every $u \in E$, $0 \le k \le n-1$,

$$
\begin{aligned}
&\int_0^\infty e^{-\mu t} S_k^{(k)}(t) u \, dt \\
&= \frac{1}{2\pi i} \int_\gamma \sum_{i=k+1}^n \lambda^{i-1} \overline{R_\lambda A_i} u \left[\int_0^\infty e^{-(\mu-\lambda)t} dt \right] d\lambda \\
&= \frac{1}{2\pi i} \int_\gamma (\mu - \lambda)^{-1} \sum_{i=k+1}^n \lambda^{i-1} \overline{R_\lambda A_i} u \, d\lambda \\
&= \sum_{i=k+1}^n \mu^{i-1} \overline{R_\mu A_i} u.
\end{aligned}
\tag{1.20}
$$

Similarly,

$$\int_0^\infty e^{-\mu t} A_k S_{n-1}^{(k-1)}(t) u \, dt = \mu^{k-1} A_k R_\mu u \quad (u \in E, \ 1 \le k \le n-1). \tag{1.21}$$

Now we justify from (1.20), (1.21) and (1.1), (1.2) that for every $1 \le k \le n-1$, $\mu^{n-1} R_\mu$, $\mu^{k-1} A_k R_\mu$, $\mu^{k-1} \overline{R_\mu A_k}$ and their derivatives satisfy the estimates in Theorem 2.2.3. So the strong wellposedness of (ACP_n) follows.

Thus the implication of (ii) $\Longrightarrow$ (i) is proved.

According to (1.16) we have (1.4) and (1.8) (noting that (1.16) holds also for $z \in \Sigma_{\theta'}^-$). (1.5) and (1.7) follow from (1.17) ((1.17) holds also for $z \in \Sigma_{\theta'}^-$). This ends the proof of Theorem 1.1.

Definition 1.2. Let $\theta \in (0, \frac{\pi}{2}]$. We say that (ACP_n) is analytically wellposed in Σ_θ, if it is strongly wellposed, the propagators $S_k(\cdot)$ $(0 \le k \le n-1)$ can be extended analytically to Σ_θ such that $A_k S_{n-1}^{(k-1)}(\cdot)$ are analytic in Σ_θ (for $1 \le k \le n-1$), and for each $\theta' \in (0, \theta)$, there are $C_{\theta'}$, $a_{\theta'} > 0$ satisfying

$$\|S_0(z)\|, \ \left\| S_k^{(k)}(z) \right\|, \ \left\| A_k S_{n-1}^{(k-1)}(z) \right\| \le C_{\theta'} e^{a_{\theta'} \operatorname{Re} z},$$

$$z \in \Sigma_{\theta'}, \ 1 \le k \le n-1,$$

$$\lim_{\substack{z \to 0 \\ z \in \Sigma_{\theta'}}} S_k^{(j)}(z)u = \delta_{kj}u, \quad u \in \bigcap_{i=0}^{k} \mathcal{D}(A_i), \ 0 \leq k, \ j \leq n-1.$$

Immediately from Theorem 1.1, we have

Theorem 1.3. *Let* $\theta \in (0, \frac{\pi}{2}]$. *Suppose that* $A_0, \cdots, A_{n-1}$ *are closed linear operators in* E. *Then* (ACP_n) *is analytically wellposed in* Σ_θ *if and only if* $\bigcap_{i=0}^{n-1} \mathcal{D}(A_i)$ *is dense in* E, *for each* $\theta' \in (0, \theta)$, *there exist* $M_{\theta'}$, $\omega_{\theta'} > 0$ *such that*

$$\omega_{\theta'} + \Sigma_{\frac{\pi}{2}+\theta'} \subset \rho_0(A_0, \cdots, A_{n-1})$$

and for $1 \leq k \leq n-1$,

$$\left\| \lambda^{n-1}R_\lambda \right\|, \ \left\| \lambda^{k-1}A_k R_\lambda \right\|, \ \left\| \lambda^{k-1}\overline{R_\lambda A_k} \right\| \leq M_{\theta'}|\lambda|^{-1},$$

$$\lambda \in \omega_{\theta'} + \Sigma_{\frac{\pi}{2}+\theta'}. \tag{1.22}$$

Definition 1.4. *Let* $\theta \in (0, \frac{\pi}{2}]$ *and let* $B_0, \cdots, B_{n-1}$ *be linear operators in* E. *We say that* $(ACP_n)_{[B_{n-1}, \cdots, B_0]}$ *is analytically wellposed in* Σ_θ *if, it is strongly wellposed (see Section 2.4), the propagators* $\widetilde{S}_k(\cdot)$ $(0 \leq k \leq n-1)$ *can be extended analytically to* Σ_θ *such that* $A_k \widetilde{S}_{n-1}^{(k-1)}(\cdot)$, $B_k \widetilde{S}_{n-1}^{(k-1)}(\cdot)$ *are analytic in* Σ_θ *(for* $1 \leq k \leq n-1$), *and for each* $\theta' \in (0, \theta)$, *there are* $C_{\theta'}$, $\omega_{\theta'} > 0$ *satisfying*

$$\left\| \widetilde{S}_0(z) \right\|, \ \left\| \widetilde{S}_k^{(k)}(z) \right\|, \ \left\| A_k\widetilde{S}_{n-1}^{(k-1)}(z) \right\|, \ \left\| B_k\widetilde{S}_{n-1}^{(k-1)}(z) \right\| \leq C_{\theta'}e^{\omega_{\theta'}\mathrm{Re}z},$$

$$z \in \Sigma_{\theta'}, \ 1 \leq k \leq n-1,$$

$$\lim_{\substack{z \to 0 \\ z \in \Sigma_{\theta'}}} \widetilde{S}_k^{(j)}(z)u = \delta_{kj}u, \quad u \in \bigcap_{i=0}^{k} \mathcal{D}(A_i + B_i), \ 0 \leq k, \ j \leq n-1.$$

Theorem 1.5. *Let* $\theta \in (0, \frac{\pi}{2}]$ *and let* $A_0, \cdots, A_{n-1}$ *be closed linear operators in* E *such that* (ACP_n) *is analytically wellposed in* Σ_θ *and* $R_\lambda A_k$ *is closable for* $1 \leq k \leq n-1$ *and*

$$\lambda \in \bigcup_{\theta' \in (0, \theta)} \left(a_{\theta'} + \Sigma_{\frac{\pi}{2}+\theta'} \right),$$

where $a_{\theta'}$ *is some constant depending on* θ'. *Assume that* $B_0, \cdots, B_{n-1}$ *are closable linear operators in* E *satisfying that for each* $0 \leq k \leq n-1$, *there is an* i_k *with* $k+1 \leq i_k \leq n$ *such that* $\mathcal{D}(B_k) \supset \mathcal{D}(A_{i_k})$; *moreover there exists*

$\lambda_k \in \rho(A_{i_k})$ such that $(\lambda_k - A_{i_k})^{-1} B_k$ has a bounded extension to E. Then $(ACP_n)_{[B_{n-1}, \cdots, B_0]}$ is analytically wellposed in Σ_θ.

Proof. By hypothesis, (1.22) holds. Accordingly, there is $\widetilde{\omega}_{\theta'} > \max\{a_{\theta'}, \omega_{\theta'}\}$ such that for $\lambda \in \widetilde{\omega}_{\theta'} + \Sigma_{\frac{\pi}{2}+\theta'}$,

$$\left\| \sum_{k=0}^{n-1} \lambda^k B_k R_\lambda \right\| \leq \sum_{k=0}^{n-1} \left\{ \left\| B_k (\lambda_k - A_{i_k})^{-1} \right\| \left(\left\| \lambda_k \lambda^k R_\lambda \right\| + \left\| \lambda^k A_{i_k} R_\lambda \right\| \right) \right\}$$

$$< \frac{1}{2},$$

$$\left\| \sum_{k=0}^{n-1} \lambda^k \overline{R_\lambda B_k} \right\| \leq \sum_{k=0}^{n-1} \left\{ \left(\left\| \lambda_k \lambda^k R_\lambda \right\| + \left\| \lambda^k \overline{R_\lambda A_{i_k}} \right\| \right) \left\| \overline{(\lambda_k - A_{i_k})^{-1} B_k} \right\| \right\}$$

$$< \frac{1}{2}.$$

Therefore, for $\lambda \in \widetilde{\omega}_{\theta'} + \Sigma_{\frac{\pi}{2}+\theta'}$, $\widetilde{P}_\lambda := \lambda^n + \sum_{i=0}^{n-1} \lambda^i (A_i + B_i)$ is boundedly invertible and

$$\widetilde{R}_\lambda := \widetilde{P}_\lambda^{-1} = R_\lambda \left[I + \sum_{k=0}^{n-1} \lambda^k B_k R_\lambda \right]^{-1}$$

$$= \left[I + \sum_{k=0}^{n-1} \lambda^k \overline{R_\lambda B_k} \right]^{-1} R_\lambda.$$

Thus we easily see that for $0 \leq k \leq n-1$, $\lambda \in \widetilde{\omega}_{\theta'} + \Sigma_{\frac{\pi}{2}+\theta'}$, $\widetilde{R}_\lambda (A_k + B_k)$ is closable and for each $1 \leq k \leq n-1$,

$$\left\| \lambda^{n-1} \widetilde{R}_\lambda \right\|, \quad \left\| \lambda^{k-1} A_k \widetilde{R}_\lambda \right\|, \quad \left\| \lambda^{k-1} A_{i_k} \widetilde{R}_\lambda \right\|, \quad \left\| \lambda^{k-1} \overline{\widetilde{R}_\lambda (A_k + B_k)} \right\| \leq \widetilde{M}_{\theta'} |\lambda|^{-1}$$

for some $\widetilde{M}_{\theta'} > M_{\theta'}$. Now arguing similarly as in the proof of Theorem 1.1 and recalling the results in Section 2.4, we verify our claim.

Corollary 1.6. Let $\theta \in (0, \frac{\pi}{2}]$, $n \geq 2$. Let $-A_{n-1}$ be the generator of an analytic semigroup of angle θ on E. Assume that $A_0, \cdots, A_{n-2}$ are closable linear operators in E with $\mathcal{D}(A_k) \supset \mathcal{D}(A_{n-1})$ $(0 \leq k \leq n-2)$; in addition, there exists $\lambda_k \in \rho(A_{n-1})$ such that $(\lambda_k - A_{n-1})^{-1} A_k$ has a bounded extension to E for any $0 \leq k \leq n-2$. Then (ACP_n) is analytically wellposed in Σ_θ.

Next, we state the well-known moment inequality (cf. Fattorini [6, p. 365], or Krasnosel'skii-Sobolevskii [1]).

Let $0 \leq \alpha < \beta < \varepsilon \leq 1$, A be a densely defined and nonnegative operator in E (see Definition 3.6.1 and the statement below it). Then there exists a constant $C = C(\alpha, \beta, \varepsilon)$ such that

$$\|A^\beta u\| \leq C \|A^\varepsilon u\|^{(\beta-\alpha)(\varepsilon-\alpha)} \|A^\alpha u\|^{(\varepsilon-\beta)(\varepsilon-\alpha)} \quad (u \in \mathcal{D}(A^\varepsilon)).$$

Theorem 1.7. *Let $\theta \in (0, \frac{\pi}{2}]$. Let $A_0, \cdots, A_{n-1}$ be densely defined and nonnegative operators in E with their resolvents commuting mutually such that (ACP_n) is analytically wellposed in Σ_θ. Suppose that $B_0, \cdots, B_{n-1}$ are closable linear operators in E such that for each $0 \leq k \leq n - 1$, $\mathcal{D}(B_k) \supset \mathcal{D}(A_k^{a_k})$ and $(I + A_k)^{-a_k} B_k$ has a bounded extension to E, for some $a_k \in [0, 1)$. Then $(ACP_n)_{[B_{n-1}, \cdots, B_0]}$ is analytically wellposed in Σ_θ.*

Proof. We make use of the same type of arguments as in the proof of Theorem 1.5, noting that, by the moment inequality,

$$\left\|\lambda^k B_k R_\lambda\right\| \leq \left\|B_k (I + A_k)^{-a_k}\right\| \left\|\lambda^k (I + A_k)^{a_k} R_\lambda\right\|$$

$$\leq \text{const } |\lambda|^k \|(I + A_k) R_\lambda\|^{a_k} \|R_\lambda\|^{1-a_k}$$

$$\leq \text{const } |\lambda|^{-(1-a_k)(n-k)}, \quad 0 \leq k \leq n - 1, \ \lambda \in \tilde{\omega}_{\theta'} + \Sigma_{\frac{\pi}{2}+\theta'},$$

$$\left\|\lambda^k \overline{R_\lambda B_k}\right\| \leq |\lambda|^k \|(I + A_k)^{a_k} R_\lambda\| \left\|\overline{(I + A_k)^{-a_k} B_k}\right\|$$

$$\leq \text{const } |\lambda|^{-(1-a_k)(n-k)}, \quad 0 \leq k \leq n - 1, \ \lambda \in \tilde{\omega}_{\theta'} + \Sigma_{\frac{\pi}{2}+\theta'}.$$

4.2 Parabolicity

Definition 2.1. Let $\theta \in (0, \frac{\pi}{2}]$. Suppose $A_0, \cdots, A_{n-1}$ are linear operators in E. We say

$$[A_{n-1}, \cdots, A_0] \in \mathcal{A}_n(\theta),$$

if for each $\theta' \in (0, \theta)$ there exist $C_{\theta'}, \omega_{\theta'} > 0$ such that P_λ is injective with $R_\lambda \in L(E)$ and

$$\left\|\lambda^{n-1} R_\lambda\right\|, \quad \left\|\lambda^{k-1} A_k R_\lambda\right\| \leq C_{\theta'} |\lambda|^{-1}, \tag{2.1}$$

whenever $\lambda \in \omega_{\theta'} + \Sigma_{\frac{\pi}{2}+\theta'}$, $1 \leq k \leq n - 1$.

Write $\mathcal{A}_n = \bigcup_{\theta \in (0, \frac{\pi}{2}]} \mathcal{A}_n(\theta)$. When $[A_{n-1}, \cdots, A_0] \in \mathcal{A}_n$, we also say that the abstract differential equation

$$u^{(n)}(t) + \sum_{i=0}^{n-1} A_i u^{(i)}(t) = 0, \quad t \geq 0$$

is parabolic.

Clearly, when $[A_{n-1}, \cdots, A_0] \in \mathcal{A}_n(\theta)$ $(\theta \in (0, \frac{\pi}{2}])$, (2.1) holds for $k = 0$ as well; $[A_0] \in \mathcal{A}_1(\theta)$ $(\theta \in (0, \frac{\pi}{2}])$ if and only if $-A_0$ is the generator of an analytic semigroup of angle θ.

Definition 2.2. Let $\theta \in (0, \frac{\pi}{2}]$. (ACP_n) is called analytically solvable in Σ_θ if it has a unique solution $u(\cdot)$ for every $u_k \in \bigcap_{i=0}^k \mathcal{D}(A_i)$, $0 \leq k \leq n-1$, and $u(\cdot)$ can be extended analytically to Σ_θ such that for each $\theta' \in (0, \theta)$, $0 \leq j \leq n-1$, $u^{(j)}(z) \to u^{(j)}(0)$ as $z \to 0$ $(z \in \Sigma_{\theta'})$.

From the proof of Theorem 1.1 combined with Theorem 2.3.2, one easily shows

Theorem 2.3. *Let $\theta \in (0, \frac{\pi}{2}]$ and let $A_0, \cdots, A_{n-1}$ be closed linear operators in E. If $[A_{n-1}, \cdots, A_0] \in \mathcal{A}_n(\theta)$, then (ACP_n) is analytically solvable in Σ_θ.*

Let S be a nonnegative operator in E. Set

$$\theta_\infty^+(S) = \inf\{\theta \in (-\pi, \pi); \quad \text{there exist } C, \ \omega > 0 \text{ such that,}$$

$$\text{for each } \lambda \text{ with } |\lambda| \geq \omega \text{ and } \theta \leq \arg \lambda \leq \pi,$$

$$\lambda \in \rho(S) \text{ and } \|\lambda(\lambda - S)^{-1}\| \leq C\},$$

$$\theta_\infty^-(S) = \sup\{\theta \in (-\pi, \pi); \quad \text{there exist } C, \ \omega > 0 \text{ such that,}$$

$$\text{for each } \lambda \text{ with } |\lambda| \geq \omega \text{ and } -\pi \leq \arg \lambda \leq \theta,$$

$$\lambda \in \rho(S) \text{ and } \|\lambda(\lambda - S)^{-1}\| \leq C\}.$$

Obviously, $\theta_\infty^+(S) \geq \theta_\infty^-(S)$;

$$[S] \in \mathcal{A}_1(\theta) \quad \left(\theta \in \left(0, \frac{\pi}{2}\right]\right) \quad \text{if and only if}$$

$$\theta_\infty^+(S) \leq \frac{\pi}{2} - \theta \quad \text{and} \quad \theta_\infty^-(S) \geq -\frac{\pi}{2} + \theta. \tag{2.2}$$

It is not difficult to verify that, for $c \in \mathbf{C}$, cS is nonnegative if and only if either

   (i) $\arg c < -\pi - \theta_\infty^+(S)$,   or

   (ii) $-\pi - \theta_\infty^-(S) < \arg c < \pi - \theta_\infty^+(S)$,   or

   (iii) $\arg c > \pi - \theta_\infty^-(S)$,

and if cS is nonnegative, we have

$$\theta_\infty^\pm(cS) = \begin{cases} \arg c + \theta_\infty^\pm(S) + 2\pi, & \text{if} \quad \arg c < -\pi - \theta_\infty^+(S), \\ \arg c + \theta_\infty^\pm(S), & \text{if} \quad -\pi - \theta_\infty^-(S) < \arg c < \pi - \theta_\infty^+(S), \\ \arg c + \theta_\infty^\pm(S) - 2\pi, & \text{if} \quad \arg c > \pi - \theta_\infty^-(S). \end{cases}$$

$$(2.3)$$

Finally, by the functional calculus of fractional powers we obtain that

$$\theta_\infty^\pm(S^\alpha) = \alpha\theta_\infty^\pm(S), \quad 0 < \alpha < 1. \tag{2.4}$$

Combining (2.2) – (2.4) together gives

Theorem 2.4. *Let S be a nonnegative operator in E, $\theta \in (0, \frac{\pi}{2}]$, $0 < \alpha < 1$, $c \in \mathbf{C}$ with $\operatorname{Re}c \geq 0$. Then $[cS^\alpha] \in \mathcal{A}_1(\theta)$ if and only if*

$$-\frac{\pi}{2} + \theta - \alpha\theta_\infty^-(S) \leq \arg c \leq \frac{\pi}{2} - \theta - \alpha\theta_\infty^+(S).$$

From now on, in this section, A will be a densely defined and nonnegative operator in E, $c_i \in \mathbf{C}$ $(1 \leq i \leq n-1)$, and

$$P_0(\lambda) := \lambda^n + \sum_{i=1}^{n-1} c_i A^{k_i} \lambda^{n-i} + A.$$

Theorem 2.5. *Let $\theta \in (0, \frac{\pi}{2}]$ and let $k_1 > k_2 - k_1 > \cdots > k_{n-1} - k_{n-2} > 1 - k_{n-1} > 0$, $c_i \neq 0$ for each $1 \leq i \leq n-1$. Then*

$$\left[c_1 A^{k_1}, \cdots, c_{n-1}A^{k_{n-1}}, A\right] \in \mathcal{A}_n(\theta)$$

if and only if, for each $1 \leq i \leq n$,

$$\left[c_{i-1}^{-1}c_i A^{k_i - k_{i-1}}\right] \in \mathcal{A}_1(\theta),$$

where $k_0 = 0$, $k_n = c_0 = c_n = 1$.

Proof. *Sufficiency.* Set $c_{i-1}^{-1}c_i = \tilde{c}_i$, $k_i - k_{i-1} = t_i$ $(1 \leq i \leq n)$,

$$P_1(\lambda) = \prod_{i=1}^n \left(\lambda + \tilde{c}_i A^{t_i}\right),$$

$$Q(\lambda) = \sum_{m=1}^{n-1} \sum_{(i_1, \cdots, i_m) \in I_m} \tilde{c}_{i_1} \cdots \tilde{c}_{i_m} A^{t_{i_1} + \cdots + t_{i_m}} \lambda^{n-m},$$

where, for each $1 \leq m \leq n-1$,

$$I_m := \{(i_1, \cdots, i_m); \ 1 \leq i_1 < \cdots < i_m \leq n, \ (i_1, \cdots, i_m) \neq (1, \cdots, m)\}.$$

Then

$$t_1 > t_2 > \cdots > t_n, \quad k_m = \sum_{i=1}^{m} t_i \quad (1 \leq m \leq n-1).$$

By hypothesis, for each $\theta' \in (0, \theta)$, there exist $C_{\theta'}, \omega_{\theta'} > 0$ such that

$$\left\| \lambda^{n-m} A^{t_{i_1} + \cdots + t_{i_m}} P_1^{-1}(\lambda) \right\| \leq C_{\theta'}, \quad \left\| \lambda^n P_1^{-1}(\lambda) \right\| \leq C_{\theta'} \qquad (2.5)$$

whenever

$$\lambda \in \omega_{\theta'} + \Sigma_{\frac{\pi}{2}+\theta'}, \ 1 \leq i_1 < \cdots < i_m \leq n, \ 1 \leq m \leq n.$$

This together with the moment inequality yields that, for each $\theta' \in (0, \theta)$, there exist $C, C_{\theta'}, \omega_{\theta'} > 0$ such that, for $\lambda \in \omega_{\theta'} + \Sigma_{\frac{\pi}{2}+\theta'}, (i_1, \cdots, i_m) \in I_m,$ $1 \leq m \leq n-1$,

$$\left\| \lambda^{n-m} A^{t_{i_1} + \cdots + t_{i_m}} P_1^{-1}(\lambda) \right\|$$

$$\leq \quad C |\lambda|^{n-m} \left\| A^{k_m} P_1^{-1}(\lambda) \right\|^{(t_{i_1} + \cdots + t_{i_m})k_m^{-1}} \left\| P_1^{-1}(\lambda) \right\|^{1-(t_{i_1} + \cdots + t_{i_m})k_m^{-1}}$$

$$\leq \quad C \left(C_{\theta'} |\lambda| \right)^{[(t_{i_1} + \cdots + t_{i_m})k_m^{-1} - 1]m},$$

which approaches 0 as $|\lambda| \to \infty$.

Therefore, for each $\theta' \in (0, \theta)$, there is $\widetilde{\omega}_{\theta'} > \omega_{\theta'}$ such that for $\lambda \in \widetilde{\omega}_{\theta'} + \Sigma_{\frac{\pi}{2}+\theta'}$

$$\left\| Q(\lambda) P_1^{-1}(\lambda) \right\| < \frac{1}{2}.$$

Thus using (2.5) again we obtain that, for each $\theta' \in (0, \theta), \lambda \in \widetilde{\omega}_{\theta'} + \Sigma_{\frac{\pi}{2}+\theta'}$, $1 \leq m \leq n$,

$$\left\| c_m \lambda^{n-m} A^{k_m} P_0^{-1}(\lambda) \right\| = \left\| c_m \lambda^{n-m} A^{k_m} P_1^{-1}(\lambda) \left[I - Q(\lambda) P_1^{-1}(\lambda) \right]^{-1} \right\|$$

$$\leq \quad 2|c_m| C_{\theta'}.$$

In conclusion, $\left[c_1 A^{k_1}, \cdots, c_{n-1} A^{k_{n-1}}, A \right] \in \mathcal{A}_n(\theta)$.

Necessity. Making use of the moment inequality as in the proof of sufficiency, we obtain that, for each $\theta' \in (0, \theta)$, there exists $\omega_{\theta'} > 0$ such that, for $\lambda \in \omega_{\theta'} + \Sigma_{\frac{\pi}{2}+\theta'}, (i_1, \cdots, i_m) \in I_m, 1 \leq m \leq n$,

$$\left\| \lambda^{n-m} A^{t_{i_1} + \cdots + t_{i_m}} P_0^{-1}(\lambda) \right\| \leq C_{\theta'} |\lambda|^{[(t_{i_1} + \cdots + t_{i_m})k_m^{-1} - 1]m},$$

and therefore there exist $\widetilde{\omega}_{\theta'} > \omega_{\theta'}$, $M_{\theta'} > 0$ such that, for $\lambda \in \widetilde{\omega}_{\theta'} + \Sigma_{\frac{\pi}{2}+\theta'}$, $1 \leq m \leq n$,

$$
\begin{cases}
\|Q(\lambda)P_0^{-1}(\lambda)\| < \dfrac{1}{2}, \\[2mm]
\|\lambda(\lambda + \widetilde{c}_m A^{tm})^{-1} P_1(\lambda) P_0^{-1}(\lambda)\| \leq M_{\theta'}.
\end{cases}
$$

Accordingly, for each $\theta' \in (0, \theta)$, $\lambda \in \widetilde{\omega}_{\theta'} + \Sigma_{\frac{\pi}{2}+\theta'}$, $1 \leq m \leq n$,

$$
\|\lambda(\lambda + \widetilde{c}_m A^{tm})^{-1}\| \ = \ \|\lambda(\lambda + \widetilde{c}_m A^{tm})^{-1} P_1(\lambda) P_0^{-1}(\lambda)[I + Q(\lambda)P_0^{-1}(\lambda)]^{-1}\|
$$

$$
\leq \ 2M_{\theta'}.
$$

This ends the proof.

Corollary 2.6. *Let* $\theta_\infty^{\pm}(A) = 0$, $c_i > 0$ $(1 \leq i \leq n-1)$, *and* $k_1 > k_2 - k_1 > \cdots > k_{n-1} - k_{n-2} > 1 - k_{n-1} > 0$. *Then*

$$
[c_1 A^{k_1}, \cdots, c_{n-1} A^{k_{n-1}}, A] \in \mathcal{A}_n\left(\frac{\pi}{2}\right).
$$

Corollary 2.7. *Let* $\theta \in (0, \frac{\pi}{2}]$, $k_1 > \dfrac{1}{2}$, *and* $c_1 \neq 0$. *Then* $[c_1 A^{k_1}, A] \in \mathcal{A}_2(\theta)$ *if and only if* $\left[c_1 A^{k_1}\right]$, $\left[c_1^{-1} A^{1-k_1}\right] \in \mathcal{A}_1(\theta)$.

Next, we consider perturbation cases.

Definition 2.8. Let $\theta \in (0, \frac{\pi}{2}]$. Suppose A_k, B_k $(0 \leq k \leq n-1)$ are linear operators in E. We say

$$
[A_{n-1} + B_{n-1}, \cdots, A_0 + B_0] \in \mathcal{A}_n(\theta)_{[B_{n-1}, \cdots, B_0]},
$$

if for each $\theta' \in (0, \theta)$ there exist $C_{\theta'}$, $\omega_{\theta'} > 0$ such that $\widetilde{P}_\lambda := \lambda^n + \sum_{i=0}^{n-1} \lambda^i (A_i + B_i)$ is injective with $\widetilde{R}_\lambda := \widetilde{P}_\lambda^{-1} \in \mathbf{L}(E)$ and

$$
\left\|\lambda^{n-1}\widetilde{R}_\lambda\right\|, \ \left\|\lambda^{k-1}A_k\widetilde{R}_\lambda\right\|, \ \left\|\lambda^{k-1}B_k\widetilde{R}_\lambda\right\| \leq C_{\theta'}|\lambda|^{-1},
$$

whenever $\lambda \in \omega_{\theta'} + \Sigma_{\frac{\pi}{2}+\theta'}$, $1 \leq k \leq n-1$.

Clearly, $[A_{n-1} + B_{n-1}, \cdots, A_0 + B_0] \in \mathcal{A}_n(\theta)_{[B_{n-1}, \cdots, B_0]}$ implies $[A_{n-1} + B_{n-1}, \cdots, A_0 + B_0] \in \mathcal{A}_n(\theta)$.

Definition 2.9. Let $\theta \in (0, \frac{\pi}{2}]$ and let B_k $(0 \leq k \leq n-1)$ be linear operators in E. $(ACP_n)_{[B_{n-1}, \cdots, B_0]}$ is called analytically solvable in Σ_θ if it has a unique solution $u(\cdot)$ (see Definition 2.4.1) for every $u_k \in \bigcap_{i=0}^k \mathcal{D}(A_i + B_i)$, $0 \leq k \leq n-1$, and $u(\cdot)$ can be extended analytically to Σ_θ such that for each $\theta' \in (0, \theta)$, $0 \leq j \leq n-1$, $u^{(j)}(z) \to u^{(j)}(0)$ as $z \to 0$ $(z \in \Sigma_{\theta'})$.

Using the arguments similarly as in the proof of Theorem 1.1 and recalling the discussion in Section 2.4, we obtain

Theorem 2.10. *Let* $\theta \in (0, \frac{\pi}{2}]$ *and let* A_k, B_k $(0 \le k \le n-1)$ *be closed linear operators in* E. *If*

$$[A_{n-1} + B_{n-1}, \cdots, A_0 + B_0] \in \mathcal{A}_n(\theta)_{[B_{n-1}, \cdots, B_0]},$$

then $(ACP_n)_{[B_{n-1}, \cdots, B_0]}$ *is analytically solvable in* Σ_θ .

Theorem 2.11. *Assume* $[A_{n-1}, \cdots, A_0] \in \mathcal{A}_n(\theta)$ *for some* $\theta \in (0, \frac{\pi}{2}]$, *and* $B_0, \cdots, B_{n-1}$ *are linear operators in* E. *If for each* $0 \le m \le n-1$, *there exist* i_m, ε_m *with* $0 \le i_m \le n-1, 0 < \varepsilon_m \le 1$ *such that* $\mathcal{D}(B_m) \supset \mathcal{D}(A_{i_m})$, *and for each* $u \in \mathcal{D}(A_{i_m})$,

$$\|B_m u\| \le C\|u\| + C\|A_{i_m} u\|^{\varepsilon_m} \|u\|^{1-\varepsilon_m}, \quad \text{for some} \quad C > 0,$$

then for each $\theta' \in (0, \theta)$, *there is* $\omega_{\theta'} > 0$ *such that for* $0 \le m \le n-1$,

$$\sup_{\lambda \in \omega_{\theta'} + \Sigma_{\frac{\pi}{2}+\theta'}} \|\lambda^m B_m R_\lambda\| < \begin{cases} \dfrac{1}{2}, & \text{if} \quad (n-i_m)\varepsilon_m < n-m, \\ \\ +\infty, & \text{if} \quad (n-i_m)\varepsilon_m = n-m. \end{cases} \tag{2.6}$$

Furthermore, when $(n-i_m)\varepsilon_m < n-m$, *for each* $0 \le m \le n-1$

$$[A_{n-1} + B_{n-1}, \cdots, A_0 + B_0] \in \mathcal{A}_n(\theta)_{[B_{n-1}, \cdots, B_0]}.$$

Proof. Observing that, for each $\theta' \in (0, \theta)$, there exist $C_{\theta'}, \omega_{\theta'} > 0$ such that, for $\lambda \in \omega_{\theta'} + \Sigma_{\frac{\pi}{2}+\theta'}$, $0 \le m \le n-1$,

$$\begin{aligned} \|\lambda^m B_m R_\lambda\| &\le C|\lambda|^m \|R_\lambda\| + C|\lambda|^m \|A_{i_m} R_\lambda\|^{\varepsilon_m} \|R_\lambda\|^{1-\varepsilon_m} \\ &\le CC_{\theta'} \left(|\lambda|^{m-n} + |\lambda|^m |\lambda|^{-i_m \varepsilon_m} |\lambda|^{-n(1-\varepsilon_m)} \right) \\ &= CC_{\theta'} \left(|\lambda|^{m-n} + |\lambda|^{(n-i_m)\varepsilon_m - (n-m)} \right), \end{aligned}$$

we obtain (2.6). The remaining part follows from the plain equality

$$\left(P_\lambda + \sum_{m=0}^{n-1} \lambda^m B_m \right)^{-1} = R_\lambda \left[I + \sum_{m=0}^{n-1} \lambda^m B_m R_\lambda \right]^{-1}.$$

Theorem 2.12. *Let* $\tilde{c}_i \in \mathbf{C}$, $0 < l_i < \frac{i}{n}$ *for each* $1 \le i \le n-1$, $\theta \in (0, \frac{\pi}{2}]$. *Then*

$$[c_1 A^{k_1}, \cdots, c_{n-1} A^{k_{n-1}}, A] \in \mathcal{A}_n(\theta)$$

if and only if

$$[c_1 A^{k_1} + \tilde{c}_1 A^{l_1}, \; \cdots, \; c_{n-1} A^{k_{n-1}} + \tilde{c}_{n-1} A^{l_{n-1}}, \; A] \in \mathcal{A}_n(\theta).$$

Proof. This theorem is an immediate consequence of Theorem 2.11 by taking $i_m = 0$, $A_0 = A$, $B_0 = 0$, $\varepsilon_m = l_{n-m}$ $(1 \le m \le n-1)$, and using the moment inequality.

Theorem 2.13. *Let $\theta \in (0, \frac{\pi}{2}]$ and let $B_1, \cdots, B_{n-1}$ be closed linear operators in E satisfying that, for each $1 \le m \le n-1$, there is l_m with $k_{m-1} < l_m < \frac{1}{2}(k_{m-1} + k_{m+1})$ such that $\mathcal{D}(B_m) \supset \mathcal{D}(A^{l_m})$. If*

$$[c_1 A^{k_1}, \; \cdots, \; c_{n-1} A^{k_{n-1}}, \; A] \in \mathcal{A}_n(\theta),$$

then

$$[c_1 A^{k_1} + B_1, \; \cdots, \; c_{n-1} A^{k_{n-1}} + B_{n-1}, \; A] \in \mathcal{A}_n(\theta)_{[B_1, \, \cdots, \, B_{n-1}, \, 0]}.$$

Proof. By hypothesis, there is $C > 0$ such that, for each $1 \le m \le n-1$, $u \in \mathcal{D}(A^{l_m})$,

$$\|B_m u\| \le C\|u\| + C \|A^{l_m} u\|.$$

So using the moment inequality yields that, for each $\theta' \in (0, \theta)$, there exist $C_{\theta'}$, $\omega_{\theta'} > 0$ such that, for each $1 \le m \le n-1$, $\lambda \in \omega_{\theta'} + \Sigma_{\frac{\pi}{2} + \theta'}$,

$$\left\| \lambda^{n-m} B_m P_0^{-1}(\lambda) \right\|$$

$$\le \;\; C|\lambda|^{n-m} \left\| P_0^{-1}(\lambda) \right\| + C|\lambda|^{n-m} \left\| A^{l_m} P_0^{-1}(\lambda) \right\|$$

$$\le \;\; C C_{\theta'} |\lambda|^{-m} + C|\lambda|^{n-m} \left\| A^{k_{m-1}} P_0^{-1}(\lambda) \right\|^{\tau} \left\| A^{k_{m+1}} P_0^{-1}(\lambda) \right\|^{1-\tau}$$

$$\le \;\; C C_{\theta'} \left(|\lambda|^{-m} + |\lambda|^{n-m} |\lambda|^{(m-n-1)\tau} |\lambda|^{(m-n+1)(1-\tau)} \right)$$

$$= \;\; C C_{\theta'} \left(|\lambda|^{-m} + |\lambda|^{1-2\tau} \right)$$

which approaches 0 as $|\lambda| \to \infty$, where

$$\tau := (l_m - k_{m-1})(k_{m+1} - k_{m-1})^{-1} < 1.$$

Consequently, for each $\theta' \in (0, \theta)$ there is $\tilde{\omega}_{\theta'} > \omega_{\theta'}$ such that, for $\lambda \in \tilde{\omega}_{\theta'} + \Sigma_{\frac{\pi}{2} + \theta'}$,

$$\left\| \sum_{m=1}^{n-1} \lambda^{n-m} B_m P_0^{-1}(\lambda) \right\| < \frac{1}{2}.$$

This leads to the result as claimed.

Corollary 2.14. *Let* $\theta \in (0, \frac{\pi}{2}]$, $0 < k_1 < \cdots < k_{n-1} < 1$ *and* $k_j < \frac{1}{2}(k_{j-1} + k_{j+1})$ *for some* $1 \leq j \leq n-1$. *Then* $[c_1 A^{k_1}, \cdots, c_{n-1} A^{k_{n-1}}, A] \in \mathcal{A}_n(\theta)$ *implies*

$$[c_1 A^{k_1}, \cdots, c_{j-1} A^{k_{j-1}}, 0, c_{j+1} A^{k_{j+1}}, \cdots, c_{n-1} A^{k_{n-1}}, A] \in \mathcal{A}_n(\theta).$$

In the sequel, we specialize to the case of $n = 3$. We assume that A is densely defined, unbounded and nonnegative operator in E with $\theta_\infty^\pm(A) = 0$.

Lemma 2.15. *For* $0 < \beta \leq 1$, $a > 0$, $\mathrm{Re}\, c > 0$, *we have:*
(i) $\theta_\infty^\pm\left(a A^\beta\right) = 0$;
(ii) $[c A^\beta] \in \mathcal{A}_1$, $[-c A^\beta] \notin \mathcal{A}_1$;
(iii) *for* $b \in \mathbb{R}$, $[b A^{\beta/2}, a A^\beta] \in \mathcal{A}_2$ *if and only if* $b > 0$;
(iv) *let* B *be a nonnegative operator in* E *and let* $\frac{1}{2} < \beta < 1$; *then* $[c B^\beta, B] \in \mathcal{A}_2$ *if and only if*

$$\begin{cases} \arg c > -\dfrac{\pi}{2} + \max\left\{(1-\beta)\theta_\infty^+(B), \ -\beta\theta_\infty^-(B)\right\}, \\[3mm] \arg c < \dfrac{\pi}{2} - \max\left\{\beta\theta_\infty^+(B), \ -(1-\beta)\theta_\infty^-(B)\right\}. \end{cases}$$

Proof. (i) – (iii) are easy to see. (iv) follows immediately from Theorem 2.4 and Corollary 2.7.

Theorem 2.16. *Let* $a_1, a_2 > 0$ *and* $0 < k_1, k_2 < 1$. *Then*

$$[a_1 A^{k_1}, 0, A], \quad [0, a_2 A^{k_2}, A], \quad [0, 0, A] \notin \mathcal{A}_3.$$

Proof. Observe that, for each $y_1 \geq 0$, the function

$$y(x) := x^{-1} + x(y_1 - x)$$

is continuous in $(0, +\infty)$, and $y \to +\infty$ as $x \to 0^+$, $y \to -\infty$ as $x \to +\infty$. Hence, for each $y_1, y_2 \geq 0$, there exists $x_1 > 0$ such that

$$y_2 = x_1^{-1} + x_1(y_1 - x_1).$$

Set $x_2 = y_1 - x_1$. If $x_2 > 0$, i.e., $y_1 > x_1$, then $y_2 > x_1^{-1}$; therefore,

$$y_1 y_2 > y_1 x_1^{-1} > 1.$$

If $x_2 \leq 0$, i.e., $y_1 \leq x_1$, then $y_2 \leq x_1^{-1}$; therefore,

$$y_1 y_2 \leq y_1 x_1^{-1} \leq 1.$$

In other words,

$$\begin{cases} x_2 > 0, & \text{if} \quad y_1 y_2 > 1, \\ x_2 \le 0, & \text{if} \quad y_1 y_2 \le 1. \end{cases}$$

So from the equality

$$\lambda^3 + y_1 A^{1/3}\lambda^2 + y_2 A^{2/3}\lambda + A = \left(\lambda + x_1 A^{1/3}\right)\left(\lambda^2 + x_2 A^{1/3}\lambda + x_1^{-1}A^{2/3}\right),$$

we see by (ii) and (iii) in Lemma 2.15,

$$\left[y_1 A^{1/3}, \; y_2 A^{2/3}, \; A\right] \in \mathcal{A}_3, \quad \text{if} \quad y_1 y_2 > 1. \tag{2.7}$$

But

$$\left[y_1 A^{1/3}, \; y_2 A^{2/3}, \; A\right] \notin \mathcal{A}_3, \quad \text{if} \quad y_1 y_2 \le 1. \tag{2.8}$$

In fact, if $\left[y_1 A^{1/3}, \; y_2 A^{2/3}, \; A\right] \in \mathcal{A}_3$ $(y_1 y_2 \le 1)$, then by virtue of (2.6) we have that there are C, $\omega > 0$, $\theta \in (0, \frac{\pi}{2}]$ such that, for $\lambda \in \omega + \Sigma_{\frac{\pi}{2}+\theta}$, $i = 1, 2, 3$,

$$\left\| \lambda^{3-i} A^{i/3} \left(\lambda^3 + y_1 A^{1/3}\lambda^2 + y_2 A^{2/3}\lambda + A\right)^{-1} \right\| \le C.$$

According to this, the equality

$$\left(\lambda^2 + x_2 A^{1/3}\lambda + x_1^{-1}A^{2/3}\right)^{-1}$$

$$= \left(\lambda + x_1 A^{1/3}\right)\left(\lambda^3 + y_1 A^{1/3}\lambda^2 + y_2 A^{2/3}\lambda + A\right)^{-1}$$

shows $[x_2 A^{1/3}, \; x_1^{-1}A^{2/3}] \in \mathcal{A}_2$, which contradicts (iii) of Lemma 2.15. So (2.8) holds. (2.8) indicates

$$\left[a_1 A^{1/3}, \; 0, \; A\right], \quad \left[0, \; a_2 A^{2/3}, \; A\right], \quad [0, \; 0, \; A] \notin \mathcal{A}_3.$$

Since $[0, \; 0, \; A] \notin \mathcal{A}_3$, using Theorem 2.12 yields that

$$\left[a_1 A^{k_1}, \; 0, \; A\right], \quad [0, \; a_2 A^{k_2}, \; A] \notin \mathcal{A}_3, \quad \text{if} \quad k_1 < \frac{1}{3}, \; k_2 < \frac{2}{3}.$$

Finally, we have that, for each $a > 0$, $0 < \beta < 1$,

$$[a A^\beta, \; a^{-1}A^{1-\beta}, \; A] \notin \mathcal{A}_3.$$

Indeed, if not, then the equality

$$\left(\lambda^2 + a^{-1}A^{1-\beta}\right)^{-1} = \left(\lambda + aA^\beta\right)\left(\lambda^3 + aA^\beta\lambda^2 + a^{-1}A^{1-\beta}\lambda + A\right)^{-1}$$

yields $\left[0,\ a^{-1}A^{1-\beta}\right] \in \mathcal{A}_2$, which contradicts (iii) of Lemma 2.15. Thus, we conclude by Theorem 2.12 again that

$$\left[aA^\beta,\ 0,\ A\right] \notin \mathcal{A}_3, \quad \text{if} \quad \beta > \frac{1}{3},$$

$$\left[0,\ a^{-1}A^{1-\beta},\ A\right] \notin \mathcal{A}_3, \quad \text{if} \quad \beta < \frac{1}{3}.$$

The proof is then complete.

Theorem 2.17. *Let* a_1, $a_2 > 0$ *and* $0 < k_1 < k_2 < 1$. *Then* $\left[a_1 A^{k_1},\ a_2 A^{k_2},\ A\right] \in \mathcal{A}_3$ *if and only if either*
  (i) $k_1 > \frac{1}{3}$, $\frac{1}{2}(1+k_1) \le k_2 \le 2k_1$,   *or*
  (ii) $k_1 = \frac{1}{3}$, $k_2 = \frac{2}{3}$, $a_1 a_2 > 1$.

Proof. Observing

$$\lambda^3 + \left(a_2 a_1^{-1} A^{(1-k_1)/2} + a_1 A^{k_1}\right)\lambda^2 + \left(a_1^{-1} A^{1-k_1} + a_2 A^{(1+k_1)/2}\right)\lambda + A$$

$$= \left(\lambda + a_1 A^{k_1}\right)\left(\lambda^2 + a_2 a_1^{-1} A^{(1-k_1)/2}\lambda + a_1^{-1} A^{1-k_1}\right),$$

we obtain

$$\left[a_2 a_1^{-1} A^{(1-k_1)/2} + a_1 A^{k_1},\ a_1^{-1} A^{1-k_1} + a_2 A^{(1+k_1)/2},\ A\right] \in \mathcal{A}_3.$$

Thus appealing to Theorem 2.12 gives

$$\left[a_1 A^{k_1},\ a_2 A^{(1+k_1)/2},\ A\right] \in \mathcal{A}_3 \quad \left(k_1 > \frac{1}{3}\right). \tag{2.9}$$

Next, let $\frac{1}{3} < k_1 < \frac{1}{2}$. Set $\tau = k_1(1-k_1)^{-1}$;

$$b_1 = \begin{cases} \dfrac{1}{2}\left[a_1 + (a_1^2 - 4a_2)^{1/2}\right], & \text{if} \quad a_1^2 \ge 4a_2, \\[2mm] re^{i\theta}, & \text{if} \quad a_1^2 < 4a_2, \end{cases}$$

$$b_2 = \begin{cases} \dfrac{1}{2}\left[a_1 - (a_1^2 - 4a_2)^{1/2}\right], & \text{if} \quad a_1^2 \ge 4a_2, \\[2mm] re^{-i\theta}, & \text{if} \quad a_1^2 < 4a_2, \end{cases}$$

where $\theta := \arccos\left(\frac{1}{2}a_1 a_2^{-1/2}\right)$, $r := a_2^{1/2}$; and set

$$B = r^{-1}e^{-i\theta}A^{1-k_2}.$$

Then $\theta_\infty^\pm(B) = -\theta$, $\frac{1}{2} < \tau < 1$, $b_1 + b_2 = a_1$, and $b_1 b_2 = a_2$. Therefore, if $a_1^2 < 4a_2$,

$$\max\left\{(1-\tau)\theta_\infty^+(B),\ -\tau\theta_\infty^-(B)\right\} = \theta\tau,$$

$$\max\left\{\tau\theta_\infty^+(B),\ -(1-\tau)\theta_\infty^-(B)\right\} = \theta(1-\tau),$$

which implies by (iv) of Lemma 2.15 that

$$\left[r^{\tau+1}e^{(\tau-1)\theta i}B^\tau, \ B\right] \in \mathcal{A}_2.$$

Consequently, using

$$\lambda^3 + a_1 A^{k_1}\lambda^2 + \left(a_2 A^{2k_1} + b_1^{-1}A^{1-k_1}\right)\lambda + A$$

$$= \ \left(\lambda + b_1 A^{k_1}\right)\left(\lambda^2 + b_2 A^{k_1}\lambda + b_1^{-1}A^{1-k_1}\right),$$

$$\lambda^2 + b_2 A^{k_1}\lambda + b_1^{-1}A^{1-k_1} = \lambda^2 + r^{\tau+1}e^{(\tau-1)\theta i}B^\tau\lambda + B, \qquad \text{if } a_1^2 < 4a_2,$$

we see by (ii) and (iii) in Lemma 2.15 that

$$\left[a_1 A^{k_1}, \ a_2 A^{2k_1} + r^{-1}e^{-i\theta}A^{1-k_1}, \ A\right] \in \mathcal{A}_3.$$

Since $1 - k_1 < \frac{2}{3}$, we claim using Theorem 2.12 again that

$$\left[a_1 A^{k_1}, \ a_2 A^{2k_1}, \ A\right] \in \mathcal{A}_3, \quad \frac{1}{3} < k_1 < \frac{1}{2}. \tag{2.10}$$

In conclusion, Corollary 2.6, combined with (2.7), (2.9) and (2.10), shows the "if" part. For the "only if" part, apply Theorems 2.16 and 2.12 and see that

$$\left[a_1 A^{k_1}, \ a_2 A^{k_2}, \ A\right] \notin \mathcal{A}_3, \qquad \text{if } k_1 < \frac{1}{3} \ \text{ or } \ k_2 > \frac{2}{3}.$$

Furthermore, Corollary 2.14, together with Theorem 2.16, gives that

$$\left[a_1 A^{k_1}, \ a_2 A^{k_2}, \ A\right] \notin \mathcal{A}_3, \qquad \text{if } k_2 < \frac{1}{2}(1+k_1) \ \text{ or } \ k_2 > 2k_1.$$

Then referring to (2.8) ends the proof.

Remark. If $a_1, a_2 > 0$, $k_1 \geq k_2$, then $\left[a_1 A^{k_1}, a_2 A^{k_2}, A\right] \notin \mathcal{A}_3$. Indeed by virtue of Theorem 2.11, $\left[a_1 A^{k_1}, a_2 A^{k_2}, A\right] \in \mathcal{A}_3$ implies $\left[a_1 A^{k_1}, 0, A\right] \in \mathcal{A}_3$, which contradicts Theorem 2.16. Again by Theorem 2.11, if $k_2 \geq 1$, then $\left[a_1 A^{k_1}, a_2 A^{k_2}, A\right] \in \mathcal{A}_3$ if and only if $\left[a_1 A^{k_1}, a_2 A^{k_2}\right] \in \mathcal{A}_2$.

4.3 The case of differential operators as coefficient operators

In this section, we assume that E is one of the Banach spaces $L^p(R^n)$ $(1 \leq p < \infty)$, $C_0(R^n)$ or $UC_b(R^n)$. Given a polynomial $p(x)$, $p(D)$ will be defined as in Section 3.5. We claim that $\mathcal{D}(p(D))$ is dense in E. Indeed, if $E = L^p(R^n)$ $(1 \leq p < \infty)$ or $C_0(R^n)$, then the Schwartz space $\mathcal{S}(R^n)$ (which is contained in $\mathcal{D}(p(D))$) is dense in E. If $E = UC_b(R^n)$, then

$$\mathcal{D}(p(D)) \supset \{J_\epsilon * f; \ \epsilon > 0, \ f(x) \in E\},$$

where $J_\epsilon \in C^\infty(R^n)$ with support in $\{x \in R^n;\ |x| \le \epsilon\}$ satisfying

$$\int_{R^n} J_\epsilon(x)dx = 1.$$

This implies that $\mathcal{D}(p(D))$ is dense in $UC_b(R^n)$ since

$$\lim_{\epsilon \to 0} J_\epsilon * f(x) = f(x)$$

uniformly in R^n, whenever $f \in UC_b(R^n)$.

We also remark that in general, $\mathcal{D}(p(D))$ is not dense in $L^\infty(R^n)$ or $C_b(R^n)$.

Theorem 3.1. *Suppose that $p(x)$, $q(x)$ are real polynomials of degrees l, m respectively on R^n such that they are strongly elliptic and $l < m < 2l$. Assume B_0, B_1 are closable linear operators in E such that*

$$\mathcal{D}(B_0) \supset \mathcal{D}\left(\Delta^{\frac{am}{2}}\right), \quad \mathcal{D}(B_1) \supset \mathcal{D}\left(\Delta^{\frac{bl}{2}}\right),$$

for some $a,\ b \in [0,\ 1)$. Let $A_0 = q(D)$, $A_1 = p(D)$. Then the Cauchy problem $(ACP_2)_{[B_1,\ B_0]}$ is analytically solvable in $\Sigma_{\frac{\pi}{2}}$; furthermore, $(ACP_2)_{[B_1,\ B_0]}$ is analytically wellposed in $\Sigma_{\frac{\pi}{2}}$ provided $(I-\Delta)^{-\frac{am}{2}}B_0$, $(I-\Delta)^{-\frac{bl}{2}}B_1$ have bounded extensions to E.

Proof. By hypothesis, there are constants L_0, $C_0 > 0$ such that

$$p(x) \ge C_0|x|^l, \quad q(x) \ge C_0|x|^m, \quad |x| > L_0.$$

Without loss of generality, we may and do assume (with $B_0 + d_1 I$, $B_1 + d_2 I$ replacing B_0, B_1 respectively for some d_1, $d_2 > 0$, if necessary) that for any $x \in R^n$,

$$\begin{cases} q(x) \ge C_1|x|^m, \\[2mm] p^2(x) - 4q(x) \ge C_1(1 + |x|)^{2l}, \\[2mm] \sqrt{p^2(x) - 4q(x)} + p(x) \ge C_1(1 + |x|)^l, \end{cases} \tag{3.1}$$

for some $C_1 > 1$. Define $\mu_\pm(x)$ as in (3.5.8). Since

$$\mu_+(x) = -\frac{2q(x)}{p(x) + \sqrt{p^2(x) - 4q(x)}}, \quad x \in R^n, \tag{3.2}$$

we have

$$\sigma_0 := \sup_{x \in R^n} \operatorname{Re}\mu_\pm(x) \le 0.$$

Also, a simple calculation shows by (3.1) that for each multiindex β,

$$|D^\beta \mu_-(x)| \le \text{const } (1 + |x|)^{l - |\beta|}, \quad x \in R^n, \tag{3.3}$$

$$|D^\beta \mu_+(x)| \leq \text{const} \, (1 + |x|)^{m-l-|\beta|}, \quad x \in R^n. \tag{3.4}$$

Now, set

$$v_0(x; \, z) = \frac{1}{\sqrt{p^2(x) - 4q(x)}} \left(e^{\mu_+(x)z} - e^{\mu_-(x)z} \right), \quad x \in R^n, \, z \in \mathbf{C},$$

$$v(x; \, z) = p(x)v_0(x; \, z), \quad x \in R^n, \, z \in \mathbf{C},$$

$$w(x; \, z) = e^{\mu_+(x)z} + e^{\mu_-(x)z}, \quad x \in R^n, \, z \in \mathbf{C}.$$

Then, (3.1) implies that for each $z \in \mathbf{C}$,

$$e^{\mu_\pm(x)z}, \, v_0(x; \, z), \, v(x; \, z), \, w(x; \, z) \in C^\infty(R^n).$$

Fix $z_0 \in \Sigma_{\frac{\pi}{2}}$. Observe that for each multiindex β,

$$\left| D^\beta \left[\mu_-(x)e^{\mu_-(x)z} \right] \right|, \, \left| D^\beta \left[\frac{p(x)\mu_-(x)}{\sqrt{p^2(x) - 4q(x)}} e^{\mu_-(x)z} \right] \right|,$$

$$\left| D^\beta \left[\frac{\mu_-(x)}{\sqrt{p^2(x) - 4q(x)}} e^{\mu_-(x)z} \right] \right|$$

$$\leq \quad \text{const} \, (1 + |x|)^{l+(l-1)|\beta|} e^{-\frac{1}{2}C_1|x|^l \text{Re} z_0}$$

valid for all $x \in R^n$, $z \in \mathbf{C}$ with $|z - z_0| < \frac{1}{2}\text{Re}z_0$, by (3.1) and (3.3);

$$\left| D^\beta \left[\mu_+(x)e^{\mu_+(x)z} \right] \right|, \, \left| D^\beta \left[\frac{p(x)\mu_+(x)}{\sqrt{p^2(x) - 4q(x)}} e^{\mu_+(x)z} \right] \right|,$$

$$\left| D^\beta \left[\frac{\mu_+(x)}{\sqrt{p^2(x) - 4q(x)}} e^{\mu_+(x)z} \right] \right|$$

$$\leq \quad \text{const} \, (1 + |x|)^{m-l+(m-l-1)|\beta|} e^{-C_2|x|^{m-l} \text{Re} z_0}$$

(where C_2 is some constant) valid for x, z as above, by (3.1), (3.2) and (3.4). Accordingly, we can see by Lemma 1.5.3 that the $\mathcal{F}L^1$-valued functions

$$z \mapsto e^{\mu_\pm(x)z}, \, z \mapsto \frac{p(x)}{\sqrt{p^2(x) - 4q(x)}} e^{\mu_\pm(x)z},$$

$$z \mapsto \frac{1}{\sqrt{p^2(x) - 4q(x)}} e^{\mu_\pm(x)z}$$

are analytic in $\Sigma_{\frac{\pi}{2}}$. Hence, letting

$$V_0(z) = \mathbf{T}\langle v_0(x; \, z)\rangle, \quad V(z) = \mathbf{T}\langle v(x; \, z)\rangle, \quad W(z) = \mathbf{T}\langle w(x; \, z)\rangle, \quad z \in \Sigma_{\frac{\pi}{2}},$$

we know that

$$V_0(z), \ V(z), \ W(z) \text{ are analytic in } \Sigma_{\frac{\pi}{2}}, \tag{3.5}$$

$$V_0(z) \subset \mathcal{D}(A_1) \text{ and } A_1 V_0(z) = V(z), \quad z \in \Sigma_{\frac{\pi}{2}}. \tag{3.6}$$

Next, we have by (3.1) and (3.2) that

$$\left| e^{\mu_-(x)z} \right| \le e^{-C_3|x|^l \operatorname{Re} z}, \quad x \in R^n, \ z \in \Sigma_{\frac{\pi}{2}}, \tag{3.7}$$

$$\left| e^{\mu_+(x)z} \right| \le e^{-C_3|x|^{m-l} \operatorname{Re} z}, \quad x \in R^n, \ z \in \Sigma_{\frac{\pi}{2}},$$

for some constant $C_3 > 0$. This combined with (3.1) – (3.4) implies that for any multiindex β with $|\beta| \ge 1$,

$$\left| D^\beta \left[e^{\mu_-(x)z} \right] \right|, \ \left| D^\beta \left[\frac{1}{\sqrt{p^2(x) - 4q(x)}} e^{\mu_-(x)z} \right] \right|,$$

$$\left| D^\beta \left[\frac{p(x)}{\sqrt{p^2(x) - 4q(x)}} e^{\mu_-(x)z} \right] \right|$$

$$\le \quad \text{const} \sum_{i=1}^{|\beta|} |z|^i (1 + |x|)^{li - |\beta|} e^{-C_3|x|^l \operatorname{Re} z}, \quad x \in R^n, \ z \in \Sigma_{\frac{\pi}{2}},$$

$$\left| D^\beta \left[e^{\mu_+(x)z} \right] \right|, \ \left| D^\beta \left[\frac{1}{\sqrt{p^2(x) - 4q(x)}} e^{\mu_+(x)z} \right] \right|,$$

$$\left| D^\beta \left[\frac{p(x)}{\sqrt{p^2(x) - 4q(x)}} e^{\mu_+(x)z} \right] \right|$$

$$\le \quad \text{const} \sum_{i=1}^{|\beta|} |z|^i (1 + |x|)^{(m-l)i - |\beta|} e^{-C_3|x|^{m-l} \operatorname{Re} z}, \quad x \in R^n, \ z \in \Sigma_{\frac{\pi}{2}}.$$

When

$$li - |\beta| < -\frac{n}{2},$$

we get

$$\left\| (1 + |x|)^{li - |\beta|} e^{-C_3|x|^l \operatorname{Re} z} \right\|_{L^2(R^n)} \le \left\| (1 + |x|)^{li - |\beta|} \right\|_{L^2(R^n)}$$

$$\le \quad \text{const}, \quad z \in \Sigma_{\frac{\pi}{2}}.$$

When

$$-\frac{n}{2} < li - |\beta| < 0,$$

we have

$$\left\| (1 + |x|)^{li - |\beta|} e^{-C_3 |x|^l \operatorname{Re} z} \right\|_{L^2(R^n)}$$

$$\leq \quad \operatorname{const} (\operatorname{Re} z)^{-i + \frac{2|\beta| - n}{2l}} \left\| |x|^{li - |\beta|} e^{-C_3 |x|^l} \right\|_{L^2(R^n)}$$

$$\leq \quad \operatorname{const} (\operatorname{Re} z)^{-i + \frac{2|\beta| - n}{2l}}, \quad z \in \Sigma_{\frac{\pi}{2}}.$$

When

$$li - |\beta| \geq 0,$$

we get

$$\left\| (1 + |x|)^{li - |\beta|} e^{-C_3 |x|^l \operatorname{Re} z} \right\|_{L^2(R^n)}$$

$$\leq \quad \operatorname{const} \left\| \left(1 + |x|^{li - |\beta|}\right) e^{-C_3 |x|^l \operatorname{Re} z} \right\|_{L^2(R^n)}$$

$$\leq \quad \operatorname{const} \left((\operatorname{Re} z)^{-\frac{n}{2l}} + (\operatorname{Re} z)^{-i + \frac{2|\beta| - n}{2l}} \right), \quad z \in \Sigma_{\frac{\pi}{2}}.$$

Keep these observations in mind. Now, we fix $\phi \in (0, \frac{\pi}{2})$. Then $|z| \leq \frac{2}{\cos \phi} \operatorname{Re} z$ for $z \in \Sigma_\phi$. Take $\beta \in N_0^n$ such that $|\beta| = \left[\frac{n}{2}\right] + 1$. Then

$$li - |\beta| \neq -\frac{n}{2}, \quad \text{for every } i \in \{1, \cdots, |\beta|\},$$

noting $l \geq 2$ by the strong ellipticity of $p(x)$. Hence

$$\left\| D^\beta \left[e^{\mu_-(x)z} \right] \right\|_{L^2(R^n)} \leq \operatorname{const} (\operatorname{Re} z)^{\frac{2|\beta| - n}{2l}} e^{\operatorname{Re} z}, \quad z \in \Sigma_\phi.$$

Moreover by (3.7),

$$\left\| e^{\mu_-(x)z} \right\|_{L^2(R^n)} \leq \operatorname{const} (\operatorname{Re} z)^{-\frac{n}{2l}}, \quad z \in \Sigma_\phi.$$

Thus, an application of the Bernstein theorem shows that

$$\left\| e^{\mu_-(x)z} \right\|_{\mathcal{F} L^1} \leq \operatorname{const} e^{\operatorname{Re} z}, \quad z \in \Sigma_\phi,$$

noting

$$-\frac{n}{2l} \left(1 - \frac{n}{2|\beta|}\right) + \frac{(2|\beta| - n)}{2l} \frac{n}{2|\beta|} = 0.$$

Similarly, we can obtain

$$\left\| e^{\mu_+(x)z} \right\|_{\mathcal{F} L^1}, \quad \left\| \frac{1}{\sqrt{p^2(x) - 4q(x)}} e^{\mu_\pm(x)z} \right\|_{\mathcal{F} L^1},$$

$$\left\| \frac{p(x)}{\sqrt{p^2(x) - 4q(x)}} e^{\mu_\pm(x)z} \right\|_{\mathcal{F} L^1} \leq \operatorname{const} e^{\operatorname{Re} z}, \quad z \in \Sigma_\phi.$$

Consequently,

$$\|V_0(z)\|, \ \|V(z)\|, \ \|W(z)\| \le \text{const } e^{\text{Re}z}, \quad z \in \Sigma_\phi. \tag{3.8}$$

Pick $\lambda_0 < 0$. Then (3.1) implies by Lemma 1.5.3 that

$$\left(\lambda_0 - p^2(x)\right)^{-1}, \ q(x)\left(\lambda_0 - p^2(x)\right)^{-1} \in \mathcal{F}L^1$$

and

$$\mathcal{R}\left(\mathbf{T}\left\langle \left(\lambda_0 - p^2(x)\right)^{-1}\right\rangle\right) = \mathcal{D}\left(A_1^2\right).$$

A simple calculation shows that for $x \in R^n$, $z \in \Sigma_{\frac{\pi}{2}}$,

$$v_0(x; \ z)(\lambda_0 - p^2(x))^{-1}$$

$$= \ \frac{1}{2}(\lambda_0 - p^2(x))^{-1} \int_0^z [w(x; \ \eta) - v(x; \ \eta)]d\eta,$$

$$w(x; \ z)(\lambda_0 - p^2(x))^{-1}$$

$$= \ -\frac{1}{2}p(x)(\lambda_0 - p^2(x))^{-1} \int_0^z [w(x; \ \eta) - v(x; \ \eta)]d\eta$$

$$-q(x)(\lambda_0 - p^2(x))^{-1} \int_0^z (z - \eta)[w(x; \ \eta) - v(x; \ \eta)]d\eta$$

$$+2(\lambda_0 - p^2(x))^{-1}.$$

It follows that for each $\phi \in (0, \frac{\pi}{2})$,

$$V_0(z)\mathbf{T}\langle(\lambda_0 - p^2(x))^{-1}\rangle \longrightarrow 0,$$

$$V(z)\mathbf{T}\langle(\lambda_0 - p^2(x))^{-1}\rangle \longrightarrow 0,$$

$$W(z)\mathbf{T}\langle(\lambda_0 - p^2(x))^{-1}\rangle \longrightarrow 2\mathbf{T}\langle(\lambda_0 - p^2(x))^{-1}\rangle,$$

as $z \to 0$ ($z \in \Sigma_\phi$). Thus referring to (3.8) and the denseness of $\mathcal{D}\left(A_1^2\right)$ yields that for each $u \in E$, $\phi \in (0, \frac{\pi}{2})$,

$$\lim_{\substack{z \to 0 \\ z \in \Sigma_\phi}} V_0(z)u = 0, \quad \lim_{\substack{z \to 0 \\ z \in \Sigma_\phi}} V(z)u = 0, \quad \lim_{\substack{z \to 0 \\ z \in \Sigma_\phi}} W(z)u = 2u. \tag{3.9}$$

Then proceeding similarly as in the proof of Theorem 3.5.5 and noting that for $\lambda > \sigma_0$, $x \in R^n$,

$$\int_0^\infty e^{-\lambda t} v_0(x; \ t)dt = \left(\lambda^2 + p(x)\lambda + q(x)\right)^{-1},$$

$$\int_0^\infty e^{-\lambda t} v(x; \ t) dt = p(x) \left(\lambda^2 + p(x)\lambda + q(x)\right)^{-1},$$

$$\int_0^\infty e^{-\lambda t} w(x; \ t) dt = (2\lambda + p(x)) \left(\lambda^2 + p(x)\lambda + q(x)\right)^{-1},$$

we obtain that for each $\lambda > \sigma_0$,

$$R_\lambda \in \mathbf{L}(E), \quad A_1 R_\lambda u = R_\lambda A_1 u \ (u \in \mathcal{D}(A_1)),$$

$$A_0 R_\lambda u = R_\lambda A_0 u \ (u \in \mathcal{D}(A_0)),$$

$$A_1 R_\lambda u = \int_0^\infty e^{-\lambda t} V(t) u dt, \quad u \in E,$$

$$2\lambda R_\lambda u = \int_0^\infty e^{-\lambda t} (W(t) - V(t)) u dt, \quad u \in E.$$

Thus Theorem 2.2.3 applies (by (3.8), (3.9)) and we see that (ACP_2) is strongly wellposed with two propagators

$$S_0(t) = \frac{1}{2}(W(t) + V(t)),$$

$$S_1(t)u = V_0(t)u = \frac{1}{2}\int_0^t [W(s) - V(s)]u ds, \quad t \geq 0, \ u \in E.$$

It follows from (3.5), (3.6), (3.8), (3.9) and Theorem 1.3 that (ACP_2) is analytically wellposed in $\Sigma_{\frac{\pi}{2}}$.

Finally, (3.1) implies that $p(D)$, $q(D)$ are nonnegative operators,

$$\mathcal{D}\left(\Delta^{\frac{bl}{2}}\right) \supset \mathcal{D}\left((p(D))^b\right), \quad \mathcal{D}\left(\Delta^{\frac{am}{2}}\right) \supset \mathcal{D}\left((q(D))^a\right);$$

also

$$(I + p(D))^{-b}(I - \Delta)^{\frac{bl}{2}}, \quad (I + q(D))^{-a}(I - \Delta)^{\frac{am}{2}} \in \mathbf{L}(E).$$

Therefore, applying Theorem 1.7 and Theorems 2.10, 2.11 establishes the results as claimed. The proof is then complete.

Example 3.2. Let $a_i(x) \in C^1(R^3)$ with $\frac{\partial a_i(x)}{\partial x_1}$, $\frac{\partial a_i(x)}{\partial x_2}$, $\frac{\partial a_i(x)}{\partial x_3} \in C_b(R^3)$, for each $i = 1, 2, 3$. We consider the Cauchy problem

$$\begin{cases} u_{tt} + \Delta^2 u_t + \sum_{i=1}^3 a_i(x)\frac{\partial}{\partial x_i} u_t - \Delta^3 u = 0, \quad t \geq 0, \ x \in R^3, \\ \\ u(0, \ x) = \phi(x), \ u_t(0, \ x) = \psi(x), \quad x \in R^3, \end{cases} \quad (3.10)$$

in $L^p(R^3)$ $(1 < p < \infty)$.

Take

$$p(x) = \left(\sum_{i=1}^{3} x_i^2\right)^2, \quad l = 4,$$

$$q(x) = \left(\sum_{i=1}^{3} x_i^2\right)^3, \quad m = 6,$$

$$B_0 = 0, \quad a = 0,$$

$$B_1 = \sum_{i=1}^{3} a_i(x)\frac{\partial}{\partial x_i}, \quad b = \frac{1}{2}.$$

Let $q = \frac{p}{p-1}$. Observing

$$\sum_{i=1}^{3} \frac{\partial}{\partial x_i} a_i(x)(I - \Delta)^{-1} \in \mathbf{L}\left(L^q(R^3)\right),$$

we have by a duality argument that $(I - \Delta)^{-1} \sum_{i=1}^{3} a_i(x)\frac{\partial}{\partial x_i}$ has a bounded extension on $L^p(R^3)$ $(1 < p < \infty)$. Thus, Theorem 3.1 tells us that (3.10) is analytically wellposed in $\Sigma_{\frac{\pi}{2}}$, and so for $\phi, \psi \in W^{6,p}(R^3)$, it has a unique solution

$$u(\cdot) \in C^2\left(R^+, L^p(R^3)\right) \bigcap C^1\left(R^+, W^{4,p}(R^3)\right) \bigcap C\left(R^+, W^{6,p}(R^3)\right),$$

which can be extended analytically to $\Sigma_{\frac{\pi}{2}}$ such that for each $\phi \in (0, \frac{\pi}{2})$,

$$\|u(z)\|_{L^p(R^3)} \leq C_\phi e^{\omega_\phi \operatorname{Re} z}\left(\|u(0)\|_{L^p(R^3)} + \|u'(0)\|_{L^p(R^3)}\right), \quad z \in \Sigma_\phi$$

for some $C_\phi, \omega_\phi > 0$.

4.4 Entire solutions

In this section, we write

$$\Upsilon_\phi(\theta, r) = \left\{z \in \mathbf{C}; \ z \neq 0, \ |\arg(e^{-i\phi}z)| \leq \frac{\pi}{2} + \theta, \ |z| \geq r\right\},$$

$$\theta \in \left(0, \frac{\pi}{2}\right), \ \phi \in (-\pi, \pi], \ r \geq 0.$$

Definition 4.1. A function $u(\cdot) \in C^n(R^+, E)$ is said to be an entire solution of (ACP_n) if $u(\cdot)$ is a solution of (ACP_n) and it can be extended analytically to the whole complex plane, and $A_i u^{(i)}(\cdot)$ $(0 \leq i \leq n-1)$ are also entire functions.

Theorem 4.2. *Suppose $A_0, \cdots, A_{n-1}$ are closed linear operators in E with $\bigcap_{i=0}^{n-1} \mathcal{D}(A_i)$ dense in E, and satisfy the following condition*
(i) *For some $\theta \in (0, \frac{\pi}{2})$, $\phi \in (-\pi, \pi]$, $r > 0$,*

$$\rho(A_0, \cdots, A_{n-1}) \supset \Upsilon_\phi(\theta, r). \tag{4.1}$$

(ii) *There exist constants $M > 0$, $h \in N$ such that for $\lambda \in \Upsilon_\phi(\theta, r)$,*

$$\|\lambda^n R_\lambda\| \leq M, \tag{4.2}$$

$$\|A_k R_\lambda\| \leq M|\lambda|^h, \quad 0 \leq k \leq n-1. \tag{4.3}$$

Then there exists a dense subset G of the product space $E^n = E \times E \times \cdots \times E$ such that for every initial value $(u_0, u_1, \cdots, u_{n-1}) \in G$, (ACP_n) has a unique entire solution.

Proof. Fix $a > r$ and b with

$$1 < b < \frac{\pi}{2}\left(\frac{\pi}{2} - \theta\right)^{-1}.$$

Let $(a - \lambda)^b$ be the branch of the power function which is holomorphic off the half-line $[a, \infty)$ and positive for $\lambda < a$. For each $z \in \mathbf{C}$, $0 \leq k \leq n-1$, $u \in \bigcap_{i=0}^{n-1} \mathcal{D}(A_i)$, $\varepsilon > 0$, set

$$W_k(z; \varepsilon)u$$

$$= \frac{e^{i\phi}}{2\pi i} \int_\Gamma \exp\left(ze^{i\phi}\lambda - \varepsilon(a - \lambda)^b\right) R_{e^{i\phi}\lambda} \sum_{j=k+1}^{n} \left((e^{i\phi}\lambda)^{j-k-1} A_j u\right) d\lambda, \tag{4.4}$$

where Γ is the boundary of $\Upsilon_0(\theta, r)$, and is oriented in a way so that Imλ increases along Γ. Clearly, if λ is in the sector

$$\Omega = \left\{z \in \mathbf{C}; \; z \neq 0, \; \frac{\pi}{2} + \theta \leq \arg z \leq \frac{3}{2}\pi - \theta\right\},$$

then $\lambda - a \in \Omega$. Since

$$\begin{aligned} \operatorname{Re}(a - \lambda)^b &= |a - \lambda|^b \cos(b \arg(a - \lambda)) \\ &\geq |a - \lambda|^b \cos\left(b\left(\frac{\pi}{2} - \theta\right)\right), \quad \lambda \in \Omega, \end{aligned} \tag{4.5}$$

we have that for every $\lambda \in \Omega$,

$$\begin{aligned} &\left|\exp\left(ze^{i\phi}\lambda - \varepsilon(a - \lambda)^b\right)\right| \\ &\leq \exp\left(|z||\lambda| - \varepsilon \cos\left(b\left(\frac{\pi}{2} - \theta\right)\right)|a - \lambda|^b\right). \end{aligned} \tag{4.6}$$

On the other hand, (4.2) implies that for each $u \in \bigcap_{i=0}^{n-1} \mathcal{D}(A_i)$,

$$R_{e^{i\phi}\lambda} \sum_{j=k+1}^{n} (e^{i\phi}\lambda)^{j-k-1} A_j u$$

is polynomially bounded (for λ). According to this fact, (4.6) and $1 < b < \frac{\pi}{2} \left(\frac{\pi}{2} - \theta \right)^{-1}$, we obtain that for each $z \in \mathbb{C}$, $0 \le k \le n-1$, $u \in \bigcap_{i=0}^{n-1} \mathcal{D}(A_i)$, $\varepsilon > 0$, the integral in (4.4) exists and it defines an entire function of z. Differentiating (4.4) in z up to l times, we get

$$W_k^{(l)}(z; \varepsilon)u$$

$$= \frac{e^{i\phi}}{2\pi i} \int_{\Gamma} \exp\left(ze^{i\phi}\lambda - \varepsilon(a-\lambda)^b\right) R_{e^{i\phi}\lambda} \sum_{j=k+1}^{n} (e^{i\phi}\lambda)^{j+l-k-1} A_j u d\lambda \tag{4.7}$$

$$\left(z \in \mathbb{C}, \ l \in N, \ 0 \le k \le n-1, \ u \in \bigcap_{i=0}^{n-1} \mathcal{D}(A_i), \ \varepsilon > 0\right).$$

By (4.3), for each $0 \le l, \ k \le n-1$, $u \in \bigcap_{i=0}^{n-1} \mathcal{D}(A_i)$,

$$A_l R_{e^{i\phi}\lambda} \sum_{j=k+1}^{n} (e^{i\phi}\lambda)^{j+l-k-1} A_j u$$

is also polynomially bounded (for λ in Γ). Hence, it follows from (4.7) and the closedness of A_l that

$$A_l W_k^{(l)}(z; \varepsilon)u$$

$$= \frac{e^{i\phi}}{2\pi i} \int_{\Gamma} \exp\left(ze^{i\phi}\lambda - \varepsilon(a-\lambda)^b\right) A_l R_{e^{i\phi}\lambda} \sum_{j=k+1}^{n} (e^{i\phi}\lambda)^{j+l-k-1} A_j u d\lambda$$

$$\left(z \in \mathbb{C}, \ 0 \le l, \ k \le n-1, \ u \in \bigcap_{i=0}^{n-1} \mathcal{D}(A_i), \ \varepsilon > 0\right). \tag{4.8}$$

Thus,

$$W_k^{(n)}(z; \varepsilon)u + \sum_{l=0}^{n-1} A_l W_k^{(l)}(z; \varepsilon)u \tag{4.9}$$

$$= \frac{e^{i\phi}}{2\pi i} \lim_{T \to \infty} \int_{\Gamma_T} \exp\left(ze^{i\phi}\lambda - \varepsilon(a-\lambda)^b\right) \sum_{j=k+1}^{n} (e^{i\phi}\lambda)^{j-k-1} A_j u d\lambda,$$

where

$$\Gamma_T := \Gamma \bigcap \{z \in \mathbb{C}; \ |z| \le T\}, \quad T > 0.$$

Since the integrand in (4.9) is analytic in Ω, we can shift the path of the integral to the arc

$$\left\{ Te^{i\alpha}; \ \frac{\pi}{2} + \theta \leq \alpha \leq \frac{3}{2}\pi - \theta \right\}$$

using the well-known Cauchy theorem. Thus, combining (4.6), we have

$$\left\| \int_{\Gamma_T} \exp\left(ze^{i\phi}\lambda - \varepsilon(a-\lambda)^b\right) \sum_{j=k+1}^{n} (e^{i\phi}\lambda)^{j-k-1} A_j u d\lambda \right\|$$

$$\leq \int_{\frac{\pi}{2}+\theta}^{\frac{3}{2}\pi-\theta} \exp\left(|z|T - \varepsilon \cos\left(b\left(\frac{\pi}{2} - \theta\right)\right)(T-a)^b\right) \sum_{j=k+1}^{n} \|A_j u\| T^{j-k} d\alpha$$

$$\longrightarrow 0, \quad \text{as } T \longrightarrow \infty.$$

$$(4.10)$$

Consequently, for any $\varepsilon > 0$, $v_k \in \bigcap_{i=0}^{n-1} \mathcal{D}(A_i)$ $(0 \leq k \leq n-1)$,

$$u_\varepsilon(t) := \sum_{k=0}^{n-1} W_k(t; \varepsilon) v_k \tag{4.11}$$

is a solution of (ACP_n) with initial value

$$u_\varepsilon^{(l)}(0) = \sum_{k=0}^{n-1} C_{\varepsilon,\,k}^l v_k,$$

where

$$C_{\varepsilon,\,k}^l u := \frac{e^{i\phi}}{2\pi i} \int_\Gamma \exp\left(-\varepsilon(a-\lambda)^b\right) (e^{i\phi}\lambda)^{l-k-1}$$

$$\cdot \left[u - \sum_{j=0}^{k} (e^{i\phi}\lambda)^j R_{e^{i\phi}\lambda} A_j u \right] d\lambda, \tag{4.12}$$

$$\varepsilon > 0, \ 0 \leq k, \ l \leq n-1, \ u \in \bigcap_{i=0}^{n-1} \mathcal{D}(A_i).$$

Now, observe that a deformation of contour as in the treatment of (4.10) shows

$$\frac{e^{i\phi}}{2\pi i} \int_\Gamma \exp\left(-\varepsilon(a-\lambda)^b\right) (e^{i\phi}\lambda)^{l-k-1} v_k d\lambda$$

$$= \begin{cases} 0, & \text{if } l-k-1 \geq 0, \\[2mm] \dfrac{(e^{i\phi})^{l-k}}{(k-l)!} \left[\dfrac{d^{k-l}}{d\lambda^{k-l}} \exp\left(-\varepsilon(a-\lambda)^b\right) v_k \right]_{\lambda=0}, & \text{if } l-k-1 < 0. \end{cases}$$

On the other hand, shifting the path Γ of integral to the arc

$$\left\{ Te^{i\alpha}; \quad -\frac{\pi}{2} - \theta \le \alpha \le \frac{\pi}{2} + \theta \right\}$$

instead and according to (4.2), we see that for $v \in E$, $j \in N$ and $j \le n - 2$,

$$\left\| \frac{e^{i\phi}}{2\pi i} \lim_{\varepsilon \to 0} \int_{\Gamma} \exp\left(-\varepsilon(a - \lambda)^b\right) (e^{i\phi}\lambda)^j R_{e^{i\phi}\lambda} v d\lambda \right\|$$

$$\le \quad \frac{1}{2\pi} \lim_{T \to \infty} \int_{-\frac{\pi}{2}-\theta}^{\frac{\pi}{2}+\theta} MT^{-1} \|v\| d\alpha$$

$$= \quad 0.$$

Hence, we have

$$\lim_{\varepsilon \to 0} C^l_{\varepsilon, \, k} v_k = \begin{cases} 0, & \text{if } k \ne l, \\ \\ v_l, & \text{if } k = l, \end{cases}$$

and therefore

$$\lim_{\varepsilon \to 0} u^{(l)}_\varepsilon(0) = v_l, \quad 0 \le l \le n - 1.$$

This ends the proof of existence, because of the denseness of $\bigcap_{i=0}^{n-1} \mathcal{D}(A_i)$.

Now, we show the uniqueness.

Let $u(\cdot)$ be an entire solution of (ACP_n) with the initial values $u_j = 0$, $0 \le j \le n - 1$. Clearly, $\lambda e^{i\phi} \in \Upsilon_\phi(\theta, r)$ for each $\lambda > r$. If we define

$$M_t = \sup\left\{ \left\| u\left(e^{-i\phi}s\right) \right\|; \ t - 1 \le s \le t \right\}, \quad t \ge 1,$$

then we have

$$\left\| \int_{t-1}^{t} \exp(\lambda(t - s - 1)) u\left(e^{-i\phi}s\right) ds \right\|$$

$$\le \quad M_t \int_{t-1}^{t} \exp(\lambda(t - s - 1)) ds \qquad (4.13)$$

$$= \quad M_t \lambda^{-1}(1 - e^{-\lambda}) \longrightarrow 0, \quad \text{as } \lambda \longrightarrow \infty.$$

On the other hand, Integrating by parts, we get that for each $t \ge 0$, $\lambda > r$, $1 \le k \le n$,

$$(e^{i\phi}\lambda)^k \int_{0}^{t} \exp(\lambda(t - s)) u\left(e^{-i\phi}s\right) ds$$

$$= \quad -\sum_{l=0}^{k-1} e^{i(k-l)\phi} \lambda^{k-l-1} u^{(l)}(e^{-i\phi}t) + \int_{0}^{t} \exp(\lambda(t - s)) u^{(k)}(e^{-i\phi}s) ds.$$

Thus, it follows from

$$u^{(n)}\left(e^{-i\phi}s\right) + \sum_{k=0}^{n-1} A_k u^{(k)}\left(e^{-i\phi}s\right) = 0 \quad (s \geq 0)$$

that for $\lambda > r$,

$$\left\| \int_0^{t-1} \exp(\lambda(t-s-1))u\left(e^{-i\phi}s\right) ds \right\|$$

$$= \left\| e^{-\lambda} R_{e^{i\phi}\lambda} P_{e^{i\phi}\lambda} \int_0^{t-1} \exp(\lambda(t-s))u\left(e^{-i\phi}s\right) ds \right\|$$

$$= \left\| -e^{-\lambda} \left\{ \sum_{k=1}^{n}\sum_{l=0}^{k-1} e^{i(k-l)\phi}\lambda^{k-l-1} R_{e^{i\phi}\lambda} A_k u^{(l)}\left(e^{-i\phi}t\right) \right. \right.$$

$$\left. \left. + \int_{t-1}^{t} \exp(\lambda(t-s))u\left(e^{-i\phi}s\right) ds \right\} \right\|.$$
(4.14)

By (4.13), (4.14) and (4.2), we obtain that for every given $t \geq 1$,

$$\lim_{\lambda\to\infty} \int_0^{t-1} \exp(\lambda s)u\left(e^{-i\phi}(t-s-1)\right) ds = 0.$$

In view of Lemma 4.1.1 of Pazy [2] (see also the statement below (2.3.1)), we have $u\left(e^{-i\phi}(t-s-1)\right) = 0$ $(0 \leq s \leq t-1)$. Since t is arbitrary, $u\left(e^{-i\phi}t\right) = 0$ for any $t \geq 0$. Therefore $u(z) = 0$, $z \in \mathbf{C}$. This ends the proof of the theorem.

Remark 4.3. (4.11) together with (4.4), (4.12) provides an explicit expression of the entire solutions of (ACP_n).

Recall that the parabolicity of the equation in (ACP_n) means that for some $\theta \in (0, \frac{\pi}{2})$, $r > 0$,

$$\rho(A_0, \cdots, A_{n-1}) \supset \Upsilon_0(\theta, r)$$

and there is $M > 0$ such that for $\lambda \in \Upsilon_0(\theta, r)$,

$$\|\lambda^n R_\lambda\|, \quad \|\lambda^k A_k R_\lambda\| \leq M \quad (0 \leq k \leq n-1).$$

As a direct consequence, we have

Corollary 4.4. *Assume that the differential equation in (ACP_n) is parabolic. Then the conclusion in Theorem 4.2 holds.*

Example 4.5. Let q, k, l, $h \in N$ with $k < 2q$, $l < 4q$, $h < 6q$, let $b > 1$, and let $a_i(\cdot)$ $(i = 0, 1, 2)$ be a bounded measurable function defined in the m-dimensional

Euclidean space R^m. Then the Cauchy problem

$$\begin{cases} \dfrac{\partial^3 u}{\partial t^3} + \left((-1)^q \Delta^q + a_2(x) \sum_{i=1}^{m} \dfrac{\partial^k}{\partial x_i^k}\right) \dfrac{\partial^2 u}{\partial t^2} + \left(b\Delta^{2q} + a_1(x) \sum_{i=1}^{m} \dfrac{\partial^l}{\partial x_i^l}\right) \dfrac{\partial u}{\partial t} \\[3mm] \qquad + \left((-1)^q \Delta^{3q} + a_0(x) \sum_{i=1}^{m} \dfrac{\partial^h}{\partial x_i^h}\right) u = 0, \quad (t, x) \in R^+ \times R^m, \\[3mm] u(0, x) = u_0(x), \quad \dfrac{\partial u}{\partial t}(0, x) = u_1(x), \quad \dfrac{\partial^2 u}{\partial t^2}(0, x) = u_2(x), \quad x \in R^m, \end{cases}$$

has a unique solution $u \in C^\infty\left(R^+, H^{6q}(R^m)\right)$, which can be extended to an entire function $: \mathbf{C} \to H^{6q}(R^m)$, for every initial value (u_0, u_1, u_2) in a dense subset of $L^2(R^m) \times L^2(R^m) \times L^2(R^m)$.

Proof. Take $E = L^2(R^m)$. Let

$$B_0 = (-1)^q \Delta^{3q} \quad \text{with } \mathcal{D}(B_0) = H^{6q}(R^m),$$

$$B_1 = b\Delta^{2q} \quad \text{with } \mathcal{D}(B_1) = H^{4q}(R^m),$$

$$B_2 = (-1)^q \Delta^q \quad \text{with } \mathcal{D}(B_2) = H^{2q}(R^m),$$

$$S_0 = a_0(x) \sum_{i=1}^{m} \frac{\partial^h}{\partial x_i^h} \quad \text{with } \mathcal{D}(S_0) = H^h(R^m),$$

$$S_1 = a_1(x) \sum_{i=1}^{m} \frac{\partial^l}{\partial x_i^l} \quad \text{with } \mathcal{D}(S_1) = H^l(R^m),$$

$$S_2 = a_2(x) \sum_{i=1}^{m} \frac{\partial^k}{\partial x_i^k} \quad \text{with } \mathcal{D}(S_2) = H^k(R^m).$$

Clearly, B_0 is a densely defined, nonnegative operator in E with $\theta_\infty^\pm(B_0) = 0$ (see Section 4.2), and

$$B_1 = bB_0^{\frac{2}{3}}, \quad B_2 = B_0^{\frac{1}{3}}.$$

Since $k < 2q$, $l < 4q$, $h < 6q$, we have by virtue of the moment inequality that there exists a constant $C > 0$ such that

$$\|S_0 u\| \le C\|u\|^{1-\frac{h}{6q}} \|B_0 u\|^{\frac{h}{6q}}, \quad \text{for } u \in \mathcal{D}(B_0),$$

$$\|S_1 u\| \le C\|u\|^{1-\frac{l}{4q}} \|B_1 u\|^{\frac{l}{4q}}, \quad \text{for } u \in \mathcal{D}(B_1),$$

$$\|S_2 u\| \le C\|u\|^{1-\frac{k}{2q}} \|B_2 u\|^{\frac{k}{2q}}, \quad \text{for } u \in \mathcal{D}(B_2).$$

Define $A_i = B_i + S_i$, $i = 0$, 1, 2. Then, Corollary 4.4 combined with Theorems 2.11 and 2.17 shows the result desired.

Example 4.6. Let $E = L^2(0, 1)$, $a(x) \in C^1[0, 1]$ with $a(x) \neq 0$ for each $x \in [0, 1]$, $\int_0^1 a^{-1}(\xi)d\xi \neq 0$, and

$$a(x) \in S_\phi(\theta) := \{z \in \mathbf{C}; \ z \neq 0, \ |\arg(e^{-i\phi}z)| \leq \theta\}, \quad \text{for any } x \in [0, 1],$$

for some $\theta \in (0, \frac{\pi}{2})$, $\phi \in (-\pi, \pi]$. Let

$$A_0 = -\frac{\partial^2}{\partial x^2}, \quad A_1 = -\frac{\partial}{\partial x}\left(a(x)\frac{\partial}{\partial x}\cdot\right)$$

with

$$\mathcal{D}(A_0) = \mathcal{D}(A_1) = \left\{u \in H^2(0, 1); \ u(x)\big|_{x=0,1} = 0\right\}.$$

We have that $0 \in \rho(A_1)$. In fact, for any $v(\cdot) \in E$,

$$u(x) := \left[\left(\int_0^1 a^{-1}(\xi)d\xi\right)^{-1}\int_0^1 a^{-1}(\xi)\int_0^\xi v(\eta)d\eta d\xi\right]\int_0^x a^{-1}(\xi)d\xi$$

$$- \int_0^x a^{-1}(\xi)\int_0^\xi v(\eta)d\eta d\xi \tag{4.15}$$

is in $\mathcal{D}(A_1)$ and it is the solution of $A_1 u = v$, and it is clear from (4.15) that

$$\|A_1^{-1}v\| \leq \text{const } \|v\|.$$

On the other hand, for each $u \in \mathcal{D}(A_1)$,

$$\langle A_1 u, \ u\rangle = -\int_0^1 \frac{\partial}{\partial x}\left(a(x)\frac{\partial}{\partial x}u(x)\right)\bar{u}(x)dx$$

$$= \int_0^1 a(x)|u'(x)|^2 \ dx \in S_\phi(\theta).$$

Now we make use of a result from Pazy [2, Theorem 3.9, p. 12]:

Let A be a densely defined closed linear operator in E, let n.r.(A) be the numerical range of A, i.e.,

$$n.r.(A) = \{\langle u^*, \ Au\rangle; \ u \in \mathcal{D}(A), \ u^* \in E^*, \ \|u\| = \|u^*\| = \langle u^*, \ u\rangle = 1\}$$

and let Σ be the complement of $\overline{n.r.(A)}$ in $\mathbf{C}$. If Σ_0 is a component of Σ satisfying $\rho(A)\bigcap\Sigma_0 \neq \emptyset$, then $\sigma(A)$ lies in the complement of Σ_0 and

$$\|R(\lambda; \ A)\| \leq \frac{1}{d\left(\lambda; \ \overline{n.r.(A)}\right)},$$

where $d\left(\lambda;\ \overline{n.r.(A)}\right)$ *is the distance of* λ *from* $\overline{n.r.(A)}$.

We have that

$$\rho(-A_1) \supset \bigcup_{\theta < \beta < \frac{\pi}{2}} \Upsilon_\phi(\frac{\pi}{2} - \beta,\ 0)$$

and there exists $C_0 > 0$ such that

$$\left\|(\lambda + A_1)^{-1}\right\| \le C_0 |\lambda|^{-1}, \quad \lambda \in \Upsilon_\phi\left(\frac{\pi}{4} - \frac{\theta}{2},\ 1\right). \tag{4.16}$$

Taking $\lambda_0 \in \rho(-A_1)$, then $A_0(\lambda_0 + A_1)^{-1} \in \mathbf{L}(E)$ since $\mathcal{D}(A_0) = \mathcal{D}(A_1)$. It follows from (4.16) that there exists $r > 1$ such that for $\lambda \in \Upsilon_\phi\left(\frac{\pi}{4} - \frac{\theta}{2},\ r\right)$,

$$\left\|\lambda^{-1} A_0 (\lambda + A_1)^{-1}\right\|$$

$$\le\ \left\|A_0(\lambda_0 + A_1)^{-1}\right\| \left\|\lambda^{-1}(\lambda_0 + A_1)(\lambda + A_1)^{-1}\right\|$$

$$<\ \frac{1}{2}.$$

Thus, from the equality

$$\left(\lambda^2 + \lambda A_1 + A_0\right)^{-1}$$

$$=\ \lambda^{-1}(\lambda + A_1)^{-1}\left[I + \lambda^{-1} A_0 (\lambda + A_1)^{-1}\right]^{-1}, \quad \lambda \in \Upsilon_\phi\left(\frac{\pi}{4} - \frac{\theta}{2},\ r\right),$$

we get that for λ as above, $\lambda \in \rho(A_0,\ A_1)$ and

$$\left\|\lambda^2 \left(\lambda^2 + \lambda A_1 + A_0\right)^{-1}\right\|, \quad \left\|\lambda A_1 \left(\lambda^2 + \lambda A_1 + A_0\right)^{-1}\right\| \le C$$

for some constant $C > 0$. Now applying Corollary 4.4, we claim that the Cauchy problem for the damped wave equation

$$\begin{cases} \dfrac{\partial^2 u(t,\ x)}{\partial t^2} - \dfrac{\partial}{\partial x}\left(a(x)\dfrac{\partial^2}{\partial t \partial x}u(t,\ x)\right) - \dfrac{\partial^2}{\partial x^2}u(t,\ x) = 0, \\[4mm] \qquad\qquad\qquad\qquad\qquad t > 0,\ 0 < x < 1, \\[4mm] u(t,\ x)\big|_{x=0,1} = 0, \quad t \ge 0, \\[4mm] u(0, x) = u_0(x), \quad \dfrac{\partial u}{\partial t}(0, x) = u_1(x), \quad 0 \le x \le 1, \end{cases}$$

has a unique solution $u \in C^\infty\left(R^+, H^2(0,\ 1)\right)$, which can be extended to an entire function: $\mathbf{C} \to H^2(0,\ 1)$, for every initial value $(u_0,\ u_1)$ in a dense subset of $L^2(0,\ 1) \times L^2(0,\ 1)$.

4.5 Notes

In the special case of $n = 1$, Theorem 1.1 is due to Hille [1]. A characterization of the generator of an analytic semigroup (i.e., the analytic propagator of (ACP_1)) in terms of estimates of its resolvent for only real values was got by Crandall-Pazy-Tartar [1]. Kato [2] also gave a different type of characterization of an analytic semigroup based on the behavior of the semigroup near its spectral radius. Other results on analytic semigroups can be found from, e.g., Beurling [1], Certain [1], Davies [1], deLaubenfels [9], Fattorini [6], Goldstein [7], Kato [3, 4], Nagel [2], Pazy [2], Stewart [1], Sinclair [1] and the references given there.

The analyticity of the solutions of higher order (ACP_n) was treated firstly by Obrécht [1]. Section 4.1 is adapted from Xiao-Liang [11, 12]. The essence of the part (ii) $\Longrightarrow$ (i) in Theorem 1.1 is due to Obrécht [1] except the continuous dependence of the solutions of (ACP_n) on the initial data.

Conditions, ensuring that the complete second order equation

$$u''(t) + A_1 u'(t) + A_0 u(t) = 0$$

with $A_1 = \rho A_0^\alpha$ ($\rho \in \mathbf{C}$, $0 < \alpha < 1$) be parabolic, were obtained by Favini-Obrécht [1]. Section 4.2 concerning equations of arbitrary order is taken from Xiao-Liang [14]. The notation $\theta_\infty^\pm(S)$ and Theorem 2.4 are due to Favini-Obrécht [1].

Section 4.3 comes from Xiao-Liang [23].

The analyticity of solutions means that they can be extended analytically to a sector in $\mathbf{C}$, but not sure the whole complex plane. deLaubenfels [4] disscussed firstly the existence and uniqueness of entire solutions, which can be extended analytically to the entire complex plane $\mathbf{C}$, for first order abstract Cauchy problems and gave some interesting criterions. Section 4.4 is from Xiao-Liang [18].

Chapter 5

Exponential growth bound and exponential stability

Summary

In Section 5.1, we characterize the exponential growth bound of the propagators of (ACP_n) in a Hilbert space in terms of the behavior of $\lambda^{n-1}R_\lambda$, $\lambda^{k-1}\overline{R_\lambda A_k}$ $(1 \leq k \leq n-1)$ on vertical lines in a half complex plane. As a consequence we show that the propagators are exponentially stable if P_λ is boundedly invertible in $\{\lambda \in \mathbf{C};\ \mathrm{Re}\lambda \geq 0\}$ with $\lambda^{n-1}R_\lambda$, $\lambda^{k-1}\overline{R_\lambda A_k}$ $(1 \leq k \leq n-1)$ uniformly bounded there.

Section 5.2 investigates the condition ensuring stability of every single solution of (ACP_n) in Banach spaces. It turns out to be a concise condition only requiring the uniform boundedness of R_λ in $\{\lambda \in \mathbf{C};\ \mathrm{Re}\lambda > -\delta\}$ for some $\delta > 0$.

5.1 Exponential growth bound of the propagators

Let $(H, \langle \cdot, \cdot \rangle)$ be a complex Hilbert space, and let A_k $(0 \leq k \leq n-1)$ be closed linear operators in H such that (ACP_n) is strongly wellposed.

Let $\rho_0(A_0, \cdots, A_{n-1})$ be as in (4.1.9). Then $\rho_0(A_0, \cdots, A_{n-1})$ is an open subset of $\mathbf{C}$ and R_λ, $\overline{R_\lambda A_k}$, $A_k R_\lambda$ $(0 \leq k \leq n-1)$ are analytic in $\rho_0(A_0, \cdots, A_{n-1})$.

Definition 1.1. Write

$$\omega_0 = \omega_0(A_0, \cdots, A_{n-1}) := \max_{0 \leq k \leq n-1} \left\{ \varlimsup_{t \to \infty} \frac{\ln \left\| S_k^{(k)}(t) \right\|}{t} \right\}.$$

The ω_0 is called exponential growth bound of the propagators of (ACP_n).

Obviously,

$$\omega_0 = \inf\{a \in R; \quad \text{there exists a constant } M \text{ with } \left\|S_k^{(k)}(t)\right\| \le Me^{at},$$

$$t \ge 0, \ 0 \le k \le n-1\};$$

in the case of $\omega_0 \ge 0$, for each $\varepsilon > 0$ there is $M_\varepsilon > 0$ such that for $0 \le k \le n-1$

$$\|S_k(t)\| \le M_\varepsilon e^{(\omega_0+\varepsilon)t}, \quad t \ge 0.$$

Lemma 1.2. *Let E be a Banach space and let $G(\cdot)$ be an E-valued function defined on R with $\|G(\cdot)\| \in L^1(R)$. Then*

$$\left\|\int_{-\infty}^{\infty} e^{ist}G(t)dt\right\| \longrightarrow 0, \quad as \ |s| \longrightarrow +\infty \ (s \in R).$$

Proof. Immediately by an application of the Riemann-Lebesgue lemma for E-valued Fourier transforms (see Proof of Theorem 1.1.8).

Lemma 1.3 (the Plancherel theorem for H-valued Fourier transforms). *There is a unique linear isometry Φ of $L^2(R, H)$ onto $L^2(R, H)$, such that*

$$\Phi(f)(x) = (2\pi)^{-\frac{1}{2}}\int_R e^{-ixy}f(y)dy, \quad x \in R,$$

whenever $f \in L^1(R, H) \bigcap L^2(R, H)$.

Proof. It is known that H is isomorphic to $L^2(\mu)$ for some measure space (X, Σ, μ). Denote by ϕ the scalar-valued Fourier transform on $L^2(R)$. We define $\Phi_0 : L^2(\mu, L^2(R)) \to L^2(\mu, L^2(R))$ by

$$\Phi_0(g)(\lambda) = \phi(g(\lambda)), \quad \lambda \in X, \ g \in L^2(\mu, L^2(R)).$$

Since ϕ is linear isometric, so is Φ_0. Observe that $L^2(\mu, L^2(R))$ is canonically isomorphic to $L^2(R, L^2(\mu)) = L^2(R, H)$. Then this isomorphism transforms Φ_0 into the linear isometry on $L^2(R, H)$ as desired.

The following is a characterization of the ω_0 which depends only on the properties of the coefficient operators of (ACP_n). By $\mathcal{H}_a$, for any $a \in R$, we will denote the half plane $\{\lambda \in C; \ \mathrm{Re}\lambda > a\}$.

Theorem 1.4. $\omega_0 = \inf\Big\{\tau_0 \in R; \ \text{for each } 1 \le k \le n, \ \text{the map } \lambda \mapsto \overline{R_\lambda A_k}$
admits an analytic continuation $\Psi_k(\lambda)$ in $\mathcal{H}_{\tau_0}$ satisfying

$$\sup_{\mu \in R} \left\|(\tau + i\mu)^{k-1}\Psi_k(\tau + i\mu)\right\| < \infty \ (\tau > \tau_0)\Big\} \overset{def.}{=} \sigma.$$

Moreover, for each $\varepsilon > 0$, there exists $M(\varepsilon)$ such that $\|S_k(t)\| \leq M(\varepsilon)e^{(\omega_0+\varepsilon)t}$, $t \geq 0, 0 \leq k \leq n-1$.

Proof. By the assumption of strong wellposedness we have from Sections 2.1 and 2.2 that $\mathcal{H}_\omega \subset \rho_0(A_0, \cdots, A_{n-1})$,

$$\left\| S_k^{(k)}(t) \right\| \leq Ce^{\omega t}, \quad t \geq 0, \ 0 \leq k \leq n-1, \tag{1.1}$$

and for $u \in H$, $\mathrm{Re}\lambda > \omega$,

$$\lambda^{n-1}R_\lambda u = \int_0^\infty e^{-\lambda t} S_{n-1}^{(n-1)}(t)u\,dt, \tag{1.2}$$

$$\lambda^{k-1}\overline{R_\lambda A_k}u = \int_0^\infty e^{-\lambda t} \left[S_{k-1}^{(k-1)}(t) - S_k^{(k)}(t) \right] u\,dt, \quad 1 \leq k \leq n-1, \tag{1.3}$$

where C, ω are some suitable constants.

First, we prove $\omega_0 \leq \sigma$.

Without loss of generality, we suppose $\sigma < \omega$. For each $\tau > \sigma$, set

$$M_\tau = \sup\left\{ \left\| \sum_{i=1}^n \left[\sum_{l=i}^n \binom{l}{i} \lambda^{l-i}\Psi_l(\lambda) \right] \right\|; \ \mathrm{Re}\lambda = \tau \right\}.$$

By hypothesis, $M_\tau < +\infty$. Let $r(\tau) = \min\left\{ \frac{1}{2M_\tau}, 1 \right\}$ (for any $\tau > \sigma$), and for $\lambda_0 \in \mathcal{H}_\sigma$, $\lambda \in \mathbf{C}$, let

$$
\begin{aligned}
K(\lambda_0; \lambda) &:= \sum_{i=1}^n \left[\sum_{l=i}^n \binom{l}{i} \lambda_0^{l-i}\Psi_l(\lambda_0) \right] (\lambda - \lambda_0)^i \\
&= \sum_{l=1}^n \left[(\lambda - \lambda_0 + \lambda_0)^l - \lambda_0^l \right] \Psi_l(\lambda_0) \tag{1.4} \\
&= \sum_{l=1}^n (\lambda^l - \lambda_0^l) \Psi_l(\lambda_0).
\end{aligned}
$$

Then

$$\|K(\lambda_0; \lambda)\| \leq \frac{1}{2}, \quad \text{for } \mathrm{Re}\lambda_0 > \sigma, \ |\lambda - \lambda_0| < r(\mathrm{Re}\lambda_0). \tag{1.5}$$

Take $\varepsilon \in (0, 1)$. Referring to the finite covering principle, we get that there exists a sequence $\{\tau_j\}_{j=1}^m$ with $\sigma + \varepsilon \leq \tau_1 < \tau_2 < \cdots < \tau_m \leq \omega + 1$ such that

$$[\sigma + \varepsilon, \ \omega + 1] \subset \bigcup_{j=1}^m (\tau_j - r(\tau_j), \ \tau_j + r(\tau_j)).$$

Noting that for $\lambda, \nu \in \rho_0(A_0, \cdots, A_{n-1})$,

$$
\begin{aligned}
R_\lambda &= R_\nu + R_\lambda(P_\nu - P_\lambda)R_\nu \\[2mm]
&= R_\nu + \left(\sum_{l=1}^{n}(\nu^l - \lambda^l)R_\lambda A_l\right) R_\nu \\[2mm]
&= R_\nu + K(\lambda;\ \nu)R_\nu,
\end{aligned}
$$

we obtain

$$
\Psi_k(\lambda) = \Psi_k(\nu) + K(\lambda;\ \nu)\Psi_k(\nu), \quad 1 \le k \le n,\ \lambda,\ \nu \in \mathcal{H}_\sigma, \tag{1.6}
$$

due to the fact that $\Psi_k(\cdot)$ is analytic in $\mathcal{H}_\sigma$, and for each fixed λ (resp. ν) $\in \mathcal{H}_\sigma$, the function $\nu \mapsto K(\lambda;\ \nu)$ (resp. $\lambda \mapsto K(\lambda;\ \nu)$) is analytic in $\mathcal{H}_\sigma$. Set

$$
F_k(\lambda) = \lambda^{k-1}\Psi_k(\lambda), \quad 1 \le k \le n,\ \lambda \in \mathcal{H}_\sigma.
$$

We obtain by (1.5) and (1.6) that for each $\tau \in (\tau_j - r(\tau_j),\ \tau_j + r(\tau_j))$ $(1 \le j \le m)$, $\mu \in R$,

$$
\begin{aligned}
F_k(\tau + i\mu) &= (\tau_j + i\mu)^{1-k}(\tau - \tau_j + \tau_j + i\mu)^{k-1} \\[4mm]
&\qquad \cdot [I + K(\tau_j + i\mu;\ \tau + i\mu)]^{-1}F_k(\tau_j + i\mu) \\[4mm]
&= \left\{\sum_{l=0}^{k-1}\binom{k-1}{l}(\tau_j + i\mu)^{-l}(\tau - \tau_j)^l\right\} \\[4mm]
&\qquad \cdot [I + K(\tau_j + i\mu;\ \tau + i\mu)]^{-1}F_k(\tau_j + i\mu) \\[4mm]
&\qquad (1 \le k \le n,\ |\mu| > 1).
\end{aligned}
$$

Hence, there exists a constant $C_0 > 0$ such that for $1 \le k \le n$, $\tau \in [\sigma + \varepsilon,\ \omega + 1]$, $\mu \in R$ with $|\mu| > 1$,

$$
\|F_k(\tau + i\mu)u\| \le C_0\|F_k(\omega + 1 + i\mu)u\|, \quad u \in H, \tag{1.7}
$$

$$
\|F_k^*(\tau + i\mu)\| \le C_0. \tag{1.8}
$$

On the other hand, letting $\lambda = \omega + 1 + i\mu$ and $\nu = \tau + i\mu$ in (1.6) yields

$$
\begin{aligned}
&F_k^*(\tau + i\mu) \\[4mm]
&= \left\{\sum_{l=0}^{k-1}\binom{k-1}{l}(\omega + 1 - i\mu)^{-l}(\tau - \omega - 1)^l\right\} F_k^*(\omega + 1 + i\mu) \\[4mm]
&\qquad - F_k^*(\tau + i\mu)K^*(\omega + 1 + i\mu;\ \tau + i\mu),
\end{aligned}
$$

for τ, μ and k as above. This together with (1.4) and (1.8) implies the existence of a constant $C_1 > 0$ such that for any $1 \le k \le n$, $\tau \in [\sigma + \varepsilon,\ \omega + 1]$, $\mu \in R$ with $|\mu| > 1$,

$$\|F_k^*(\tau + i\mu)u\| \le C_1 \sum_{i=1}^{n} \|F_i^*(\omega + 1 + i\mu)u\|. \tag{1.9}$$

But using (1.1) – (1.3) and referring to Lemmas 1.2 and 1.3, we find that for any $1 \le k \le n$, $u \in H$,

$$\lim_{|\mu| \to \infty} \|F_k(\omega + 1 + i\mu)u\| = 0, \tag{1.10}$$

$$\|F_k(\omega + 1 + i\mu)u\|,\quad \|F_k^*(\omega + 1 + i\mu)u\| \in L^2(R) \tag{1.11}$$

(each as a function of μ),

noting that for every $u \in H$ and $0 \le k \le n - 1$,

$$t \longmapsto \left(S_k^{(k)}(t)\right)^* u$$

is weakly continuous and therefore strongly measurable. Thus, (1.7) and (1.10) together show that, for any $1 \le k \le n$, $u \in H$,

$$\lim_{|\mu| \to \infty} \|F_k(\tau + i\mu)u\| = 0, \tag{1.12}$$

uniformly for $\tau \in [\sigma + \varepsilon,\ \omega + 1]$; also, (1.7) and (1.9) combined with (1.11) give that for any $1 \le k \le n$, $u \in H$,

$$\|F_k(\sigma + \varepsilon + i\mu)u\|,\quad \|F_k^*(\sigma + \varepsilon + i\mu)u\| \in L^2(R) \tag{1.13}$$

(each as a function of μ).

It follows from (1.1) – (1.3) and Theorem 1.1.8 that for any $1 \le k \le n - 1$,

$$S_{n-1}^{(n-1)}(t)u = \frac{1}{2\pi i} \int_{\omega+1-i\infty}^{\omega+1+i\infty} e^{\lambda t} F_n(\lambda)u\,d\lambda,\quad u \in \bigcap_{i=0}^{n-1} \mathcal{D}(A_i),\ t > 0,$$

$$S_{k-1}^{(k-1)}(t)u - S_k^{(k)}(t)u$$

$$= \frac{1}{2\pi i} \int_{\omega+1-i\infty}^{\omega+1+i\infty} e^{\lambda t} F_k(\lambda)u\,d\lambda,\quad u \in \bigcap_{i=0}^{n-1} \mathcal{D}(A_i),\ t > 0.$$

Shifting the path of integrals and using (1.12) gives that for any $1 \le k \le n - 1$,

$$S_{n-1}^{(n-1)}(t)u = \frac{1}{2\pi i} \int_{\sigma+\varepsilon-i\infty}^{\sigma+\varepsilon+i\infty} e^{\lambda t} F_n(\lambda)u\,d\lambda,\quad u \in \bigcap_{i=0}^{n-1} \mathcal{D}(A_i),\ t > 0, \tag{1.14}$$

$$S_{k-1}^{(k-1)}(t)u - S_k^{(k)}(t)u$$

$$= \frac{1}{2\pi i} \int_{\sigma+\epsilon-i\infty}^{\sigma+\epsilon+i\infty} e^{\lambda t} F_k(\lambda)u \, d\lambda, \quad u \in \bigcap_{i=0}^{n-1} \mathcal{D}(A_i), \ t > 0. \tag{1.15}$$

For any $u \in H$, $t > 0$, $m \in N$, $1 \leq k \leq n$, let

$$\tilde{S}_k(t;\, m)u = \frac{1}{2\pi i} \int_{\sigma+\epsilon-im}^{\sigma+\epsilon+im} e^{\lambda t} F_k(\lambda)u \, d\lambda.$$

Integrating by parts and using (1.12), we obtain that for $u,\, v \in H$, $t \geq 1$, $m \in N$, $1 \leq k \leq n$,

$$\left| \left\langle \tilde{S}_k(t;\, m)u,\, v \right\rangle \right|$$

$$\leq \frac{1}{2\pi} \left\| \frac{1}{t} e^{\lambda t} F_k(\lambda) \Big|_{\sigma+\epsilon-im}^{\sigma+\epsilon+im} \right\| \|u\| \|v\| + \frac{1}{2\pi} \left| \left\langle \frac{1}{t} \int_{\sigma+\epsilon-im}^{\sigma+\epsilon+im} e^{\lambda t} F_k'(\lambda)u \, d\lambda,\, v \right\rangle \right|$$

$$\leq \text{const } e^{(\sigma+\epsilon)t} \|u\| \|v\| + \frac{1}{2\pi t} \left| \int_{\sigma+\epsilon-im}^{\sigma+\epsilon+im} e^{\lambda t} \left\{ (k-1) \left\langle F_k(\lambda)u,\, (\bar{\lambda})^{-1}v \right\rangle \right. \right.$$

$$\left. \left. - \sum_{j=1}^{n} j \left\langle F_k(\lambda)u,\, F_j^*(\lambda)v \right\rangle \right\} d\lambda \right|,$$

noting the easily verified equalities

$$\frac{d}{d\lambda} \left(R_\lambda A_k \right) = - \left(\sum_{j=1}^{n} j\lambda^{j-1} \overline{R_\lambda A_j} \right) \overline{R_\lambda A_k}, \tag{1.16}$$

$$1 \leq k \leq n, \ \lambda \in \rho_0(A_0,\, \cdots,\, A_{n-1}).$$

Accordingly, making use of (1.13) yields that for $u,\, v,\, t,\, m,\, k$ as above,

$$\left| \left\langle \tilde{S}_k(t;\, m)u,\, v \right\rangle \right| \leq C_1(u,\, v) e^{(\sigma+\epsilon)t},$$

with some constant $C_1(u,\, v)$ independent of $m,\, t$. Hence,

$$\|\tilde{S}_k(t;\, m)\| \leq C_2 e^{(\sigma+\epsilon)t}, \quad t \geq 1, \ m \in N, \ 1 \leq k \leq n, \tag{1.17}$$

for some constant C_2. (1.14) and (1.15) tell us that

$$\lim_{m \to \infty} \tilde{S}_n(t;\, m)u = S_{n-1}^{(n-1)}(t)u, \quad u \in \bigcap_{i=0}^{n-1} \mathcal{D}(A_i), \ t \geq 1,$$

$$\lim_{m \to \infty} \widetilde{S}_k(t; \, m)u = S_{k-1}^{(k-1)}(t)u - S_k^{(k)}(t)u,$$

$$u \in \bigcap_{i=0}^{n-1} \mathcal{D}(A_i), \ t \geq 1, \ 1 \leq k \leq n - 1.$$

Thus, in view of (1.17) and the density of $\bigcap_{i=0}^{n-1} \mathcal{D}(A_i)$, we obtain that there exists a constant C such that

$$\left\| S_k^{(k)}(t) \right\| \leq C e^{(\sigma + \varepsilon)t}, \qquad t \geq 0, \ 0 \leq k \leq n - 1,$$

that is $\omega_0 \leq \sigma + \varepsilon$. The arbitrariness of ε implies $\omega_0 \leq \sigma$.

Moreover, by Theorem 2.2.2

$$S_k(t)u = \frac{1}{2\pi i} \int_{\omega + 1 - i\infty}^{\omega + 1 + i\infty} e^{\lambda t} \sum_{i=k+1}^{n} \lambda^{i-k-1} \overline{R_\lambda A_i} u d\lambda,$$

$$u \in H, \ t > 0, \ 0 \leq k \leq n - 1.$$

We have

$$\|S_k(t)\| \leq M_1 e^{(\sigma + \varepsilon)t}, \qquad t \geq 0, \ M_1 \text{ is a constant},$$

by arguing similarly as above.

Now, we prove $\sigma \leq \omega_0$. For any $\varepsilon > 0$, by the definition of ω_0, there is a constant C such that for each $0 \leq k \leq n - 1$,

$$\left\| S_k^{(k)}(t) \right\| \leq C e^{(\omega_0 + \varepsilon)t}, \qquad t \geq 0.$$

Accordingly, the integral on the right-hand side of (1.2) or (1.3) exists and it is an analytic function of λ for $\mathrm{Re}\lambda > \omega_0 + \varepsilon$. Thus, for any $1 \leq k \leq n$,

$$F_k(\lambda) := \lambda^{k-1} \overline{R_\lambda A_k}$$

can be extended analytically to $\mathcal{H}_{\omega_0 + \varepsilon}$, and

$$\|F_k(\lambda)\| \leq 2C(\mathrm{Re}\lambda - \omega_0 - \varepsilon)^{-1} \qquad (\lambda \in \mathcal{H}_{\omega_0 + \varepsilon}). \tag{1.18}$$

Clearly, for any $1 \leq k \leq n$, $\overline{R_\lambda A_k}$ can be extended analytically to $\mathcal{H}_{\omega_0 + \varepsilon} \setminus \{0\}$. Write

$$\Psi_k(\lambda) = \lambda^{1-k} F_k(\lambda), \qquad 1 \leq k \leq n, \ \lambda \in \mathcal{H}_{\omega_0 + \varepsilon} \setminus \{0\}.$$

In the case of $\omega_0 + \varepsilon < 0$, it remains to show that for any $1 \leq k \leq n$, $\|\Psi_k(\cdot)\|$ is bounded on a neighbourhood of zero. To this end, we observe by (1.18) that

$$\|F_k(\lambda)\| \leq \text{const} \qquad (1 \leq k \leq n, \ \mathrm{Re}\lambda > 0).$$

Therefore, there exists $\delta > 0$ such that for each $\tau \in [-\delta, \, \delta]$ and $\mu \in R$,

$$\|K(\delta + i\mu; \, \tau + i\mu)\| < \frac{1}{2},$$

noting

$$K(\delta + i\mu; \ \tau + i\mu) = \sum_{l=1}^{n} \left[(\tau - \delta) \sum_{i=0}^{l-1} \left((\delta + i\mu)^{-1}(\tau + i\mu) \right)^i \right] F_l(\delta + i\mu)$$

and

$$\left\| (\delta + i\mu)^{-1}(\tau + i\mu) \right\| \le 1.$$

Arguing similarly as in the proof of (1.6), we obtain that for each $1 \le k \le n$, $\lambda \neq 0$, $|\lambda| \le \delta$,

$$\Psi_k(\lambda) = [1 + K(\delta + i \ \mathrm{Im}\lambda; \ \lambda)]^{-1} \Psi_k(\delta + i \ \mathrm{Im}\lambda)$$

and so $\|\Psi_k(\lambda)\| \le$ const. Consequently $\sigma \le \omega_0 + \varepsilon$, and therefore $\sigma \le \omega_0$ by virtue of the arbitrariness of ε.

To sum up, we see that the conclusions of Theorem 1.4 hold. The proof is then complete.

As an immediate consequence, we have

Corollary 1.5. $\omega_0 \le \inf \Big\{ \tau_0 \in R; \ \text{for each } \tau > \tau_0, \ \tau + iR \subset \rho_0(A_0, \ \cdots, \ A_{n-1})$

and

$$\sup_{\mu \in R} \left\| (\tau + i\mu)^{k-1} \overline{R_{\tau + i\mu} A_k} \right\| < \infty, \ 1 \le k \le n \Big\}.$$

A function $f : R^+ \to E$ (or $L(E)$), for a Banach space E, is called exponentially stable if there exist constants M, $\sigma > 0$ such that

$$\|f(t)\| \le M e^{-\sigma t}, \quad t \ge 0.$$

Corollary 1.6. *Assume that for each λ with $\mathrm{Re}\lambda \ge 0$, $\lambda \in \rho_0(A_0, \ \cdots, \ A_{n-1})$, and for each $1 \le k \le n$, $\tau \ge 0$,*

$$\sup \left\{ \left\| \lambda^{k-1} \overline{R_\lambda A_k} \right\|; \ \mathrm{Re}\lambda = \tau \right\} < +\infty.$$

Then, $S_k^{(k)}(\cdot)$, $S_k(\cdot)$ $(0 \le k \le n-1)$ are exponentially stable.

Proof. Arguing similarly as in the proof of (1.5), we obtain that there is $\tau_0 > 0$ such that

$$\|K(\lambda_0; \ \lambda)\| < \frac{1}{2}, \quad \text{for all} \quad \mathrm{Re}\lambda_0 = 0, \ |\lambda - \lambda_0| \le \tau_0.$$

Hence, for any $\tau \in [-\tau_0, \ \tau_0]$, $\mu \in R$,

$$R_{\tau + i\mu} = [I + K(i\mu; \ \tau + i\mu)]^{-1} R_{i\mu}.$$

We thus see that for any $1 \leq k \leq n$, $\overline{R_\lambda A_k}$ can be extended analytically to $\mathcal{H}_{-\frac{r_0}{2}}$, and for any $1 \leq k \leq n$, $\tau > -r_0$,

$$\sup \left\{ \left\| \lambda^{k-1} \Psi_k(\lambda) \right\|; \ \mathrm{Re}\lambda = \tau \right\} < +\infty,$$

where $\Psi_k(\cdot)$ denotes the extension of $\overline{R_\lambda A_k}$. According to Theorem 1.4, $\omega_0 \leq -r_0 < 0$. This ends the proof.

Example 1.7. Let A_0 and A_1 be self-adjoint and strictly positive operators in H such that $\mathcal{D}(A_0) \subset \mathcal{D}(A_1)$ and

$$\langle A_0 u, \ A_1 u \rangle \geq 0, \quad u \in \mathcal{D}(A_0). \tag{1.19}$$

Then (ACP_2) is strongly wellposed and each of $S_0(t)$, $S_1(t)$ and $S_1'(t)$ is exponentially stable.

Proof. Fix $\lambda \in \mathbf{C}$ with $\mathrm{Re}\lambda \geq 0$. From (1.19) we obtain

$$\langle A_1 A_0^{-1} u, \ u \rangle \geq 0, \quad u \in H, \tag{1.20}$$

which indicates that $-\lambda A_1 A_0^{-1}$ is dissipative. Notice $(0, \ \infty) \bigcap \rho(-\lambda A_1 A_0^{-1}) \neq \emptyset$ due to $A_1 A_0^{-1} \in \mathbf{L}(H)$. We infer that $I + \lambda A_1 A_0^{-1}$ is boundedly invertible by the Lumer-Phillips theorem. It follows that

$$\lambda A_1 + A_0 = (I + \lambda A_1 A_0^{-1}) A_0 \text{ is boundedly invertible.} \tag{1.21}$$

This implies

$$(\lambda A_1 + A_0)^* = \overline{\lambda} A_1 + A_0. \tag{1.22}$$

Noting

$$A_0^{-1} A_1 \subset \left(A_1 A_0^{-1} \right)^* \in \mathbf{L}(H),$$

we deduce that $A_0^{-1} A_1$ is closable and $\overline{A_0^{-1} A_1} \in \mathbf{L}(H)$. Evidently, there exists $\eta_0 > 0$ such that for any $\lambda \in \mathbf{C}$ with $|\lambda| \leq \eta_0$

$$\left\| \lambda^2 A_0^{-1} + \lambda A_1 A_0^{-1} \right\|, \ \left\| \lambda^2 A_0^{-1} + \lambda \overline{A_0^{-1} A_1} \right\| < \frac{1}{2}.$$

Therefore

$$P_\lambda = \left[\lambda^2 A_0^{-1} + \lambda A_1 A_0^{-1} + I \right] A_0$$

is boundedly invertible and

$$\| R_\lambda \| \leq \text{const}, \quad |\lambda| \leq \eta_0; \tag{1.23}$$

$$\left\| \overline{R_\lambda A_1} \right\| \ \leq \ \left\| \left[\lambda^2 A_0^{-1} + \lambda \overline{A_0^{-1} A_1} + I \right]^{-1} \right\| \left\| \overline{A_0^{-1} A_1} \right\| \tag{1.24}$$

$$\leq \ \text{const}, \quad |\lambda| \leq \eta_0.$$

Recalling that A_0 and A_1 are strictly positive, we have

$$\langle A_0 u, \ u \rangle \geq \alpha \|u\|^2, \quad u \in \mathcal{D}(A_0),$$

$$\langle A_1 u, \ u \rangle \geq \alpha \|u\|^2, \quad u \in \mathcal{D}(A_1),$$

for some $\alpha > 0$. So

$$\begin{aligned} \|P_\lambda u\| \|\lambda u\| &\geq |\langle P_\lambda u, \ \lambda u \rangle| \\ &\geq |\lambda|^2 (\mathrm{Re}\lambda) \|u\|^2 + \alpha |\lambda|^2 \|u\|^2 + \alpha (\mathrm{Re}\lambda) \|u\|^2 \\ &\geq \alpha |\lambda|^2 \|u\|^2, \quad u \in \mathcal{D}(A_0), \ \mathrm{Re}\lambda \geq 0. \end{aligned}$$

Thus, we obtain

$$\|P_\lambda u\| \geq \alpha |\lambda| \|u\|, \quad u \in \mathcal{D}(A_0), \ \mathrm{Re}\lambda \geq 0. \tag{1.25}$$

Noting $P_\lambda^* = P_{\overline{\lambda}}$ by (1.22), we get also

$$\|P_\lambda^* u\| \geq \alpha |\lambda| \|u\|, \quad u \in \mathcal{D}(A_0), \ \mathrm{Re}\lambda \geq 0. \tag{1.26}$$

Thus, combining (1.25) and (1.26) we deduce that for $\mathrm{Re}\lambda \geq 0$, P_λ is boundedly invertible and

$$\|\lambda R_\lambda\| \leq \mathrm{const}, \quad \mathrm{Re}\lambda \geq 0, \ |\lambda| > \eta_0. \tag{1.27}$$

Moreover, we observe by (1.19) that for $\mathrm{Re}\lambda \geq 0$, $u \in H$,

$$\begin{aligned} |\lambda| \, \|A_1^{-1} P_\lambda u\| \, \|u\| &\geq |\langle A_1^{-1} P_\lambda u, \ \lambda u \rangle| \\ &\geq |\lambda|^2 (\mathrm{Re}\lambda) \langle A_1^{-1} u, \ u \rangle \\ &\qquad + |\lambda|^2 \langle u, \ u \rangle + (\mathrm{Re}\lambda) \langle A_1^{-1} A_0 u, \ u \rangle \\ &\geq |\lambda|^2 \|u\|^2. \end{aligned}$$

Hence,

$$\left\| \overline{R_\lambda A_1} \right\| \leq \frac{1}{|\lambda|}, \quad \mathrm{Re}\lambda \geq 0, \ |\lambda| > \eta_0. \tag{1.28}$$

Next, let us look at the operator matrix

$$\mathcal{A} = \begin{pmatrix} 0 & I \\ -A_0 & -A_1 \end{pmatrix}, \quad \mathcal{D}(\mathcal{A}) := \mathcal{D}(A_0) \times \mathcal{D}(A_0).$$

It is easy to verify that for $\lambda > 0$, $\lambda - \mathcal{A}$ is injective and

$$(\lambda - \mathcal{A})^{-1} = \begin{pmatrix} \lambda^{-1}(I - R_\lambda A_0) & R_\lambda \\ -R_\lambda A_0 & \lambda R_\lambda \end{pmatrix}, \tag{1.29}$$

$$\mathcal{D}\left((\lambda - \mathcal{A})^{-1}\right) := \mathcal{D}(A_0) \times H.$$

Define H_1, H_2 by

$$H_1 := \left(\mathcal{D}\left(A_0^{\frac{1}{2}} \right), \langle \cdot, \cdot \rangle_1 \right), \quad \langle u, v \rangle_1 := \left\langle A_0^{\frac{1}{2}} u, A_0^{\frac{1}{2}} v \right\rangle,$$

$$H_2 := \text{the completion of } (H, \langle \cdot, \cdot \rangle_2),$$

$$\langle u, v \rangle_2 := \left\langle A_0^{-\frac{1}{2}} u, A_0^{-\frac{1}{2}} v \right\rangle.$$

Then we have that in the product space $H_1 \times H$,

$$\text{Re} \left\langle \mathcal{A} \begin{pmatrix} u \\ v \end{pmatrix}, \begin{pmatrix} u \\ v \end{pmatrix} \right\rangle_{H_1 \times H} = \text{Re} \left\langle \begin{pmatrix} v \\ -A_0 u - A_1 v \end{pmatrix}, \begin{pmatrix} u \\ v \end{pmatrix} \right\rangle_{H_1 \times H}$$

$$= \text{Re} \left\langle A_0^{\frac{1}{2}} v, A_0^{\frac{1}{2}} u \right\rangle - \text{Re} \left\langle A_0 u, v \right\rangle - \langle A_1 v, v \rangle$$

$$= -\langle A_1 v, v \rangle \leq 0, \quad u, v \in \mathcal{D}(A_0).$$

Thus an application of the Lumer-Phillips theorem shows that $\overline{\mathcal{A}}$ generates a strongly continuous semigroup of contractions on $H_1 \times H$, since $\mathcal{A}$ has dense range. As a consequence, using (1.29) gives

$$\lambda \longmapsto \lambda R_\lambda \in LT - \mathbf{L}(H). \tag{1.30}$$

On the other hand, we get that in the product space $H \times H_2$,

$$\text{Re} \left\langle \mathcal{A} \begin{pmatrix} u \\ v \end{pmatrix}, \begin{pmatrix} u \\ v \end{pmatrix} \right\rangle_{H \times H_2}$$

$$= \text{Re} \langle v, u \rangle - \text{Re} \left\langle A_0^{\frac{1}{2}} u, A_0^{-\frac{1}{2}} v \right\rangle - \left\langle A_0^{-\frac{1}{2}} A_1 v, A_0^{-\frac{1}{2}} v \right\rangle$$

$$= -\langle A_1 A_0^{-1} v, v \rangle \leq 0, \quad u, v \in \mathcal{D}(A_0), \quad \text{by (1.20)}.$$

Hence $\overline{\mathcal{A}}$ is the generator of a strongly continuous semigroup of contractions on $H \times H_2$, and so is the adjoint $\left(\overline{\mathcal{A}} \right)^*$ in $H \times H_2$. Accordingly, making use of (1.29) and the equality

$$\left[\lambda^{-1} \left(I - \overline{R_\lambda A_0} \right) \right]^* = \lambda^{-1}(I - A_0 R_\lambda), \quad \lambda > 0,$$

yields that

$$\lambda \longmapsto \lambda^{-1} \overline{R_\lambda A_0}, \quad \lambda \longmapsto \lambda^{-1} A_0 R_\lambda \in LT - \mathbf{L}(H).$$

From this and (1.30), we see by means of Theorem 2.2.3 that (ACP_2) is strongly wellposed.

Now, combining (1.23), (1.24), (1.27) and (1.28) together enables us to apply Corollary 1.6 and obtain the required result. The proof is complete.

5.2   Exponential stability of solutions

In this section, E is a Banach space.

First, we show a general theorem concerning the growth estimate of the determining function of a Laplace transform.

Theorem 2.1. *Given an E-valued function $f(t)$ defined in R^+ with $f(t) = O(e^{\tau t})$ $(t \to +\infty)$ for some $\tau > 0$, let*

$$F(\lambda) := \mathcal{L}[f(t)](\lambda) \quad (\operatorname{Re}\lambda > \tau). \tag{2.1}$$

Assume that $F(\lambda)$ admits an analytic continuation $\widetilde{F}(\lambda)$ to the half plane $\mathcal{H}_\sigma$ for some $\sigma < \tau$, satisfying

$$\left\|\widetilde{F}(\lambda)\right\| \le \text{const } (1 + |\lambda|)^{-1} \quad (\sigma < \operatorname{Re}\lambda \le \omega) \tag{2.2}$$

for some $\omega > \tau$; moreover, on the line: $\omega + iR$, $F(\lambda)$ can be decomposed as

$$F(\omega + i\mu) = \sum_{p=1}^{m} a_p(\mu) G_p(\omega + i\mu), \quad \mu \in R, \tag{2.3}$$

where, $\{a_p(\cdot)\}_{p=1}^{m}$ is a family of complex functions, and $\{G_p(\cdot)\}_{p=1}^{m}$ is a family of E-valued functions, satisfying

$$|a_p(\mu)| \le \text{const } (1 + |\mu|)^{-1} \quad (\mu \in R), \tag{2.4}$$

$$G_p(\lambda) = \mathcal{L}[g_p(t)](\lambda) \quad (\operatorname{Re}\lambda > \tau), \tag{2.5}$$

for some $g_p(t) = O(e^{\tau t})$ $(t \to +\infty)$. Then

$$\|f(t)\| \le \text{const } (1 + t)e^{\sigma t}, \quad t \ge 0.$$

Proof. Take $\sigma_0 \in (\sigma, \tau)$. Let $u^* \in E^*$. It is clear that

$$h_p(t) := \begin{cases} \langle u^*, \ e^{-\omega t} g_p(t) \rangle, & \text{if } t \ge 0 \\ \\ 0, & \text{if } t < 0 \end{cases}$$

$\in L^2(R, \ \mathbf{C})$. Hence, by Plancherel's theorem, we obtain

$$\mu \ \longmapsto \ \langle u^*, \ G_p(\omega + i\mu) \rangle$$

$$= \left\langle u^*, \ \int_0^\infty e^{-i\mu t} \left(e^{-\omega t} g_p(t) \right) dt \right\rangle$$

$$\in L^2(R, \ \mathbf{C}),$$

and

$$\left(\frac{1}{2\pi} \int_{-\infty}^{\infty} |\langle u^*, \; G_p(\omega + i\mu)\rangle|^2 \, d\mu\right)^{\frac{1}{2}}$$

$$= \left(\int_0^{\infty} |\langle u^*, \; e^{-\omega t} g_p(t)\rangle|^2 \, dt\right)^{\frac{1}{2}} \tag{2.6}$$

$$\le \quad \text{const } \|u^*\|.$$

Therefore, using (2.3), (2.4), (2.6) and Hölder's inequality, we have

$$\mu \longmapsto \langle u^*, \; F(\omega + i\mu)\rangle \in L^1(R, \; C).$$

Moreover, it follows from (2.1) that

$$\int_0^{\infty} e^{-i\mu t} \langle u^*, \; e^{-\omega t} f(t)\rangle \, dt = \langle u^*, \; F(\omega + i\mu)\rangle, \quad \mu \in R.$$

Thus, by virtue of the inversion formula of Fourier transforms, we obtain

$$\frac{1}{2\pi i} \int_{\omega - i\infty}^{\omega + i\infty} e^{\lambda t} \langle u^*, \; F(\lambda)\rangle \, d\lambda = \frac{e^{\omega t}}{2\pi} \int_{-\infty}^{\infty} e^{it\mu} \langle u^*, \; F(\omega + i\mu)\rangle \, d\mu \tag{2.7}$$

$$= \langle u^*, \; f(t)\rangle, \quad t \ge 0.$$

This combined with Cauchy's formula yields that for each $t \ge 0$,

$$\langle u^*, \; f(t)\rangle = \frac{1}{2\pi i} \int_{\Gamma} e^{\lambda t} \langle u^*, \; F(\lambda)\rangle \, d\lambda,$$

where $\Gamma = \Gamma_1 \bigcup \Gamma_2 \bigcup \Gamma_3 \bigcup \Gamma_4 \bigcup \Gamma_5$, and

$$\Gamma_1 = \{\lambda \in C; \; \text{Re}\lambda = \omega, \; \text{Im}\lambda \le -b\},$$

$$\Gamma_2 = \{\lambda \in C; \; \sigma_0 \le \text{Re}\lambda \le \omega, \; \text{Im}\lambda = -b\},$$

$$\Gamma_3 = \{\lambda \in C; \; \text{Re}\lambda = \sigma_0, \; -b \le \text{Im}\lambda \le b\},$$

$$\Gamma_4 = \{\lambda \in C; \; \sigma_0 \le \text{Re}\lambda \le \omega, \; \text{Im}\lambda = b\},$$

$$\Gamma_5 = \{\lambda \in C; \; \text{Re}\lambda = \omega, \; \text{Im}\lambda \ge b\},$$

for any given $b > 0$. By (2.2), we get that for $t \ge 0$,

$$\left|\int_{\Gamma_2} e^{\lambda t} \langle u^*, \; F(\lambda)\rangle \, d\lambda\right|, \quad \left|\int_{\Gamma_4} e^{\lambda t} \langle u^*, \; F(\lambda)\rangle \, d\lambda\right|$$

$$\le \quad \text{const } e^{\omega t} b^{-1} \|u^*\|,$$

$$\left| \int_{\Gamma_s} e^{\lambda t} \langle u^*, \, F(\lambda) \rangle \, d\lambda \right| \quad \leq \quad \text{const } e^{\sigma_0 t} \|u^*\| \int_{-b}^{b} (1 + |\eta|)^{-1} d\eta$$

$$\leq \quad \text{const } e^{\sigma_0 t} \ln(1+b) \|u^*\|$$

$$\longrightarrow \quad \text{const } e^{\sigma t} \ln(1+b) \|u^*\|, \quad \text{as } \sigma_0 \to \sigma.$$

Moreover, by (2.3), (2.4), (2.6) and Hölder's inequality, we have that for $t \geq 0$,

$$\left| \int_{\Gamma_1} e^{\lambda t} \langle u^*, \, F(\lambda) \rangle \, d\lambda \right|, \quad \left| \int_{\Gamma_s} e^{\lambda t} \langle u^*, \, F(\lambda) \rangle \, d\lambda \right|$$

$$\leq \quad \text{const } e^{\omega t} \|u^*\| \left\{ \int_b^\infty \frac{1}{\mu^2} d\mu \right\}^{\frac{1}{2}}$$

$$\leq \quad \text{const } b^{-\frac{1}{2}} e^{\omega t} \|u^*\|.$$

Hence, taking $b = e^{4t \max\{\omega, |\sigma|\}}$ gives

$$| \langle u^*, \, f(t) \rangle |$$

$$\leq \quad \text{const } \|u^*\| \left(e^{-3t \max\{\omega, |\sigma|\}} + e^{-t \max\{\omega, |\sigma|\}} + e^{\sigma t}(1+t) \right), \quad t \geq 0.$$

Consequently,

$$\|f(t)\| \leq \text{const } (1+t) e^{\sigma t}, \quad t \geq 0.$$

The proof is then complete.

The following is a general criterion for existence and uniqueness of solutions of (ACP_n), from Sections 2.2 and 2.3 with a slight generalization(cf. Proof of Lemma 2.4.4).

Theorem 2.2. *Assume that each A_j $(0 \leq j \leq n-1)$ is closable in E with*

$$\bigcap_{j=0}^{n-1} \mathcal{D}\left(\overline{A}_j\right) = \bigcap_{j=0}^{n-1} \mathcal{D}(A_j),$$

and there exists $\omega_0 > 0$ such that R_λ exists and $\mathcal{D}(R_\lambda) = E$ for $\lambda > \omega_0$ with $R_\lambda u = O\left(e^{\omega_0 \lambda}\right) (\lambda \to +\infty)$ for each $u \in E$. Let $u_k \in \bigcap_{p=0}^{n-1} \mathcal{D}(A_p) (0 \leq k \leq n-1)$. If there exist operators $\tilde{A}_j$ $(0 \leq j \leq n-1)$ in E, with $\mathcal{D}\left(\tilde{A}_j\right) \subset \mathcal{D}(A_j)$, such that A_j is $\tilde{A}_j$-bounded and

$$\lambda \longmapsto \lambda^j \tilde{A}_j R_\lambda u_{n-1}, \quad \lambda \longmapsto \lambda^j \tilde{A}_j \sum_{p=0}^{k} \lambda^{p-k-1} R_\lambda u_k \in LT - E$$

for $0 \leq j \leq n-1$ and $0 \leq k \leq n-2$, then (ACP_n) has a unique solution $u(t)$ with initial data $u^{(k)}(0) = u_k$, $0 \leq k \leq n-1$, satisfying that for each $0 \leq j \leq n-1$, $u^{(j)}(t) = O\left(e^{\omega_1 t}\right)$ $(t \to +\infty)$ for some $\omega_1 > 0$ and

$$\mathcal{L}\left[u^{(j)}(t)\right](\lambda)$$

$$= \sum_{k=j}^{n-1} \lambda^{j-k-1} u_k - \sum_{k=0}^{n-1} \sum_{p=0}^{k} \lambda^{j+p-k-1} R_\lambda A_p u_k, \quad \mathrm{Re}\lambda > \omega_1. \tag{2.8}$$

Theorem 2.3. *Let the hypothesis of Theorem 2.2 hold. Fix $\delta > 0$ and $l \in \{0, 1, \cdots, n-1\}$. If R_λ exists, belongs to $\mathbf{L}(E)$ and is analytic in $\{\lambda \in \mathbf{C}; \mathrm{Re}\lambda > -\delta\}$ with $\lambda^l R_\lambda \in LT - \mathbf{L}(E)$ and uniformly bounded in $\{\lambda \in \mathbf{C}; \mathrm{Re}\lambda > -\delta\}$, then the solution $u(t)$ satisfies*

$$\left\| u^{(j)}(t) \right\| \leq \text{const } (1+t)e^{-\delta t} \qquad (t \geq 0, \ 0 \leq j \leq l). \tag{2.9}$$

Proof. Set, for $\mathrm{Re}\lambda > -\delta$,

$$U_{1j}(\lambda) := -\sum_{k=0}^{j-1} \sum_{p=0}^{k} \lambda^{j+p-k-1} R_\lambda A_p u_k, \quad 1 \leq j \leq n-1, \tag{2.10}$$

$$U_{2j}(\lambda) := \sum_{k=j}^{n-1} \sum_{p=k+1}^{n} \lambda^{j+p-k-1} R_\lambda A_p u_k, \quad 0 \leq j \leq n-1, \tag{2.11}$$

$$U_j(\lambda) := \begin{cases} U_{1j}(\lambda) + U_{2j}(\lambda), & 1 \leq j \leq n-1, \\[2mm] U_{2j}(\lambda), & j = 0. \end{cases} \tag{2.12}$$

Then by (2.8) and the identity

$$R_\lambda \sum_{p=k+1}^{n} \lambda^p A_p = I - R_\lambda \sum_{p=0}^{k} \lambda^p A_p, \quad \mathrm{Re}\lambda > -\delta, \tag{2.13}$$

we get

$$\mathcal{L}\left[u^{(j)}(t)\right](\lambda) = U_j(\lambda), \quad \mathrm{Re}\lambda > \omega_1. \tag{2.14}$$

Now, we take an observation on $U_j(\lambda)$ $(0 \leq j \leq l)$.

Since $j + p - k - 1 \geq 0$ and $R_\lambda(A_p u_k)$ is analytic in $\{\lambda \in \mathbf{C}; \mathrm{Re}\lambda > -\delta\}$, we see that

(i) for each $0 \leq j \leq n-1$, $U_j(\lambda)$ is analytic in $\{\lambda \in \mathbf{C}; \mathrm{Re}\lambda > -\delta\}$;

(ii)

$$\left. \begin{array}{l} \|U_{1j}(\lambda)\| \quad (1 \leq j \leq l) \\[3mm] \|U_{2j}(\lambda)\| \quad (0 \leq j \leq l) \end{array} \right\} \leq \text{const}, \quad |\lambda| \leq \frac{1}{2}\delta. \tag{2.15}$$

In the sequel, we will use (2.15) freely. By hypothesis, $\lambda^l R_\lambda(A_p u_k)$ is bounded in $\{\lambda \in \mathbf{C}; \operatorname{Re}\lambda > -\delta\}$. Therefore, we have that
(a) for any $1 \le j \le l$,

$$\|U_{1j}(\lambda)\| \le \text{const}\,(1+|\lambda|)^{-1}, \qquad \operatorname{Re}\lambda > -\delta, \tag{2.16}$$

noting

$$U_{1j}(\lambda) = -\sum_{k=0}^{j-1}\sum_{p=0}^{k} \lambda^{j+p-k-l-1}\lambda^l R_\lambda A_p u_k \tag{2.17}$$

$$(1 \le j \le l,\ \operatorname{Re}\lambda > -\delta,\ \lambda \ne 0),$$

and $j+p-k \le l$;
(b) for $k=j$,

$$\left\| \sum_{p=j+1}^{n} \lambda^{p-1} R_\lambda A_p u_j - (\lambda+\delta+1)^{-1} u_j \right\| \tag{2.18}$$

$$\le \quad \text{const}\,(1+|\lambda|)^{-1}, \qquad \operatorname{Re}\lambda > -\delta,$$

since by (2.13),

$$\sum_{p=j+1}^{n} \lambda^{p-1} R_\lambda A_p u_j - (\lambda+\delta+1)^{-1} u_j$$

$$= \quad (\lambda+\delta+1)^{-1} \sum_{p=j+1}^{n} \lambda^{p} R_\lambda A_p u_j$$

$$+ (\delta+1)(\lambda+\delta+1)^{-1}\sum_{p=j+1}^{n}\lambda^{p-1}R_\lambda A_p u_j - (\lambda+\delta+1)^{-1}u_j$$

$$= \quad -(\lambda+\delta+1)^{-1}\sum_{p=0}^{j}\lambda^{p-l}\left(\lambda^l R_\lambda A_p u_j\right) + (\delta+1)(\lambda+\delta+1)^{-1}\lambda^{-1}u_j$$

$$-(\delta+1)(\lambda+\delta+1)^{-1}\lambda^{-1}\sum_{p=0}^{j}\lambda^{p-l}\left(\lambda^l R_\lambda A_p u_j\right)$$

$$(\operatorname{Re}\lambda > -\delta,\ \lambda \ne 0),$$

$$\tag{2.19}$$

and $0 \le p \le j \le l$;
(c) for $j+1 \le k \le n-1$,

$$\left\| \sum_{k=j+1}^{n-1}\sum_{p=k+1}^{n} \lambda^{j+p-k-1} R_\lambda A_p u_k \right\| \le \text{const}\,(1+|\lambda|)^{-1}, \qquad \operatorname{Re}\lambda > -\delta, \tag{2.20}$$

since by (2.13),

$$\sum_{k=j+1}^{n-1} \sum_{p=k+1}^{n} \lambda^{j+p-k-1} R_\lambda A_p u_k$$

$$= \sum_{k=j+1}^{n-1} \left[\lambda^{j-k-1} u_k - \lambda^{j-k-1} \sum_{p=0}^{k} \lambda^p R_\lambda A_p u_k \right] \qquad (2.21)$$

$$= \sum_{k=j+1}^{n-1} \left[\lambda^{j-k-1} u_k - \lambda^{-1} \sum_{p=0}^{k} \lambda^{p-k+j-l} \lambda^l R_\lambda A_p u_k \right]$$

$$(\mathrm{Re}\lambda > -\delta, \ \lambda \neq 0),$$

and $j - k - 1 \leq -2$, $p - k + j - l \leq 0$.

Fixing $0 \leq j \leq l$, let

$$f(t) = u^{(j)}(t) - e^{-(\delta+1)t} u_j,$$

$$F(\lambda) = U_j(\lambda) - (\lambda + \delta + 1)^{-1} u_j \quad (\mathrm{Re}\lambda > \omega_1).$$

Then, Theorem 2.2 implies

$$\|f(t)\| = O(e^{\omega_1 t}) \ (t \to +\infty),$$

and by (2.14),

$$F(\lambda) = \mathcal{L}[f(t)](\lambda) \quad (\mathrm{Re}\lambda > \omega_1).$$

From (i), (2.16), (2.18) and (2.20), it follows that $F(\lambda)$ has an analytic extension to the half plane $\{\lambda \in \mathbf{C}; \ \mathrm{Re}\lambda > -\delta\}$ and satisfies (2.2) for any $\mathrm{Re}\lambda > -\delta$. Moreover, making use of the fact that $\lambda \mapsto \lambda^l R_\lambda u_k \in LT - E$ and for any $q \leq 0$,

$$\left| (\omega_1 + i\mu)^{q-1} \right| \leq \mathrm{const} \ (1 + |\mu|)^{-1} \quad (\mu \in R),$$

$$\lambda^{-1} = \mathcal{L}[1](\lambda), \quad \mathrm{Re}\lambda > 0,$$

we find from (2.17), (2.19) and (2.21) that $F(\lambda)$ takes the form as in (2.3) for some ω sufficiently large. Thus, by virtue of Theorem 2.1, we obtain (2.9). The proof is then complete.

Example 2.4. Assume that B_0 and B_1 are self-adjoint and positive operators in a Hilbert space $(H, \langle \cdot, \cdot \rangle)$, satisfying $\mathcal{D}(B_0) \subset \mathcal{D}(B_1)$ and $\langle B_0 u, B_1 u \rangle \geq 0$ for $u \in \mathcal{D}(B_0)$. Let B_{00} be a symmetric operator in H such that $\mathcal{D}(B_{00}) \supset \mathcal{D}\left(B_0^{\frac{1}{2}}\right)$ and $\langle B_{00} u, u \rangle \geq \beta \|u\|^2$ with some $\beta > 0$ for $u \in \mathcal{D}(B_0)$. Let B_{11} be a bounded operator on H such that

$$\langle B_{11} u, u \rangle \geq \alpha \|u\|^2, \quad u \in H. \qquad (2.22)$$

If $A_0 = B_0 + B_{00}$ and $A_1 = B_1 + B_{11}$, then for every $u_0, u_1 \in \mathcal{D}(A_0)$, (ACP_2) has a unique solution

$$u(t) \in C^2(R^+, H) \bigcap C^1(R^+, [\mathcal{D}(B_1)]) \bigcap C(R^+, [\mathcal{D}(B_0)])$$

satisfying

$$\|u(t)\|, \ \|u'(t)\| \leq \text{const } e^{-\delta t}, \quad t \geq 0$$

for some $\delta > 0$.

Proof. Without loss of generality, we may and do assume that B_0 and B_1 are strictly positive. Write $\eta = \left(2 \|B_1 B_0^{-1}\| + 1\right)^{-1}$. Then for $\lambda \in \mathbf{C}$ with $\mathrm{Re}\lambda \in [-\eta, 0)$,

$$\mathrm{Re} \left\langle \left(\lambda B_1 B_0^{-1} + \frac{1}{2}\right) u, \ u \right\rangle = \frac{1}{2}\|u\|^2 + (\mathrm{Re}\lambda) \left\langle B_1 B_0^{-1}, \ u \right\rangle$$

$$\geq \left(\frac{1}{2} - \eta \|B_1 B_0^{-1}\|\right) \|u\|^2 > 0, \quad u \in H;$$

it follows that

$$I + \lambda B_1 B_0^{-1} = \frac{1}{2}I + \left(\lambda B_1 B_0^{-1} + \frac{1}{2}I\right)$$

is boundedly invertible by the Lumer-Phillips theorem. Therefore, $\lambda B_1 + B_0$ is boundedly invertible and $(\lambda B_1 + B_0)^* = \overline{\lambda} B_1 + B_0$ for λ as above. Actually, this statement holds true for all $\lambda \in \mathbf{C}$ with $\mathrm{Re}\lambda \geq -\eta$ by (1.21) and (1.22).

Let $\lambda \in \mathbf{C}$ with $\mathrm{Re}\lambda \geq -\eta$ and fixed. Observe by Theorem A1.9 that for all $u \in \mathcal{D}(B_0)$, $a > 0$,

$$\|(\lambda B_{11} + B_{00})u\| \leq \text{const } \left\|B_0^{\frac{1}{2}} u\right\|$$

$$\leq \text{const } \left(a^{\frac{1}{2}}\|u\| + a^{-\frac{1}{2}}\|B_0 u\|\right)$$

$$\leq \text{const } \left(a^{\frac{1}{2}}\|u\| + a^{-\frac{1}{2}}\|(\lambda B_1 + B_0)u\|\right),$$

noting $B_0(\lambda B_1 + B_0)^{-1} \in \mathbf{L}(H)$. This implies that $\lambda B_{11} + B_{00}$ is $(\lambda B_1 + B_0)$-bounded with $(\lambda B_1 + B_0)$-bound less than 1, due to the arbitrariness of a. Accordingly we obtain (cf., e.g., Weidmann [1, Theorems 5.5 and 5.27]) that $\lambda A_1 + A_0$ is closed and

$$(\lambda A_1 + A_0)^* = \overline{\lambda} A_1 + A_0, \quad \mathrm{Re}\lambda \geq -\eta. \tag{2.23}$$

A similar reasoning as in the proof of Example 1.7 establishes that for $\lambda \in \mathbf{C}$ satisfying $\mathrm{Re}\lambda \geq 0$ or $|\lambda| \leq \eta_0$ (with some $\eta_0 > 0$), P_λ is boundedly invertible and

$$\|\lambda R_\lambda\| \leq \text{const} \quad (\mathrm{Re}\lambda \geq 0), \qquad \|R_\lambda\| \leq \text{const} \quad (|\lambda| \leq \eta_0). \tag{2.24}$$

Let $\sigma = \min\left\{\eta, \frac{\alpha}{4}, \frac{\eta_0}{2}\right\}$. Then for each $\lambda \in \mathbf{C}$ with $-\sigma \le \text{Re}\lambda < 0$, $|\lambda| > \eta_0$,

$$\|P_\lambda u\| \|u\| \ge |\langle (\lambda^2 + \lambda A_1 + A_0) u, \, u \rangle|$$

$$\ge |2(\text{Re}\lambda)(\text{Im}\lambda)\langle u, \, u \rangle + (\text{Im}\lambda)\langle A_1 u, \, u \rangle|$$

$$\ge |\text{Im}\lambda|(\alpha - 2|\text{Re}\lambda|)\|u\|^2$$

$$\ge \frac{\sqrt{3}}{4}|\lambda| \alpha \|u\|^2, \quad u \in \mathcal{D}(A_0);$$

namely

$$\|P_\lambda u\| \ge \frac{\sqrt{3}}{4}|\lambda| \alpha \|u\|, \quad u \in \mathcal{D}(A_0).$$

This is also true with P_λ^* instead of P_λ since $P_\lambda^* = P_{\bar\lambda}$ by (2.23). Therefore we conclude that P_λ is boundedly invertible and

$$\|\lambda R_\lambda\| \le \text{const}, \quad -\delta \le \text{Re}\lambda < 0, \ |\lambda| > \eta_0. \tag{2.25}$$

Moreover, note that for any λ, $\lambda_0 \in \{\lambda \in \mathbf{C}; \ \text{Re}\lambda > -\sigma\}$,

$$R_\lambda u - R_{\lambda_0} u = R_\lambda (P_{\lambda_0} - P_\lambda) R_{\lambda_0} u$$

$$= R_\lambda (\lambda_0^2 - \lambda^2 + (\lambda_0 - \lambda)A_1) R_{\lambda_0} u, \quad u \in H;$$

also $A_1 R_{\lambda_0} \in L(H)$ due to $\mathcal{D}(A_0) \subset \mathcal{D}(A_1)$. From this, we easily see that R_λ is analytic in $\{\lambda \in \mathbf{C}; \ \text{Re}\lambda > -\sigma\}$.

In view of Theorem 1.7, the Cauchy problem for

$$u''(t) + B_1 u'(t) + B_0 u(t) = 0 \quad (t \ge 0)$$

is strongly wellposed. Also from the proof of Theorem 1.7, it follows that

$$\lambda \longmapsto B_0^{\frac{1}{2}} \left(\lambda^2 + \lambda B_1 + B_0 \right)^{-1} \in LT - L(H),$$

$$\lambda \longmapsto \overline{\left(\lambda^2 + \lambda B_1 + B_0 \right)^{-1} B_0^{\frac{1}{2}}} \in LT - L(H).$$

Consequently, we claim by an application of Theorem 2.4.7 that (ACP_2) is strongly wellposed, and so the conditions of Theorem 2.2 are satisfied.

Recall (2.24) and (2.25). An appeal to Theorem 2.3 now completes the proof.

Example 2.5. Suppose that $a(x)$ and $b(x)$ are real valued and bounded measurable functions in R^3 with ess.sup$a(x) > 0$, ess.sup$b(x) > 0$. Consider the following Cauchy problem arising from the theory of linear viscoelastic materials

$$\begin{cases} u_{tt} - \nabla(\text{div}u_t) + b(x)u_t - \Delta u + a(x)u = 0, \\ \\ \qquad\qquad\qquad\qquad\qquad t \ge 0, \ x \in R^3, \tag{2.26} \\ \\ u(0, \, x) = \phi(x), \ u_t(0, \, x) = \psi(x), \ x \in R^3. \end{cases}$$

Let

$$H = L^2\left(R^3,\ \mathbf{C}^3\right),$$

$$B_0 u = -\Delta u, \quad u \in \mathcal{D}(B_0) := H^2\left(R^3,\ \mathbf{C}^3\right),$$

$$B_1 u = -\nabla\left(\operatorname{div} u\right), \quad u \in \mathcal{D}(B_1) := \left\{u \in H;\ \operatorname{div} u \in H^1\left(R^3,\ \mathbf{C}\right)\right\},$$

$$B_{00} u = a(\cdot) u, \quad u \in H,$$

$$B_{11} u = b(\cdot) u, \quad u \in H.$$

It can be verified (cf. Clément-Prüss [1, p. 636]) that the hypotheses in Example 2.4 are satisfied. Therefore, we claim that for every $\phi(x),\ \psi(x) \in H^2\left(R^3,\ \mathbf{C}^3\right)$, (2.26) has a unique solution

$$u(t,\ \cdot) \in C^2\left(R^+,\ L^2\left(R^3,\ \mathbf{C}^3\right)\right) \bigcap C\left(R^+,\ H^2\left(R^3,\ \mathbf{C}^3\right)\right)$$

satisfying

$$\|u(t,\ \cdot)\|_{L^2(R^3,\ \mathbf{C}^3)},\ \|u_t(t,\ \cdot)\|_{L^2(R^3,\ \mathbf{C}^3)} \le \operatorname{const} e^{-\sigma t}, \quad t \ge 0,$$

for a suitable $\sigma > 0$.

5.3 Notes

The notion of type of a semigroup $\{T(t)\}_{t \ge 0}$, which is defined as

$$\omega_0 := \inf_{t > 0} \frac{\ln \|T(t)\|}{t} = \lim_{t \to \infty} \frac{\ln \|T(t)\|}{t},$$

is due to Hille-Phillips [1]. ω_0 is also called growth bound (see, e.g., Nagle [2]), or growth abscissa (see, e.g., Prüss [1]) of $\{T(t)\}_{t \ge 0}$, etc. A characterization of the type of a strongly continuous semigroup in a Hilbert space determined by (roughly speaking) the spectrum of the generator appears in F. L. Huang [4] and Prüss [1]. It is also implied in the papers of Herbst [1] and Howland [1]. These works are significant generalizations of the interesting characterization of the $\sigma(T(t))$, for semigroups of contractions in Hilbert spaces, obtained by Gearhart [1]. A further simplification of the proof of the Gearhart type theorem was given by Greiner [3] (see, e.g., Nagel [2]). For other extensions and the related works please see, e.g., Arendt-Batty [1, 2], Batty [2-4], Nagel [2], Voigt [1], G. Weiss [1, 2] and the references cited in.

Definition 1.1 (which is an analogue of the type of semigroups), Theorem 1.4 (a characterization of the exponential growth bound of the propagators for

(ACP_n)), Corollaries 1.5 and 1.6, Example 1.7 are taken from Xiao-Liang [25]. The special case of $n = 2$ for Theorem 1.4 and Corollary 1.6 are implied in Xiao-Liang [6]. Lemma 1.3 is adapted from Greiner-Nagel [1]. Some more information about the wellposedness discussion in the situation like that in Example 1.7 may be found in Clément-Prüss [1] and Engel [6].

Theorems 2.1 – 2.3 comes from Xiao-Liang [22]. Theorem 2.1 is motivated by the idea in van Neerven [3].

Chapter 6

Differentiability and norm continuity

Summary

Section 6.1 is intended to give a characterization of the infinitely differentiable propagators of (ACP_n) in Banach spaces, which depends only on the properties of $\lambda^{k-1}\overline{R_\lambda A_k}$ $(1 \leq k \leq n)$. As a corollary, a concise sufficient condition is also presented.

Section 6.2 explores the characterization of the norm continuity (i.e., continuity in the uniform operator topology) for $t > 0$ of the propagators of (ACP_n) in Hilbert spaces. Following a general discussion on Laplace transforms in this respect, we obtain a succinct characterization (Theorem 2.1).

In Section 6.3, we restrict to (ACP_2) in a Banach space with $A_1 \in \mathbf{L}(E)$; see also Section 2.5. We show that $S_0(t)$ or $S_1'(t)$ is norm continuous for $t > 0$ if and only if A_0 is bounded. This leads to an interesting consequence for strongly continuous cosine operator functions or operator groups.

Section 6.4 is concerned with the operator matrix

$$\mathcal{A}_B = \begin{pmatrix} 0 & A^{\frac{1}{2}} \\ -A^{\frac{1}{2}} & -B \end{pmatrix},$$

where A is a positive self-adjoint operator in a Hilbert space and B subordinated to A in various ways. One can see that the semigroup generated by $\mathcal{A}_B$ (or $\overline{\mathcal{A}_B}$) may possess norm continuity, differentiability, analyticity, or exponential stability, respectively, as B changes.

6.1 Differentiability

In this section, $A_0, \cdots, A_{n-1}$ are closed linear operators in a Banach space E such that (ACP_n) is strongly wellposed. From Chapter 2 (Lemma 2.1.4, Theorems

2.2.2 and 2.2.3, Remark 2.2.5), we know that for all $t \geq 0$,

$$S_{n-1}(t)A_k u = S_{k-1}(t)u - S_k'(t)u, \quad 1 \leq k \leq n-1, \ u \in \bigcap_{i=0}^{k} \mathcal{D}(A_i), \quad (1.1)$$

$$S_{n-1}^{(n)}(t)u + \sum_{k=0}^{n-1} S_{n-1}^{(k)}(t)A_k u = 0, \quad u \in \bigcap_{i=0}^{n-1} \mathcal{D}(A_i); \quad (1.2)$$

there exist $C, \omega > 0$ such that

$$\{\lambda \in \mathbf{C}; \ \mathrm{Re}\lambda > \omega\} \subset \rho_0(A_0, \cdots, A_{n-1}); \quad (1.3)$$

and for $t \geq 0$,

$$\|S_k(t)\|, \ \left\|S_k^{(k)}(t)\right\| \leq Ce^{\omega t}, \quad 0 \leq k \leq n-1, \quad (1.4)$$

which in combination with (1.1) implies

$$\left\|\overline{S_{n-1}^{(k-1)}(t)A_k}\right\| \leq Ce^{\omega t}, \quad 1 \leq k \leq n-1; \quad (1.5)$$

that for $\lambda > \omega, \ u \in E$,

$$\int_0^\infty e^{-\lambda t} S_k(t)u\,dt = \sum_{i=k+1}^{n} \lambda^{i-k-1}\overline{R_\lambda A_i}u, \quad 0 \leq k \leq n-1. \quad (1.6)$$

We see by (1.3) (cf. the arguments below (4.1.9)) that for each $1 \leq k \leq n$, the map $\lambda \mapsto \lambda^{k-1}\overline{R_\lambda A_k}$ is analytic in $\{\lambda \in \mathbf{C}; \ \mathrm{Re}\lambda > \omega\}$.

By $\int_0^t F(s)ds$, for a strongly continuous $\mathbf{L}(E)$-valued function $F(t)$ on $[0, \ t]$, we will denote the bounded operators on E defined by

$$\left(\int_0^t F(s)ds\right)u = \int_0^t F(s)u\,ds, \quad \text{for every} \ u \in E.$$

Theorem 1.1. *The following two assertions are equivalent*

    (i) *For any $0 \leq k \leq n-1$, $S_k(t)$ is infinitely differentiable in $\mathbf{L}(E)$ for $t > 0$.*

    (ii) *To each $b > 0$ there correspond $a_b \in R$, $C_b > 0$ such that for each $1 \leq k \leq n$, the map $\lambda \mapsto \lambda^{k-1}\overline{R_\lambda A_k}$ admits an analytic continuation $L_k(\lambda)$ in the region*

$$\Sigma_b := \{\lambda \in \mathbf{C}; \ \omega + 1 > \mathrm{Re}\lambda > a_b - b\ln|\mathrm{Im}\lambda|\}$$

satisfying

$$\|L_k(\lambda)\| \leq C_b|\mathrm{Im}\lambda| \quad (1 \leq k \leq n, \ \lambda \in \Sigma_b).$$

Proof. (i) $\Longrightarrow$ (ii). Let $t > 0$, $\lambda \in \mathbf{C}$. Since for any $u \in E$,

$$S_{n-1}^{(k)}(0)u = \begin{cases} 0, & 0 \le k \le n-2, \\ \\ u, & k = n-1, \end{cases} \tag{1.7}$$

we have that for any $u \in \bigcap_{i=0}^{n-1} \mathcal{D}(A_i)$,

$$\lambda \int_0^t e^{\lambda(t-s)} S_{n-1}(s) u\, ds = -e^{\lambda(t-s)} S_{n-1}(s) u \Big|_0^t + \int_0^t e^{\lambda(t-s)} S_{n-1}'(s) u\, ds$$

$$= -S_{n-1}(t)u + \int_0^t e^{\lambda(t-s)} S_{n-1}'(s) u\, ds,$$

$$\lambda^2 \int_0^t e^{\lambda(t-s)} S_{n-1}(s) u\, ds$$

$$= -\lambda S_{n-1}(t)u - S_{n-1}'(t)u + \int_0^t e^{\lambda(t-s)} S_{n-1}''(s) u\, ds,$$

$$\lambda^k \int_0^t e^{\lambda(t-s)} S_{n-1}(s) u\, ds$$

$$= -\sum_{i=0}^{k-1} \lambda^{k-i-1} S_{n-1}^{(i)}(t)u + \int_0^t e^{\lambda(t-s)} S_{n-1}^{(k)}(s) u\, ds, \quad 2 < k \le n-1,$$

and

$$\lambda^n \int_0^t e^{\lambda(t-s)} S_{n-1}(s) u\, ds$$

$$= -\sum_{i=0}^{n-2} \lambda^{n-i-1} S_{n-1}^{(i)}(t)u + e^{\lambda t}u - S_{n-1}^{(n-1)}(t)u + \int_0^t e^{\lambda(t-s)} S_{n-1}^{(n)}(s) u\, ds$$

$$= e^{\lambda t}u - \sum_{i=0}^{n-1} \lambda^{n-i-1} S_{n-1}^{(i)}(t)u + \int_0^t e^{\lambda(t-s)} S_{n-1}^{(n)}(s) u\, ds.$$

Therefore, for all $u \in \bigcap_{i=0}^{n-1} \mathcal{D}(A_i)$,

$$\int_0^t e^{\lambda s} S_{n-1}(t-s) P_\lambda u\, ds$$

$$= \int_0^t e^{\lambda(t-s)} S_{n-1}(s) P_\lambda u\, ds$$

$$= e^{\lambda t}u - \sum_{i=0}^{n-1}\lambda^{n-i-1}S_{n-1}^{(i)}(t)u - \sum_{k=1}^{n-1}\sum_{j=0}^{k-1}\lambda^{k-j-1}S_{n-1}^{(j)}(t)A_k u$$

$$+ \int_0^t e^{\lambda(t-s)}\sum_{k=0}^{n}S_{n-1}^{(k)}(s)A_k u\,ds.$$

Thus, using (1.1) and (1.2) gives that for any $u \in \bigcap_{i=0}^{n-1}\mathcal{D}(A_i)$,

$$\int_0^t e^{\lambda s}S_{n-1}(t-s)P_\lambda u\,ds = e^{\lambda t}u - \sum_{i=0}^{n-1}\lambda^{n-i-1}S_{n-1}^{(i)}(t)u$$

$$- \sum_{k=1}^{n-1}\sum_{j=0}^{k-1}\lambda^{k-j-1}\left[S_{k-1}^{(j)}(t)u - S_k^{(j+1)}(t)u\right].$$

(1.8)

From (1.7), it follows that

$$\frac{d^{n-1}}{dt^{n-1}}\left[\int_0^t e^{\lambda s}S_{n-1}(t-s)P_\lambda u\,ds\right]$$

$$= \int_0^t e^{\lambda s}S_{n-1}^{(n-1)}(t-s)P_\lambda u\,ds$$

$$= \int_0^t e^{\lambda(t-s)}S_{n-1}^{(n-1)}(s)P_\lambda u\,ds, \quad u \in \bigcap_{i=0}^{n-1}\mathcal{D}(A_i),$$

and therefore

$$\frac{d^n}{dt^n}\left[\int_0^t e^{\lambda s}S_{n-1}(t-s)P_\lambda u\,ds\right]$$

$$= \frac{d}{dt}\left[\int_0^t e^{\lambda(t-s)}S_{n-1}^{(n-1)}(s)P_\lambda u\,ds\right]$$

(1.9)

$$= S_{n-1}^{(n-1)}(t)P_\lambda u + \lambda\int_0^t e^{\lambda(t-s)}S_{n-1}^{(n-1)}(s)P_\lambda u\,ds, \quad u \in \bigcap_{i=0}^{n-1}\mathcal{D}(A_i).$$

Hence, if we set

$$G(\lambda, t) := \sum_{i=0}^{n-1}\lambda^{n-i-1}S_{n-1}^{(n+i)}(t) + \sum_{k=1}^{n-1}\sum_{j=0}^{k-1}\lambda^{k-j-1}\left[S_{k-1}^{(j+n)}(t) - S_k^{(j+n+1)}(t)\right],$$

then combining (1.8) and (1.9) shows that for all $u \in \bigcap_{i=0}^{n-1}\mathcal{D}(A_i)$,

$$\lambda^{1-n}S_{n-1}^{(n-1)}(t)P_\lambda u + \lambda^{2-n}\int_0^t e^{\lambda(t-s)}S_{n-1}^{(n-1)}(s)P_\lambda u\,ds$$

$$= \lambda e^{\lambda t}u - \lambda^{1-n}G(\lambda, t)u,$$

and for any $|\lambda| \geq 1$,

$$\left\| \lambda^{1-n} G(\lambda, \, t) \right\| \leq \frac{a}{2}, \tag{1.10}$$

where

$$a := 2 \max \left\{ \sum_{i=0}^{n-1} \left\| S_{n-1}^{(n+i)}(t) \right\| + \sum_{k=1}^{n-1} \sum_{j=0}^{k-1} \left\| S_{k-1}^{(j+n)}(t) - S_{k}^{(j+n+1)}(t) \right\|, \quad e^t \right\}.$$

Denote

$$\mathcal{E} := \left\{ \lambda \in \mathbf{C}; \ \omega + 1 > \mathrm{Re}\lambda > t^{-1} \ln a - t^{-1} \ln |\mathrm{Im}\lambda| \right\}.$$

It is easy to see that

$$\begin{aligned} \mathcal{E} &= \left\{ \lambda \in \mathbf{C}; \ |\mathrm{Im}\lambda| > a e^{-t\mathrm{Re}\lambda}, \ \mathrm{Re}\lambda < \omega + 1 \right\} \\[2mm] &\subset \left\{ \lambda \in \mathbf{C}; \ |\lambda| > 1 \right\}, \end{aligned} \tag{1.11}$$

since $a > e^t$. Let $\lambda \in \mathcal{E}$. Then (1.10) implies that $\lambda e^{\lambda t} I - \lambda^{1-n} G(\lambda, \, t)$ is injective and

$$\begin{aligned} &\left\| \left[\lambda e^{\lambda t} I - \lambda^{1-n} G(\lambda, \, t) \right]^{-1} \right\| \\[2mm] &\leq \ \left| \lambda e^{\lambda t} \right|^{-1} \left\| \left[I - (\lambda e^{\lambda t})^{-1} \lambda^{1-n} G(\lambda, \, t) \right]^{-1} \right\| \\[2mm] &\leq \ \left| \lambda e^{\lambda t} \right|^{-1} \frac{1}{1 - \left| \lambda e^{\lambda t} \right|^{-1} \frac{a}{2}} \\[2mm] &\leq \ \left| \lambda e^{\lambda t} \right|^{-1} \frac{1}{1 - \frac{1}{2}} \\[2mm] &= \ 2 \left| \lambda e^{\lambda t} \right|^{-1}. \end{aligned} \tag{1.12}$$

Therefore P_λ is injective and

$$\begin{aligned} P_\lambda^{-1} u \ = \ &\left[\lambda e^{\lambda t} - \lambda^{1-n} G(\lambda, \, t) \right]^{-1} \left\{ \lambda^{1-n} S_{n-1}^{(n-1)}(t) \right. \\[2mm] &\left. + \lambda^{2-n} \int_0^t e^{\lambda(t-s)} S_{n-1}^{(n-1)}(s) ds \right\} u, \quad (u \in \mathcal{R}(P_\lambda), \ \lambda \in \mathcal{E}). \end{aligned} \tag{1.13}$$

For each $t > 0$ fixed, define

$$\begin{aligned} Q_t(\lambda) \ = \ &\left[\lambda e^{\lambda t} - \lambda^{1-n} G(\lambda, \, t) \right]^{-1} \left\{ \lambda^{1-n} S_{n-1}^{(n-1)}(t) \right. \\[2mm] &\left. + \lambda^{2-n} \int_0^t e^{\lambda(t-s)} S_{n-1}^{(n-1)}(s) ds \right\}, \quad \lambda \in \mathcal{E}. \end{aligned} \tag{1.14}$$

Then it is not difficult to verify by (1.12) and (1.4) that $\lambda \mapsto Q_t(\lambda)$ is an $\mathbf{L}(E)$-valued function, analytic in $\mathcal{E}$. Moreover, according to (1.4), (1.5) and (1.11), (1.12), we obtain that for $\lambda \in \mathcal{E}$,

$$
\begin{aligned}
\left\| \lambda^{n-1} Q_t(\lambda) \right\| &\leq 2 \left[a^{-1} \left\| S_{n-1}^{(n-1)}(t) \right\| + \left\| \int_0^t e^{-\lambda s} S_{n-1}^{(n-1)}(s)\,ds \right\| \right] \\
&\leq M_t \left(1 + e^{-\operatorname{Re}\lambda t} \right) \\
&\leq C_t |\operatorname{Im}\lambda|,
\end{aligned} \tag{1.15}
$$

and for $u \in \bigcap_{i=0}^{k} \mathcal{D}(A_i),\ 1 \leq k \leq n-1,\ \lambda \in \mathcal{E}$,

$$
\begin{aligned}
& \left\| \lambda^{k-1} Q_t(\lambda) A_k u \right\| \\
&\leq 2 \left[|\lambda|^{k-n} \left| \lambda e^{-\lambda t} \right|^{-1} \left\| S_{n-1}^{(n-1)}(t) A_k u \right\| \right. \\
&\qquad\qquad \left. + \left\| \lambda^{k-n} \int_0^t e^{-\lambda s} S_{n-1}^{(n-1)}(s) A_k u\,ds \right\| \right] \\
&= 2 \left[a^{-1} \left\| S_{k-1}^{(n-1)}(t)u - S_k^{(n)}(t)u \right\| \right. \\
&\qquad + \left\| \sum_{i=0}^{n-k-1} \lambda^{k-n+i} e^{-\lambda t} \left[S_{k-1}^{(n-i-2)}(t)u - S_k^{(n-i-1)}(t)u \right] \right. \\
&\qquad\qquad \left.\left. + \int_0^t e^{-\lambda s} S_{n-1}^{(k-1)}(s) A_k u\,ds \right\| \right] \\
&\leq M_t \left(1 + e^{-\operatorname{Re}\lambda t} \right) \|u\| \\
&\leq C_t |\operatorname{Im}\lambda| \|u\|,
\end{aligned} \tag{1.16}
$$

by (1.1) and (1.5), where M_t, C_t are suitable constants dependent only on t.

On the other hand, (1.13) indicates that for each $1 \leq k \leq n$, $\lambda^{k-1}\overline{Q_t(\lambda)A_k}$ coincides with $\lambda^{k-1}\overline{R_\lambda A_k}$ in $\mathcal{E} \bigcap \{\lambda \in \mathbb{C};\ \operatorname{Re}\lambda > \omega\}$. Consequently, given $b > 0$ we claim by (1.15) and (1.16) that

$$
\mathbf{L}_k(\lambda) := \lambda^{k-1}\overline{Q_{\frac{1}{b}}(\lambda)A_k}
$$

is just the analytic continuation of $\lambda^{k-1}\overline{R_\lambda A_k}$, as desired. The proof of the implication (i) $\Longrightarrow$ (ii) is then complete.

(ii) $\Longrightarrow$ (i). For any $0 \leq k \leq n-1$, $t > 0$, set

$$T_k(t) := \frac{1}{2\pi i} \int_\Gamma e^{\lambda t} \sum_{i=k+1}^{n} \lambda^{-k-3} L_i(\lambda) d\lambda,$$

where Γ is a path in Σ_b composed of three parts Γ_1, Γ_2 and Γ_3:

$$\Gamma_1 := \left\{ \lambda \in \mathbf{C}; \ -\infty < \mathrm{Re}\lambda \leq \omega + 1, \ \mathrm{Im}\lambda = -e^{\frac{2a_b - \mathrm{Re}\lambda}{b}} \right\},$$

$$\Gamma_2 := \left\{ \lambda \in \mathbf{C}; \ \mathrm{Re}\lambda = \omega + 1, \ -e^{\frac{2a_b - \omega - 1}{b}} \leq \mathrm{Im}\lambda \leq e^{\frac{2a_b - \omega - 1}{b}} \right\},$$

$$\Gamma_3 := \left\{ \lambda \in \mathbf{C}; \ -\infty < \mathrm{Re}\lambda \leq \omega + 1, \ \mathrm{Im}\lambda = e^{\frac{2a_b - \mathrm{Re}\lambda}{b}} \right\}.$$

Γ is oriented such that $\mathrm{Im}\lambda$ increases along Γ. Clearly, $T_k(t)$ is independent of the choice of $b > 0$.

For each fixed $l \in N$, $t_0 > 0$, taking $b = \frac{2l}{t_0}$ in (ii), then (ii) tells us that

$$\left\| \lambda^l e^{\lambda t} \sum_{i=k+1}^{n} \lambda^{-k-3} L_i(\lambda) \right\| \leq \text{const } |\lambda|^{l-3} e^{2a_b t} |\mathrm{Im}\lambda|^{-bt} \cdot C_b |\mathrm{Im}\lambda|$$

$$\leq \text{const } e^{2a_b t} |\mathrm{Im}\lambda|^{l-bt-2}$$

$$\leq M_t |\mathrm{Im}\lambda|^{-2} \quad (t > \frac{t_0}{2}, \ \lambda \in \Gamma_1 \bigcup \Gamma_3),$$

for some M_t dependent on t. Therefore for each $0 \leq k \leq n-1$, $t > \frac{t_0}{2}$, $T_k(t)$ is well-defined and is l-times differentiable in the uniform operator topology, and

$$T_k^{(j)}(t) = \frac{1}{2\pi i} \int_\Gamma e^{\lambda t} \lambda^j \sum_{i=k+1}^{n} \lambda^{-k-3} L_i(\lambda) d\lambda$$

$$(0 \leq k \leq n-1, \ t > \frac{t_0}{2}, \ 1 \leq j \leq l).$$

The arbitrariness of l and t_0 implies that $T_k(t)$ is well-defined for all $t > 0$ and is infinitely differentiable in the uniform operator topology for $t > 0$. On the other hand, for $\tau > \omega + 1$, $u \in E$, we have

$$\int_0^\infty e^{-\tau t} T_k(t) u dt = \frac{1}{2\pi i} \int_\Gamma \sum_{i=k+1}^{n} \lambda^{-k-3} L_i(\lambda) \left(\int_0^\infty e^{-(\tau - \lambda)t} dt \right) u d\lambda$$

$$= \frac{1}{2\pi i} \int_\Gamma \left(\sum_{i=k+1}^{n} \lambda^{-k-3} L_i(\lambda) \right) \frac{u}{\tau - \lambda} d\lambda$$

$$= \sum_{i=k+1}^{n} \tau^{-k-3} L_i(\tau) u$$

$$= \sum_{i=k+1}^{n} \tau^{i-k-4} \overline{R_\tau A_i} u$$

$$= \int_0^\infty e^{-\tau t} \left[\int_0^t \frac{(t-s)^2}{2} S_k(s) u \, ds \right] dt,$$

by (1.6). Hence,

$$T_k(t) u = \int_0^t \frac{(t-s)^2}{2} S_k(s) u \, ds.$$

Thus for each $0 \le k \le n-1$, $S_k(t)$ is infinitely differentiable in $\mathbf{L}(E)$ for $t > 0$. The proof of the implication (ii)$\Longrightarrow$(i) is then complete.

Corollary 1.2. *Let $\tau \in R$ such that $\tau + i\mu \in \rho_0(A_0, \cdots, A_{n-1})$ for $\mu \in R$ with $|\mu|$ sufficiently large, and*

$$\overline{\lim_{|\mu| \to \infty}} \ln |\mu| \, \|(\tau + i\mu)^{k-1} \overline{R_{\tau+i\mu} A_k}\| = 0, \quad 1 \le k \le n. \qquad (1.17)$$

Then for each $0 \le k \le n-1$, $S_k(t)$ is infinitely differentiable in $\mathbf{L}(E)$ for $t > 0$.

Proof. By virtue of (1.17), we know that for any given $\varepsilon > 0$, there exists a $\mu_0 > 1$ such that for all $|\mu| > \mu_0$,

$$\|(\tau + i\mu)^{k-1} \overline{R_{\tau+i\mu} A_k}\| \le \frac{\varepsilon}{\ln |\mu|}, \quad 1 \le k \le n, \qquad (1.18)$$

$$(n2^{n+1} \varepsilon)^{-1} \frac{\ln |\mu|}{|\tau + i\mu|} \le 1. \qquad (1.19)$$

Therefore

$$\left\| \sum_{l=i}^{n} \binom{l}{i} (\tau + i\mu)^{l-1} \overline{R_{\tau+i\mu} A_l} \right\| \le \frac{2^n \varepsilon}{\ln |\mu|} \quad (1 \le i \le n, \ |\mu| > \mu_0). \qquad (1.20)$$

Set, for $\lambda_0 \in \rho_0(A_0, \cdots, A_{n-1})$, $\lambda \in \mathbf{C}$,

$$U(\lambda_0; \lambda) = \sum_{i=1}^{n} \left[\sum_{l=i}^{n} \binom{l}{i} \lambda_0^{l-i} \overline{R_{\lambda_0} A_l} \right] (\lambda - \lambda_0)^i$$

$$= \sum_{l=1}^{n} \left[(\lambda - \lambda_0 + \lambda_0)^l - \lambda_0^l \right] \overline{R_{\lambda_0} A_l} \qquad (1.21)$$

$$= \sum_{l=1}^{n} (\lambda^l - \lambda_0^l) \overline{R_{\lambda_0} A_l}.$$

Then, by (1.19) and (1.20) we have that, for any σ with $|\sigma - \tau| < \left(n2^{n+1}\varepsilon\right)^{-1}\ln|\mu|$, $|\mu| > \mu_0$,

$$
\begin{aligned}
\|U(\tau + i\mu;\ \sigma + i\mu)\| &= \left\| \sum_{i=1}^{n} \left[\sum_{l=i}^{n} \binom{l}{i}(\tau + i\mu)^{l-i}\overline{R_{\tau+i\mu}A_l} \right] (\sigma - \tau)^i \right\| \\
&\leq \frac{2^n \varepsilon}{\ln|\mu|} \sum_{i=1}^{n} \left| \frac{\sigma - \tau}{\tau + i\mu} \right|^{i-1} |\sigma - \tau| \\
&= \frac{n2^n \varepsilon}{\ln|\mu|} \left(n2^{n+1}\varepsilon\right)^{-1} \ln|\mu| \\
&= \frac{1}{2}.
\end{aligned}
\tag{1.22}
$$

Now let $\mu_1 \geq \mu_0$ such that

$$
|\omega + 1 - \tau| < \left(n2^{n+1}\varepsilon\right)^{-1}\ln\mu_1,
$$

and let $\sigma, \mu \in R$ such that

$$
\omega + 1 + \left(n2^{n+1}\varepsilon\right)^{-1}\ln\mu_1 - \left(n2^{n+1}\varepsilon\right)^{-1}\ln|\mu| < \sigma < \omega + 1,
\tag{1.23}
$$

which implies that

$$
|\mu| > \mu_1, \quad \text{and} \quad |\sigma - \tau| < \left(n2^{n+1}\varepsilon\right)^{-1}\ln|\mu|.
$$

Observing

$$
P_{\sigma+i\mu} = P_{\tau+i\mu}\left\{ I + \sum_{l=1}^{n} \left[(\sigma + i\mu)^l - (\tau + i\mu)^l\right] R_{\tau+i\mu}A_l \right\},
$$

it follows from (1.21) and (1.22) that $P_{\sigma+i\mu}$ is injective and

$$
P_{\sigma+i\mu}^{-1} = \left[I + U(\tau + i\mu;\ \sigma + i\mu)\right]^{-1} R_{\tau+i\mu}, \quad \text{on} \quad \mathcal{R}(P_{\sigma+i\mu}).
\tag{1.24}
$$

Write, for $1 \leq k \leq n$,

$$
L_k(\lambda) := \left[I + U(\tau + i\mu;\ \lambda)\right]^{-1} \lambda^{k-1}\overline{R_{\tau+i\mu}A_k}, \quad \lambda = \sigma + i\mu.
$$

We see easily that $L_k(\cdot)$ is analytic in the region

$$
\Omega_0 := \Big\{ \lambda \in \mathbf{C};
$$

$$
\omega + 1 + \left(n2^{n+1}\varepsilon\right)^{-1}\ln\mu_1 - \left(n2^{n+1}\varepsilon\right)^{-1}\ln|\mathrm{Im}\lambda| < \mathrm{Re}\lambda < \omega + 1 \Big\},
$$

and (1.18), (1.19), (1.24) together imply that for all $1 \le k \le n$ and $\lambda \in \Omega_0$,

$$\|L_k(\lambda)\| \;\le\; 2 \left| \frac{\lambda}{\tau + i\, \mathrm{Im}\lambda} \right|^{k-1} \frac{\varepsilon}{\ln |\mathrm{Im}\lambda|}$$

$$\le\; 2^n \frac{\varepsilon}{\ln |\mathrm{Im}\lambda|}$$

$$\le\; C_\varepsilon |\mathrm{Im}\lambda|,$$

where C_ε depends only on ε. On the other hand, (1.24) also indicates that for each $1 \le k \le n-1$, $L_k(\lambda)$ coincides with $\lambda^{k-1}\overline{R_\lambda A_k}$ in $\Omega_0 \bigcap \{\lambda \in \mathbf{C};\ \mathrm{Re}\lambda > \omega\}$. Thus, according to Theorem 1.1, we obtain that for any $0 \le k \le n-1$, $S_k(t)$ is infinitely differentiable in $\mathbf{L}(E)$ for $t > 0$. The proof is then complete.

Theorem 1.3. *Let a_0, $b_0 > 0$ and let p_0, p_1, q_0, q_1 be real coefficient polynomials with*

$$\deg(p_0) < 2n_0, \quad \deg(p_1) = n_1, \quad \deg(q_0) < 2m_0, \quad \deg(q_1) = m_1$$

(n_0, n_1, m_0, $m_1 \in N_0$) satisfying

$$m_0 > 2n_0, \quad m_0 > n_1, \quad m_1 < m_0 + 2n_0.$$

Assume that A is an unbounded self-adjoint operator in a Hilbert space E, and B is a closable linear operator on E with $\mathcal{D}(B) \supset \mathcal{D}\left(A^{2n_0}\right)$ satisfying that there is a $\lambda_0 \in \rho\left(A^{2n_0}\right)$ such that $\left(\lambda_0 - A^{2n_0}\right)^{-1} B$ has a bounded extension to E. Set

$$p(s) = a_0 s^{2n_0} + p_0(s) + i p_1(s),$$

$$q(s) = b_0 s^{2m_0} + q_0(s) + i q_1(s).$$

Then the following Cauchy problem

$$\begin{cases} u''(t) + p(A)u'(t) + [q(A) + B]u(t) = 0, \quad t \ge 0, \\ u(0) = u_0, \quad u'(0) = u_1, \end{cases} \qquad (ACP_2)_{[0,\ B]}$$

is strongly wellposed and the two propagators $S_0(t)$, $S_1(t)$ are infinitely differentiable in $\mathbf{L}(E)$ for $t > 0$.

Proof. Set

$$r_0(s) = 4\left(b_0 s^{2m_0} + q_0(s)\right) - \left(a_0 s^{2n_0} + p_0(s)\right)^2 + p_1^2(s), \quad s \in \sigma(A), \qquad (1.25)$$

$$r_1(s) = 2p_1(s)\left(a_0 s^{2n_0} + p_0(s)\right) - 4q_1(s), \quad s \in \sigma(A). \qquad (1.26)$$

Then

$$\deg(r_0(s)) = 2m_0, \quad \deg(r_1(s)) \le 2n_0 + m_0.$$

We observe that for $|s|$ sufficiently large, $r_0(s) > 0$. Without loss of generality, we may and do assume (with $B - aI$ replacing B for some $a > 0$, if necessary) that

$$r_0(s) > 0 \quad \text{(for each } s \in \sigma(A)).$$

Write

$$\tilde{r}_0(s) = \frac{\sqrt{2}}{2} r_1(s) \left[r_0(s) + \sqrt{r_0^2(s) + r_1^2(s)} \right]^{-\frac{1}{2}}, \quad s \in \sigma(A),$$

$$\tilde{r}_1(s) = \frac{\sqrt{2}}{2} \left[r_0(s) + \sqrt{r_0^2(s) + r_1^2(s)} \right]^{\frac{1}{2}}, \quad s \in \sigma(A).$$

Then there exist constants M, $M_1 > 0$ such that for s sufficiently large,

$$|\tilde{r}_0(s)| \leq M|s|^{2n_0}, \quad |\tilde{r}_1(s)| \geq M_1|s|^{m_0}, \quad s \in \sigma(A).$$

Clearly, in the case of $r_1(s) = 0$,

$$[\tilde{r}_0(s) + i\tilde{r}_1(s)]^2 = -r_0(s),$$

and in the case of $r_1(s) \neq 0$,

$$[\tilde{r}_0(s) + i\tilde{r}_1(s)]^2$$

$$= \frac{1}{2} \left\{ r_1^2(s) \left[r_0(s) + \sqrt{r_0^2(s) + r_1^2(s)} \right]^{-1} \right.$$

$$\left. + 2ir_1(s) - \left[r_0(s) + \sqrt{r_0^2(s) + r_1^2(s)} \right] \right\}$$

$$= \frac{1}{2} \left\{ \left[\sqrt{r_0^2(s) + r_1^2(s)} - r_0(s) \right] + 2ir_1(s) - \left[r_0(s) + \sqrt{r_0^2(s) + r_1^2(s)} \right] \right\}$$

$$= -r_0(s) + ir_1(s).$$

That is

$$[-r_0(s) + ir_1(s)]^{\frac{1}{2}} = \pm [\tilde{r}_0(s) + i\tilde{r}_1(s)], \quad s \in \sigma(A). \tag{1.27}$$

For $s \in \sigma(A)$, define

$$\tau_\pm(s) = \frac{1}{2} \left[-(a_0 s^{2n_0} + p_0(s)) \pm \tilde{r}_0(s) \right] + \frac{1}{2} i \left[-p_1(s) \pm \tilde{r}_1(s) \right].$$

Then there exists $M_2 > 0$ such that for $|s|$ sufficiently large,

$$|\text{Im}(\tau_\pm(s))| \geq M_2|s|^{m_0}, \quad s \in \sigma(A). \tag{1.28}$$

Also

$$\begin{cases} |\text{Re}(\tau_\pm(s))| \le C_0(1+|s|)^{2n_0} \quad (s \in \sigma(A)), \\[2mm] |\text{Im}(\tau_\pm(s))| \le C_0(1+|s|)^{m_0} \quad (s \in \sigma(A)), \end{cases} \tag{1.29}$$

for some $C_0 > 0$;

$$(1+|s|)^{-2n_0}\text{Re}(\tau_\pm(s)) \longrightarrow -\frac{1}{2}a_0, \quad \text{as} \quad |s| \to \infty \quad (s \in \sigma(A)),$$

so that for $s \in \sigma(A)$ with $|s| > s_0$ (for some $s_0 > 0$),

$$\begin{aligned} \text{Re}(\tau_\pm(s)) &\le -\frac{1}{4}a_0(1+|s|)^{2n_0} \\[2mm] &\le -\frac{1}{4}a_0(1+s_0)^{2n_0}. \end{aligned} \tag{1.30}$$

Let $\mu \in R$ with $|\mu| > 2C_0(1+s_0)^{m_0}$. Then (1.30) yields that

$$\begin{aligned} |i\mu - \tau_\pm(s)| &\ge |\text{Re}(\tau_\pm(s))| \\[2mm] &\ge \frac{1}{4}a_0\left(\frac{1}{2C_0}\right)^{\frac{2n_0}{m_0}}|\mu|^{\frac{2n_0}{m_0}} \end{aligned}$$

whenever $|\mu| < 2C_0(1+|s|)^{m_0}$ (which implies $|s| > s_0$), $s \in \sigma(A)$. On the other hand, we get from (1.29) that

$$\begin{aligned} |i\mu - \tau_\pm(s)| &\ge |\mu| - |\text{Im}(\tau_\pm(s))| \\[2mm] &\ge |\mu| - C_0(1+|s|)^{m_0} \\[2mm] &\ge |\mu| - \frac{1}{2}|\mu| \\[2mm] &= \frac{1}{2}|\mu| \end{aligned}$$

for all $|\mu| \ge 2C_0(1+|s|)^{m_0}$, $s \in \sigma(A)$. It follows that there is a constant $C_1 > 0$ such that

$$|(i\mu - \tau_\pm(s))^{-1}| \le C_1|\mu|^{-\frac{2n_0}{m_0}}, \tag{1.31}$$

for all $s \in \sigma(A)$, $\mu \in R$ with $|\mu| > 2C_0(1+s_0)^{m_0}$. Thus equality (1.27), combined with (1.25) and (1.26), gives that for any $s \in \sigma(A)$, $\lambda \in \mathbf{C} \setminus \{\tau_\pm(s); \ s \in \sigma(A)\}$,

$$\begin{aligned} &\lambda\left(\lambda^2 + p(s)\lambda + q(s)\right)^{-1} \\[2mm] &= (\tilde{r}_0(s) + i\tilde{r}_1(s))^{-1}\left[\tau_+(s)(\lambda - \tau_+(s))^{-1} - \tau_-(s)(\lambda - \tau_-(s))^{-1}\right], \end{aligned} \tag{1.32}$$

$$p(s)\left(\lambda^2 + p(s)\lambda + q(s)\right)^{-1}$$

$$= p(s)(\widetilde{r}_0(s) + i\widetilde{r}_1(s))^{-1}\left[(\lambda - \tau_-(s))^{-1} - (\lambda - \tau_+(s))^{-1}\right], \tag{1.33}$$

$$s^{2n_0}\left(\lambda^2 + p(s)\lambda + q(s)\right)^{-1}$$

$$= s^{2n_0}(\widetilde{r}_0(s) + i\widetilde{r}_1(s))^{-1}\left[(\lambda - \tau_-(s))^{-1} - (\lambda - \tau_+(s))^{-1}\right]. \tag{1.34}$$

Since for $|s|$ sufficiently large, we have by (1.28) that

$$|\widetilde{r}_0(s) + i\widetilde{r}_1(s)| \ge C|s|^{m_0}$$

for some $C > 0$, it follows from (1.29) that

$$\sup_{s \in \sigma(A)} \left|\tau_\pm(s)\left[\widetilde{r}_0(s) + i\widetilde{r}_1(s)\right]^{-1}\right| < \infty.$$

Similarly, we can deduce that

$$\sup_{s \in \sigma(A)} \left|p(s)\left[\widetilde{r}_0(s) + i\widetilde{r}_1(s)\right]^{-1}\right| < \infty,$$

$$\sup_{s \in \sigma(A)} \left|s^{2n_0}\left[\widetilde{r}_0(s) + i\widetilde{r}_1(s)\right]^{-1}\right| < \infty.$$

Thus, employing the self-adjoint calculus, we obtain by (1.31) – (1.34) that

$$\lim_{|\mu| \to \infty} \ln|\mu| \left\|i\mu\left((i\mu)^2 + p(A)(i\mu) + q(A)\right)^{-1}\right\| = 0, \tag{1.35}$$

$$\lim_{|\mu| \to \infty} \ln|\mu| \left\|p(A)\left((i\mu)^2 + p(A)(i\mu) + q(A)\right)^{-1}\right\| = 0, \tag{1.36}$$

$$\lim_{|\mu| \to \infty} \ln|\mu| \left\|A^{2n_0}\left((i\mu)^2 + p(A)(i\mu) + q(A)\right)^{-1}\right\| = 0; \tag{1.37}$$

that there exists a constant $C_2 > 0$ such that for $\mathrm{Re}\lambda > \mu_0$, $m \in N_0$,

$$\left\|\left[\lambda\left(\lambda^2 + p(A)\lambda + q(A)\right)^{-1}\right]^{(m)}\right\|, \quad \left\|\left[p(A)\left(\lambda^2 + p(A)\lambda + q(A)\right)^{-1}\right]^{(m)}\right\|,$$

$$\left\|\left[A^{2n_0}\left(\lambda^2 + p(A)\lambda + q(A)\right)^{-1}\right]^{(m)}\right\| \le C_2 m!(\mathrm{Re}\lambda - \mu_0)^{-m-1}, \tag{1.38}$$

where

$$\mu_0 := \sup\{\mathrm{Re}(\tau_\pm(s)); \ s \in \sigma(A)\},$$

which is finite by (1.30). Now, we recall the hypothesis on the operator B. Making use of Theorems 2.2.3 and 2.4.6, we infer by (1.38) that $(ACP_2)_{[0, B]}$ is strongly wellposed. Moreover, we obtain by (1.35), (1.37) and (1.38) that for λ with either $\mathrm{Re}\lambda = 0$, $\mathrm{Im}\lambda$ sufficiently large or $\mathrm{Re}\lambda$ sufficiently large,

$$\left\| B \left(\lambda^2 + p(A)\lambda + q(A)\right)^{-1} \right\|$$

$$\leq \left\| B \left(\lambda_0 - A^{2n_0}\right)^{-1} \right\| \left\| \left(\lambda_0 - A^{2n_0}\right) \left(\lambda^2 + p(A)\lambda + q(A)\right)^{-1} \right\|$$

$$\leq \frac{1}{2},$$

$$\left\| \left(\lambda^2 + p(A)\lambda + q(A)\right)^{-1} B \right\|$$

$$\leq \left\| \left(\lambda_0 - A^{2n_0}\right) \left(\lambda^2 + p(A)\lambda + q(A)\right)^{-1} \right\| \left\| \overline{\left(\lambda_0 - A^{2n_0}\right)^{-1} B} \right\|$$

$$\leq \frac{1}{2}.$$

Thus, observing that for λ as above,

$$\lambda^2 + p(A)\lambda + q(A) + B$$

$$= \left[I + B \left(\lambda^2 + p(A)\lambda + q(A)\right)^{-1} \right] \left(\lambda^2 + p(A)\lambda + q(A)\right)$$

$$= \left(\lambda^2 + p(A)\lambda + q(A)\right) \left[I + \left(\lambda^2 + p(A)\lambda + q(A)\right)^{-1} B \right],$$

we have by (1.35) and (1.36) that

$$\lim_{|\mu| \to \infty} \ln |\mu| \left\| i\mu \left((i\mu)^2 + p(A)(i\mu) + q(A) + B\right)^{-1} \right\| = 0,$$

$$\lim_{|\mu| \to \infty} \ln |\mu| \left\| \overline{\left((i\mu)^2 + p(A)(i\mu) + q(A) + B\right)^{-1} p(A)} \right\| = 0.$$

This implies the desired conclusion by an application of Corollary 1.2. The proof is then complete.

Example 1.4. Let Ω is a smooth bounded domain of R^n, Γ be the boundary of Ω, and let $\xi \in C$ with $\mathrm{Re}\xi > 0$, $\eta > 0$, $\eta_1 \in R$. We consider the following initial-boundary value problem in $L^2(\Omega)$:

$$\begin{cases} u_{tt} - \xi\Delta u_t + \left(\eta\Delta^4 + i\eta_1\Delta^2\right) u = 0, & (t,\, x) \in R^+ \times \Omega, \\[2mm] u(0, x) = u_0, \quad u_t(0, x) = u_1, \quad x \in \Omega, \\[2mm] u\Big|_{\Gamma} = \Delta u\Big|_{\Gamma} = \Delta^2 u\Big|_{\Gamma} = \Delta^3 u\Big|_{\Gamma} = 0, \end{cases} \tag{1.39}$$

where Δ denotes the Laplacian.

Set

$$A_0 u = -\Delta u, \quad \mathcal{D}(A_0) = \left\{ u \in H^2(\Omega); \ u\big|_\Gamma = 0 \right\}.$$

It is known that A_0 is a positive, self-adjoint operator in $L^2(\Omega)$. Let $A = A_0^{\frac{1}{2}}$. Then the abstract version of problem (1.39) is

$$\begin{cases} u''(t) + p(A)u'(t) + q(A)u(t) = 0, \quad t \geq 0, \\[2mm] u(0) = u_0, \quad u'(0) = u_1, \end{cases}$$

where

$$p(A) := \xi A^2, \quad q(A) := \eta A^8 + i\eta_1 A^4,$$

and

$$\mathcal{D}\left(A^8\right) = \left\{ u \in H^8(\Omega); \ u\big|_\Gamma = \Delta u\big|_\Gamma = \Delta^2 u\big|_\Gamma = \Delta^3 u\big|_\Gamma = 0 \right\},$$

$$\mathcal{D}\left(A^2\right) = \mathcal{D}(A_0) = \left\{ u \in H^2(\Omega); \ u\big|_\Gamma = 0 \right\}.$$

Making use of Theorem 1.3 with $m_0 = 4$, $n_0 = 1$, $n_1 = 2$, $m_1 = 4$, we obtain that for every

$$u_0, \ u_1 \in \left\{ u \in H^8(\Omega); \ u\big|_\Gamma = \Delta u\big|_\Gamma = \Delta^2 u\big|_\Gamma = \Delta^3 u\big|_\Gamma = 0 \right\},$$

(1.39) has a unique solution u in

$$C^2\left(R^+, L^2(\Omega)\right) \bigcap C^1\left(R^+, H^2(\Omega)\right) \bigcap C\left(R^+, H^8(\Omega)\right) \bigcap C^\infty\left((0, \infty), L^2(\Omega)\right).$$

Example 1.5. Let $k \in N$ be odd, let p_1, q_1 be real coefficient polynomials with $\deg p_1 \leq k$ and $\deg q_1 \leq 2k - 1$ respectively, and let $\mathcal{A}\left(x, \frac{\partial}{\partial x}\right)$ be a linear differential operator in R with bounded smooth coefficients of degree $\leq k - 1$. Consider the following initial value problem in $L^2(R)$:

$$\begin{cases} \dfrac{\partial^2 u(t, x)}{\partial t^2} + \left((-1)^{\frac{k-1}{2}} \dfrac{\partial^{k-1}}{\partial x^{k-1}} + ip_1\left(i\dfrac{\partial}{\partial x}\right) \right) \dfrac{\partial u(t, x)}{\partial t} + \\[4mm] \left(\dfrac{\partial^{2(k+1)}}{\partial x^{2(k+1)}} + iq_1\left(i\dfrac{\partial}{\partial x}\right) + \mathcal{A}\left(x, \dfrac{\partial}{\partial x}\right) \right) u(t, x) = 0, \qquad (1.40) \\[4mm] \qquad\qquad t \geq 0, \ x \in R, \\[4mm] u(0, x) = u_0(x), \quad u_t(0, x) = u_1(x), \quad x \in R. \end{cases}$$

Take $E = L^2(R)$. Let A, B be the $L^2(R)$ realization of $i\frac{\partial}{\partial x}$, $\mathcal{A}\left(x, \frac{\partial}{\partial x}\right)$ respectively. Clearly, A is a self-adjoint operator in E; moreover,

$$\mathcal{D}(B) \bigcap \mathcal{D}(B^*) \supset \mathcal{D}(A^{k-1})$$

which implies that for each $\lambda_0 \in \rho(A^{k-1})$, $\left(\lambda_0 - A^{k-1}\right)^{-1} B$ has a bounded extension. Then applying Theorem 1.3 with $2n_0 = k - 1$, $m_0 = k + 1$, $n_1 \leq k$, $m_1 \leq 2k - 1$, we have that for every $u_0, u_1 \in H^{2(k+1)}(R)$, (1.40) has a unique solution u in

$$C^2 \left(R^+, L^2(R)\right) \bigcap C^1 \left(R^+, H^{k-1}(R)\right)$$

$$\bigcap C \left(R^+, H^{2(k+1)}(R)\right) \bigcap C^\infty \left((0, \infty), L^2(R)\right).$$

6.2   Norm continuity (general case)

In this section, we assume that $(H, \langle \cdot, \cdot \rangle)$ is a Hilbert space, A_k $(0 \leq k \leq n-1)$ are closed linear operators in H such that (ACP_n) is strongly wellposed.

A function $U(\cdot) : R^+ \to L(H)$ is called norm continuous (or continuous in the uniform operator topology) for $t > 0$ if

$$\lim_{h \to 0} \|U(t + h) - U(t)\| = 0, \quad \text{for any } t > 0.$$

Theorem 2.1. *For each* $0 \leq k \leq n - 1$, $S_k^{(k)}(t)$ *is norm continuous for* $t > 0$ *if and only if there is a* $\tau_0 \in R$ *such that* $\tau_0 + i\mu \in \rho_0(A_0, \cdots, A_{n-1})$ *for* $\mu \in R$ *with* $|\mu|$ *large enough and*

$$\lim_{|\mu| \to \infty} \left\|(\tau_0 + i\mu)^{k-1} \overline{R_{\tau_0 + i\mu} A_k}\right\| = 0, \quad 1 \leq k \leq n. \tag{2.1}$$

In this case, (2.1) holds actually for any $\tau_0 \in R$ *with* $\tau_0 + iR \subset \rho_0(A_0, \cdots, A_{n-1})$.

In order to prove this characterization, we consider the corresponding problem for Laplace transforms.

Let $l \in N_0$. For each $0 \leq m \leq l$, $\{U_m(t)\}_{t \geq 0}$ will be an $L(H)$-valued function, strongly continuous for $t > 0$ such that

$$\|U_m(t)\| \leq \text{const } e^{\omega t}, \quad t \geq 0, \tag{2.2}$$

for some $\omega > 0$. Write

$$G_m(\lambda)u := \int_0^\infty e^{-\lambda t} U_m(t)u\,dt, \quad u \in H, \ \text{Re}\lambda > \omega, \ 0 \leq m \leq l. \tag{2.3}$$

Theorem 2.2. *Let* $\eta > \omega$, $p \in N_0$ *and* $M > 0$ *fixed such that for any* $\mu \in R$ *with* $|\mu| \geq M$,

$$G_0^{(p)}(\eta + i\mu)u = \sum_{0 \leq j, \ k \leq l} G_j(\eta + i\mu)F_{jk}(\eta + i\mu)G_k(\eta + i\mu)u, \quad u \in H, \tag{2.4}$$

where for each $0 \leq j$, $k \leq l$, $F_{jk}(\eta + i\mu)$ is a strongly continuous $\mathbf{L}(H)$-valued function for $\mu \in \{\mu \in R; \ |\mu| \geq M\}$ satisfying

$$\lim_{|\mu| \to \infty} \|F_{jk}(\eta + i\mu)\| = 0, \qquad 0 \leq j, \ k \leq l. \qquad (2.5)$$

Then $\{U_0(t)\}_{t \geq 0}$ is norm continuous for $t > 0$.

Proof. Since

$$\langle U_m^*(t)v, \ u \rangle = \langle v, \ U_m(t)u \rangle, \qquad u, \ v \in H, \ 0 \leq m \leq l,$$

we know that for each $v \in H$, $U_m^*(t)v$ is weakly continuous for $t \geq 0$ and therefore strongly measurable for $t \geq 0$. Accordingly, it follows from (2.2) and (2.3) that

$$\|U_m^*(t)\| \leq \text{const } e^{\omega t}, \qquad t \geq 0, \ 0 \leq m \leq l,$$

$$G_m^*(\lambda)v = \int_0^\infty e^{-\bar{\lambda}t} U_m^*(t)v\,dt, \qquad v \in H, \ \text{Re}\lambda > \omega, \ 0 \leq m \leq l.$$

Thus, by virtue of Lemma 5.1.3 we obtain that for $u \in H$, $0 \leq m \leq l$,

$$
\begin{aligned}
\|G_m(\eta + i\mu)u\|_{L^2(R)} &= \left\{ \int_{-\infty}^\infty \left\| \int_0^\infty e^{-i\mu t} e^{-\eta t} U_m(t)u\,dt \right\|^2 d\mu \right\}^{\frac{1}{2}} \\
&= \sqrt{2\pi} \left\{ \int_0^\infty \left\| e^{-\eta t} U_m(t)u \right\|^2 dt \right\}^{\frac{1}{2}} \\
&\leq \text{const } \|u\|,
\end{aligned}
\qquad (2.6)
$$

and for $v \in H$, $0 \leq m \leq l$,

$$\|G_m^*(\eta + i\mu)v\|_{L^2(R)} \leq \text{const } \|v\|. \qquad (2.7)$$

Making use of (2.4) – (2.7) together yields that for any $W \geq M$, $u, \ v \in H$,

$$
\begin{aligned}
&\int_{|\mu| \geq W} \left| \left\langle G_0^{(p)}(\eta + i\mu)u, \ v \right\rangle \right| d\mu \\
&\leq \sum_{0 \leq j, \ k \leq l} \sup_{|\mu| \geq W} \|F_{jk}(\eta + i\mu)\| \left\{ \int_{|\mu| \geq W} \|G_k(\eta + i\mu)u\|^2 d\mu \right\}^{\frac{1}{2}} \\
&\qquad\qquad \cdot \left\{ \int_{|\mu| \geq W} \|G_j^*(\eta + i\mu)v\|^2 d\mu \right\}^{\frac{1}{2}} \\
&\leq \sum_{0 \leq j, \ k \leq l} \sup_{|\mu| \geq W} \|F_{jk}(\eta + i\mu)\| \text{ const } \|u\|\|v\|.
\end{aligned}
\qquad (2.8)
$$

This implies by (2.5) again that for any $\varepsilon > 0$, there exists $W_0 > 1$ such that

$$\int_{|\mu| \geq W_0} \left| \left\langle G_0^{(p)}(\eta + i\mu)u,\ v \right\rangle \right| d\mu \leq \varepsilon \|u\| \|v\|, \qquad u,\ v \in H. \tag{2.9}$$

On the other hand, for each $u,\ v \in H$, $\operatorname{Re}\lambda > \omega$,

$$\langle G_0(\lambda)u,\ v \rangle = \int_0^\infty e^{-\lambda t} \langle U_0(t)u,\ v \rangle\, dt,$$

and so for any $\operatorname{Re}\lambda > \omega$, $u,\ v \in H$,

$$\left\langle G_0^{(p)}(\lambda)u,\ v \right\rangle = \int_{-\infty}^\infty e^{-i\,\operatorname{Im}\lambda t} e^{-\operatorname{Re}\lambda t} \langle X(t)u,\ v \rangle\, dt, \tag{2.10}$$

in which

$$X(t) := \begin{cases} (-1)^p t^p U_0(t), & \text{if} \quad t \geq 0, \\[2mm] 0, & \text{if} \quad t < 0. \end{cases}$$

This, together with (2.2), gives that for each $u,\ v \in H$,

$$\left| \left\langle G_0^{(p)}(\eta + i\mu)u,\ v \right\rangle \right| \leq \text{const } \|u\| \|v\|. \tag{2.11}$$

According to this, it follows from (2.9) that for any $u,\ v \in H$,

$$\int_{-\infty}^\infty \left| \left\langle G_0^{(p)}(\eta + i\mu)u,\ v \right\rangle \right| d\mu$$

$$= \int_{|\mu| \geq W_0} \left| \left\langle G_0^{(p)}(\eta + i\mu)u,\ v \right\rangle \right| d\mu$$

$$+ \int_{|\mu| \leq W_0} \left| \left\langle G_0^{(p)}(\eta + i\mu)u,\ v \right\rangle \right| d\mu \tag{2.12}$$

$$\leq \quad \text{const } \|u\| \|v\|.$$

Applying (2.10), (2.12) and the elementary properties of Fourier transforms, we obtain
   (i) for each $t \geq 0$,

$$\frac{1}{2\pi} \int_{-\infty}^\infty e^{i\mu t} \left\langle G_0^{(p)}(\eta + i\mu)u,\ v \right\rangle d\mu$$

exists and equals $\langle e^{-\eta t}(-1)^p t^p U_0(t)u,\ v \rangle$;
moreover, by (2.11),
   (ii) for any $\varepsilon > 0$, $t,\ s \in (0,\ \infty)$, there exists $\delta(\varepsilon) > 0$ such that

$$\left| e^{i\mu t} - e^{i\mu s} \right| \left| \left\langle G_0^{(p)}(\eta + i\mu)u,\ v \right\rangle \right| < \frac{\varepsilon}{W_0} \|u\| \|v\|, \qquad u,\ v \in H,\ |\mu| \leq W_0,$$

whenever $|t - s| < \delta(\varepsilon)$.

Hence, combining (i), (ii) and (2.9) shows that for all $\varepsilon > 0$, t, $s \in (0, \infty)$ with $|t - s| < \delta(\varepsilon)$,

$$|\langle e^{-\eta t} t^p U_0(t)u - e^{-\eta s} s^p U_0(s)u, \ v\rangle|$$

$$\leq \ \frac{1}{2\pi} \left| \int_{|\mu| \geq W_0} e^{i\mu t} \left\langle G_0^{(p)}(\eta + i\mu)u, \ v \right\rangle d\mu \right|$$

$$+ \frac{1}{2\pi} \left| \int_{|\mu| \geq W_0} e^{i\mu s} \left\langle G_0^{(p)}(\eta + i\mu)u, \ v \right\rangle d\mu \right|$$

$$+ \frac{1}{2\pi} \int_{|\mu| \leq W_0} \left| e^{i\mu t} - e^{i\mu s} \right| \left| \left\langle G_0^{(p)}(\eta + i\mu)u, \ v \right\rangle \right| d\mu$$

$$< \ \frac{\varepsilon}{\pi} \|u\| \|v\| + \frac{\varepsilon}{\pi} \|u\| \|v\|$$

$$< \ \varepsilon \|u\| \|v\|, \qquad u, \ v \in H.$$

One thus sees that the operator family $\{e^{-\eta t} t^p U_0(t)\}_{t \geq 0}$ is norm continuous for $t > 0$, and so is $\{U_0(t)\}_{t \geq 0}$. The proof is then complete.

An examination of the steps of the above proof shows immediately

Theorem 2.3. *Let $\eta > \omega$, $p \in N_0$, $M > 0$ fixed such that for any $\mu \in R$ with $|\mu| \geq M$ and $u \in H$,*

$$G_0^{(p)}(\eta + i\mu)u \ = \ \sum_{0 \leq j, \ k \leq l} H_j(\eta + i\mu) F_{jk}(\eta + i\mu) H_k(\eta + i\mu)u$$

$$+ \sum_{0 \leq j \leq l} J_j(\eta + i\mu) K_j(\eta + i\mu)u,$$

where $F_{jk}(\eta + i\mu)$ $(0 \leq j, \ k \leq l)$, $H_j(\eta + i\mu)$, $J_j(\eta + i\mu)$, $K_j(\eta + i\mu)$ $(0 \leq j \leq l)$ are strongly continuous $L(H)$-valued functions for $\mu \in \{\mu \in R; \ |\mu| \geq M\}$ satisfying that for each $0 \leq j, \ k \leq l$, $u \in H$,

$$\lim_{|\mu| \to \infty} \|F_{jk}(\eta + i\mu)\| = \lim_{|\mu| \to \infty} \|J_j(\eta + i\mu)\| = 0,$$

$$\|K_j(\eta + i\mu)u\|_{L^1(R)}, \quad \|H_j(\eta + i\mu)u\|_{L^2(R)},$$

$$\left\| H_j^*(\eta + i\mu)u \right\|_{L^2(R)} \leq \text{const } \|u\|.$$

Then $\{U_0(t)\}_{t \geq 0}$ is norm continuous for $t > 0$.

Corollary 2.4. *Assume that for each $0 \le m \le l$, $u \in H$, $\mathrm{Re}\lambda > \omega$,*

$$G'_m(\lambda)u = \sum_{0 \le j_m,\ k_m \le l} a_{j_m k_m} G_{j_m}(\lambda) G_{k_m}(\lambda)u,$$

for some $a_{j_m k_m} \in \mathbf{C}$ $(0 \le j_m,\ k_m \le l)$. Then for each $0 \le m \le l$, $\{U_m(t)\}_{t \ge 0}$ is norm continuous for $t > 0$ if and only if

$$\lim_{|\mu| \to \infty} \|G_m(\eta + i\mu)\| = 0, \qquad 0 \le m \le l$$

for some $\eta > \omega$.

Proof. Combining Theorem 2.2 and Lemma 5.1.2 leads to the desired conclusion.

The proof of Theorem 2.1. We let ω be as in $(5.1.1) - (5.1.3)$. Take

$$l = n,$$

$$G_0(\lambda) = \frac{1}{\lambda}, \quad G_m(\lambda) = \lambda^{m-1}\overline{R_\lambda A_m} \quad (1 \le m \le l),$$

$$U_0(t) = I, \quad U_m(t) = S_{m-1}^{(m-1)}(t) - S_m^{(m)}(t) \quad (1 \le m \le l-1),$$

$$U_l(t) = S_{l-1}^{(l-1)}(t).$$

Then we have by (5.1.16) that for any $\mathrm{Re}\lambda > \omega$,

$$G'_m(\lambda) = (m-1)G_0(\lambda)G_m(\lambda) + \sum_{1 \le j \le l} j G_j(\lambda) G_m(\lambda), \quad 1 \le m \le l.$$

Also,

$$G'_0(\lambda) = -G_0(\lambda)G_0(\lambda), \quad \mathrm{Re}\lambda > \omega.$$

Note that if (2.1) holds then it holds also for any $\eta \in R$ with

$$\eta + iR \subset \rho_0(A_0, \cdots, A_{n-1}),$$

in place of τ_0; this can be verified by the fact that for $\mu \in R$ with $|\mu|$ large enough

$$R_{\eta + i\mu} = (I + U(\tau_0 + i\mu,\ \eta + i\mu))^{-1} R_{\tau_0 + i\mu},$$

where $U(\lambda_0,\ \lambda)$ (for each $\lambda_0 \in \rho_0(A_0, \cdots, A_{n-1})$, $\lambda \in \mathbf{C}$) is defined as in (1.21). From these observations, Corollary 2.4 applies and derives the conclusion as desired.

Corollary 2.5. *Let the characteristic condition of Theorem 2.1 be satisfied. Then for each $0 \le k \le n-1$, $S_k(t)$ is norm continuous for $t > 0$.*

Proof. This assertion is a direct consequence of Theorem 2.1 and the identity

$$S_k(t) = \int_0^t \frac{(t-s)^{k-1}}{(k-1)!} S_k^{(k)}(s)ds \quad (1 \le k \le n-1).$$

Theorem 2.6 (Perturbation). *Let the hypotheses in Theorem 2.4.6 hold. If $S_k^{(k)}(t)$ (for each $0 \le k \le n-1$) is norm continuous for $t > 0$, then so does $\tilde{S}_k^{(k)}(t)$ ($0 \le k \le n-1$), where $\tilde{S}_0(t), \cdots, \tilde{S}_{n-1}(t)$ denote the n propagators of $(ACP_n)_{[B_{n-1}, \cdots, B_0]}$.*

Proof. By hypothesis, (2.1) is true for any $\tau_0 \in R$ with

$$\tau_0 + iR \subset \rho_0(A_0, \cdots, A_{n-1}).$$

Therefore, defining $\tilde{P}_\lambda$, $\tilde{R}_\lambda$ as in (2.4.4) and (2.4.5), we have

$$\tilde{P}_{\tau_0+i\mu} = P_{\tau_0+i\mu}\left(I + \sum_{k=0}^{n-1}(\tau_0+i\mu)^k R_{\tau_0+i\mu}B_k\right), \quad \mu \in R, \tag{2.13}$$

with

$$\left\|\sum_{k=0}^{n-1}(\tau_0+i\mu)^k \overline{R_{\tau_0+i\mu}B_k}\right\|$$

$$\le \sum_{k=0}^{n-1}\left\{\left(\|\lambda_k(\tau_0+i\mu)^k R_{\tau_0+i\mu}\| + \left\|(\tau_0+i\mu)^k \overline{R_{\tau_0+i\mu}A_{i_k}}\right\|\right)\right.$$

$$\left.\left\|\overline{(\lambda_k - A_{i_k})^{-1}B_k}\right\|\right\}$$

$$< \frac{1}{2},$$

for $|\mu|$ large enough; so it can be easily seen that, for τ_0 sufficiently large,

$$\lim_{|\mu|\to\infty}\left\|(\tau_0+i\mu)^{k-1}\overline{\tilde{R}_{\tau_0+i\mu}(A_k+B_k)}\right\| = 0, \quad 1 \le k \le n.$$

This leads to the desired result by the arguments similar to those in the proof of Theorem 2.1.

Example 2.7. Let $n \ge 2$ and $a_1 > a_2 > \cdots > a_{n-1} > 0$. Suppose that $-A$ is the generator of a strongly continuous semigroup on H which is norm continuous for $t > 0$, with $(-\infty, 0) \subset \rho(A)$. Consider the Cauchy problem

$$\begin{cases} \displaystyle\prod_{i=1}^{n-1}\left(\frac{d}{dt} + a_i A^{\frac{1}{2}}\right)(u'(t) + Au(t)) = 0, & t \ge 0, \\ \\ u^{(k)}(0) = u_k, & 0 \le k \le n-1. \end{cases} \tag{2.14}$$

It is known that $-A^{\frac{1}{2}}$ (and so each $-a_i A^{\frac{1}{2}}$) generates an analytic semigroup. Hence, there exist constants C_0, $b_0 > 0$ such that for every λ with $\operatorname{Re}\lambda > b_0$, we have $\lambda \in \rho\left(-a_i A^{\frac{1}{2}}\right)$ and

$$\left\| \left(\lambda + a_i A^{\frac{1}{2}}\right)^{-1} \right\| \le C_0 |\lambda|^{-1}, \quad \operatorname{Re}\lambda > b_0, \quad 1 \le i \le n-1. \tag{2.15}$$

Moreover, Theorem 2.1 (with $n = 1$) gives that there is a constant $b_1 > b_0$ such that for all $\tau > b_1$, $\mu \in R$, $\tau + i\mu \in \rho(-A)$ and

$$\lim_{|\mu| \to \infty} \left\| (\tau + i\mu + A)^{-1} \right\| = 0. \tag{2.16}$$

Note that for $\operatorname{Re}\lambda > b_1$, $1 \le i \le j \le n-1$,

$$\lambda \left(\lambda + a_i A^{\frac{1}{2}}\right)^{-1} \left(\lambda + a_j A^{\frac{1}{2}}\right)^{-1}$$

$$= (a_i - a_j)^{-1} \left[a_i \left(\lambda + a_i A^{\frac{1}{2}}\right)^{-1} - a_j \left(\lambda + a_j A^{\frac{1}{2}}\right)^{-1} \right],$$

$$A^{\frac{1}{2}} \left(\lambda + a_i A^{\frac{1}{2}}\right)^{-1} \left(\lambda + a_j A^{\frac{1}{2}}\right)^{-1}$$

$$= (a_i - a_j)^{-1} \left[\left(\lambda + a_j A^{\frac{1}{2}}\right)^{-1} - \left(\lambda + a_i A^{\frac{1}{2}}\right)^{-1} \right].$$

We get that for each $0 \le k \le n-2$ there exist constants $C_1(k), \cdots, C_{n-1}(k)$ such that for $\operatorname{Re}\lambda > b_1$,

$$\lambda^k A^{\frac{1}{2}(n-k-2)} \prod_{i=1}^{n-1} \left(\lambda + a_i A^{\frac{1}{2}}\right)^{-1} = \sum_{i=1}^{n-1} C_i(k) \left(\lambda + a_i A^{\frac{1}{2}}\right)^{-1}. \tag{2.17}$$

Take $\mu_i \in \rho\left(a_i A^{\frac{1}{2}} - A\right)$ for each $1 \le i \le n-1$. We have that for $\operatorname{Re}\lambda > b_1$, $1 \le i \le n-1$,

$$\lambda \left(\lambda + a_i A^{\frac{1}{2}}\right)^{-1} (\lambda + A)^{-1}$$

$$= (\lambda + A)^{-1} - a_i A^{\frac{1}{2}} \left(\lambda + a_i A^{\frac{1}{2}}\right)^{-1} (\lambda + A)^{-1}$$

$$= \left[I + a_i A^{\frac{1}{2}} \left(\mu_i + A - a_i A^{\frac{1}{2}}\right)^{-1} \right] (\lambda + A)^{-1} \tag{2.18}$$

$$- \left[a_i A^{\frac{1}{2}} \left(\mu_i + A - a_i A^{\frac{1}{2}}\right)^{-1} \right]$$

$$\cdot \left\{ \mu_i \left(\lambda + a_i A^{\frac{1}{2}}\right)^{-1} (\lambda + A)^{-1} + \left(\lambda + a_i A^{\frac{1}{2}}\right)^{-1} \right\},$$

$$A \left(\lambda + a_i A^{\frac{1}{2}} \right)^{-1} (\lambda + A)^{-1}$$

$$= \left(\lambda + a_i A^{\frac{1}{2}} \right)^{-1} - \lambda (\lambda + a_i A)^{-1} (\lambda + A)^{-1}.$$

(2.19)

Now setting

$$P_0(\lambda) = (\lambda + A) \prod_{i=1}^{n-1} \left(\lambda + a_i A^{\frac{1}{2}} \right),$$

then (2.17), (2.18) and (2.19) together indicate that there exists C, $b > 0$ such that for $\mathrm{Re}\lambda > b$, $0 \le k \le n - 2$, $m = 0, 1, 2, \cdots$,

$$\left\| \left[\lambda^k A^{\frac{1}{2}(n-k)} P_0^{-1}(\lambda) \right]^{(m)} \right\|, \quad \left\| \left[\lambda^{n-1} P_0^{-1}(\lambda) \right]^{(m)} \right\| \le C m! (\mathrm{Re}\lambda - b)^{-m-1}.$$

Thus, strong wellposedness of (2.14) follows immediately from Theorem 2.2.3. On the other hand, making use of (2.17), (2.18) and (2.19) again, we find that for $0 \le k \le n - 2$, $\tau > b_1$,

$$\lim_{|\mu| \to \infty} \left\| (\tau + i\mu)^k A^{\frac{1}{2}(n-k)} P_0^{-1}(\tau + i\mu) \right\| = 0,$$

$$\lim_{|\mu| \to \infty} \left\| (\tau + i\mu)^{n-1} P_0^{-1}(\tau + i\mu) \right\| = 0,$$

by virtue of (2.15) and (2.16). Thus, we conclude that the hypothesis in Theorem 2.1 is satisfied. Therefore, the propagators $S_0(t)$, $S_1(t)$, $\cdots$, $S_{n-1}(t)$ of (2.14), as well as $S_k^{(k)}(t)$ (for all $1 \le k \le n - 1$), are norm continuous for $t > 0$.

6.3 Norm continuity (a special case)

Let A_0, A_1, E be as in Section 2.5. We consider the (ACP_2) in the case of $A_1 \in \mathbf{L}(E)$.

Throughout this section, (ACP_2) is assumed to be strongly wellposed, or equivalently (by Theorem 2.5.1), $-A_0$ is assumed to be the generator of a strongly continuous cosine function. If $A_1 = 0$, $S_0(t)$ is just the cosine function generated by $-A_0$.

We will show that $S_0(t)$ or $S_1'(t)$ is norm continuous for $t > 0$ if and only if A_0 is bounded. As a byproduct, we obtain that for a strongly continuous cosine function or for a strongly continuous group, it is norm continuous for $t > 0$ if and only if its generator is bounded. It is interesting to compare this with the case of general strongly continuous semigroups; it is known that many unbounded operators generate strongly continuous semigroups which are norm continuous for $t > 0$, such as analytic semigroups.

Lemma 3.1. *Let L, $a > 0$. Let A be a linear operator in E such that*

$$\sigma(A) \subset \{z \in \mathbf{C}; \ |z| \le L\}$$

and

$$\|R(z;\ A)\| \le \text{const } |z|^a, \qquad |z| > L.$$

Then $A \in L(E)$.

Proof. Copy the latter part of the proof of Theorem 3.6.5.

Theorem 3.2. $S_0(t)$ *or* $S_1'(t)$ *is norm continuous for* $t > 0$ *if and only if* $A_0 \in L(E)$.

Proof. Since (ACP_2) is strongly wellposed, it follows from Theorem 2.1.4 that for each $u \in \mathcal{D}(A_0) \bigcap \mathcal{D}(A_1) = \mathcal{D}(A_0)$

$$S_1'(t)u = S_0(t)u - S_1(t)A_1 u, \qquad t \ge 0. \tag{3.1}$$

By the boundedness of A_1 and the denseness of $\mathcal{D}(A_0)$, (3.1) holds for each $u \in E$. Therefore

$$S_1(t)u = \int_0^t [S_0(s) - S_1(s)A_1]u\,ds, \qquad t \ge 0,\ u \in E, \tag{3.2}$$

which implies that $S_1(t)$ is norm continuous for $t \ge 0$. Thus we see by (3.1) that $S_1'(t)$ is norm continuous for $t > 0$ if and only if $S_0(t)$ is norm continuous for $t > 0$.

  Necessity. The strong wellposedness implies that there exists $\omega > 0$ such that for $\text{Re}\lambda > \omega$, $\lambda \in \rho_0(A_0,\ A_1)$ and

$$\|\lambda R_\lambda\| \le \frac{C}{\text{Re}\lambda - \omega}. \tag{3.3}$$

Also by Theorem 2.1,

$$\lim_{|\text{Im}\lambda| \to \infty} \|\lambda R_\lambda\| = 0, \qquad \text{Re}\lambda > \omega. \tag{3.4}$$

Hence, there is $\omega_1 > \omega$ such that for $\text{Re}\lambda \ge \omega_1$,

$$\|\lambda A_1 R_\lambda\| \le \frac{1}{2},$$

since $A_1 \in L(E)$. Thus, for $\text{Re}\lambda \ge \omega_1$,

$$\left(\lambda^2 + A_0\right)^{-1} = R_\lambda\left(I - \lambda A_1 R_\lambda\right)^{-1} \in L(E); \tag{3.5}$$

namely,

$$\{\lambda^2;\ \ \lambda \in \mathbf{C},\ \ |\text{Re}\lambda| \ge \omega_1\} \subset \rho(-A_0). \tag{3.6}$$

On the other hand, from (3.4) and (3.5) we see that for $\text{Re}\lambda \ge \omega_1$,

$$\lim_{|\text{Im}\lambda| \to \infty} \left\|\lambda\left(\lambda^2 + A_0\right)^{-1}\right\| = 0. \tag{3.7}$$

Furthermore, for ν, $\lambda \in \mathbf{C}$ with $\mathrm{Re}\lambda \geq \omega_1$, we have that

$$\nu^2 + A_0$$

$$= \nu^2 - \lambda^2 + \lambda^2 + A_0$$

$$= (\lambda^2 + A_0)\left[I + (\nu^2 - \lambda^2)(\lambda^2 + A_0)^{-1}\right] \tag{3.8}$$

$$= (\lambda^2 + A_0)\left\{I + [(\nu - \lambda)^2 + 2\lambda(\nu - \lambda)](\lambda^2 + A_0)^{-1}\right\}.$$

Taking

$$\lambda = \omega_1 + 1 + iq, \quad \nu = p + iq$$

($p \in (-\omega_1 - 1, \omega_1 + 1)$, $q \in R$) we obtain from (3.7) that there exists $q_0 \geq \omega_1$ such that for $|q| \geq q_0$,

$$\left\|[(\nu - \lambda)^2 + 2\lambda(\nu - \lambda)](\lambda^2 + A_0)^{-1}\right\| \leq \frac{1}{2}. \tag{3.9}$$

Accordingly, $\lambda^2 \in \rho(-A_0)$ implies that

$$\{\nu^2; \quad \nu = p + iq, \quad |q| \geq q_0, \quad |p| \leq \omega_1 + 1\} \subset \rho(-A_0). \tag{3.10}$$

Combining (3.10) with (3.6), we know that

$$\sigma(-A_0) \quad \subset \quad \{\lambda^2; \ |\mathrm{Re}\lambda|, \ |\mathrm{Im}\lambda| \leq q_0\}$$

$$\subset \quad \{\lambda^2; \ |\lambda| \leq \sqrt{2}q_0\} \tag{3.11}$$

$$\subset \quad \{\lambda; \ |\lambda| \leq 2q_0^2\}.$$

Clearly, for every $\lambda \in \{\lambda; \ |\lambda| \geq 4q_0^2\}$, there is a $\nu \in \{\lambda; \ |\lambda| \geq 2q_0\}$ such that $\nu^2 = \lambda$ and $\mathrm{Re}\nu \geq 0$. Moreover,
  (i) if $\mathrm{Re}\nu \geq \sqrt{2}q_0$, then from (3.3) and (3.5) it follows that

$$\left\|(\nu^2 + A_0)^{-1}\right\| \leq \frac{C}{|\nu|}\frac{1}{q_0 - \omega + 1};$$

  (ii) if $\mathrm{Re}\nu < \sqrt{2}q_0$, then $|\mathrm{Im}\nu| > \sqrt{2}q_0$. Thus, (3.8), (3.9), (3.3) and (3.5) together show that there exists a constant C_1 such that

$$\left\|(\nu^2 + A_0)^{-1}\right\| \leq C_1 \|R_{\omega_1 + 1 + i \, \mathrm{Im}\nu}\|$$

$$\leq \frac{CC_1}{((\omega_1 + 1)^2 + (\mathrm{Im}\nu)^2)^{\frac{1}{2}}}$$

$$\leq \frac{CC_1}{((\omega_1 + 1)^2 + (q_0)^2)^{\frac{1}{2}}}.$$

Therefore, for every $\lambda \in \mathbf{C}$ with $|\lambda| \geq 4q_0^2$,

$$\left\|(\lambda + A_0)^{-1}\right\| = O(1), \quad \text{as } |\lambda| \to \infty.$$

This, together with (3.11), enables us to apply Lemma 3.1 and obtain that $A_0 \in \mathbf{L}(E)$.

Sufficiency. From Theorem 2.1.4 again, we have

$$S_0(t)u = u - \int_0^t S_1(s)A_0 u\, ds, \quad u \in \mathcal{D}(A_0).$$

Hence by the boundedness of A_0, $S_0(t)$ is norm continuous for $t > 0$, so is $S_1'(t)$ by recalling the arguments below (3.1). The proof is then complete.

The following is an immediate consequence of Theorem 3.2.

Corollary 3.3. *Both $S_0(t)$ and $S_1(t)$ are norm continuous for $t > 0$ if and only if $A_0 \in \mathbf{L}(E)$.*

Corollary 3.4. *Let $\{S_0(t)\}_{t \in R}$ be a strongly continuous cosine function on E. Then the following statements are equivalent.*
  (i) *$S_0(t)$ is norm continuous for $t > 0$.*
  (ii) *$S_0(t)$ is norm continuous at $t = 0$.*
  (iii) *$S_0(t)$ is norm continuous for $t \in R$.*
  (iv) *the generator A_0 of $S_0(t)$ is bounded.*

Proof. Letting $A_1 = 0$ in Theorem 3.2, we have that (i) is equivalent to (iv).

By Theorem A2.10 we know that A_0 is also the generator of a strongly continuous semigroup $T(t)$, defined by

$$T(t)u := \frac{2}{\sqrt{\pi}} \int_0^\infty e^{-s^2} S_0\left(2t^{\frac{1}{2}}s\right) u\, ds, \quad u \in E, \ t \geq 0.$$

Thus, if $S_0(t)$ is norm continuous at $t = 0$, so is $T(t)$. It follows from Theorem A2.8 that (ii) implies (iv). Finally the cosine function, generated by a bounded operator A_0, takes the form

$$S_0(t) = \sum_{j=0}^\infty \frac{t^{2j} A_0^j}{(2j)!}, \quad t \in R;$$

the series converges (in the uniform operator topology) uniformly on compacts of R. We thus see that (iv) implies (i) – (iii).

The remaining part is clear.

Corollary 3.5. *Let $\{T(t)\}_{t \in R}$ be a strongly continuous group on E. Then the following statements are equivalent.*
  (i) *$T(t)$ is norm continuous for $t > 0$.*

(ii) $T(t)$ *is norm continuous at $t = 0$.*
(iii) $T(t)$ *is norm continuous for $t \in R$.*
(iv) *the generator A of $T(t)$ is bounded.*

Proof. From the fact that if A generates a strongly continuous group $T(t)$, then A^2 generates a strongly continuous cosine function $S_0(\cdot)$ given by

$$S_0(t) = \frac{1}{2}[T(t) + T(-t)],$$

we know by Corollary 3.4 that each of (i) – (iii) implies (iv). The rest is obvious, since the group $T(t)$, generated by a bounded operator A, takes the form

$$T(t) = \sum_{j=0}^{\infty} \frac{t^j A^j}{j!}.$$

6.4 Operator matrices generating various semigroups

Let $(H, \langle \cdot, \cdot \rangle)$ be a Hilbert space. We consider in this section the operator matrix

$$\mathcal{A}_B = \begin{pmatrix} 0 & A^{\frac{1}{2}} \\ -A^{\frac{1}{2}} & -B \end{pmatrix},$$

$$\mathcal{D}(\mathcal{A}_B) = \mathcal{D}\left(A^{\frac{1}{2}}\right) \times \left(\mathcal{D}\left(A^{\frac{1}{2}}\right) \bigcap \mathcal{D}(B)\right)$$

in the product space $H \times H$, corresponding to the elastic model $u''(t) + Bu'(t) + Au(t) = 0$ written as a first order system. Here A (the elastic operator) is a self-adjoint and strictly positive operator in H with $\sigma(A) \subset [\sigma_0, +\infty)$ for some $\sigma_0 > 0$, and B (the dissipation operator) is a closed operator in H with $\mathcal{D}(B) \supset \mathcal{D}(A)$; B will be assumed to be 'subordinated' to A in various ways.

It can be verified that $\mathcal{A}_B$ is dissipative and closable, $0 \in \rho\left(\overline{\mathcal{A}_B}\right)$ and

$$\left(\overline{\mathcal{A}_B}\right)^{-1} = \begin{pmatrix} -\overline{A^{-\frac{1}{2}}BA^{-\frac{1}{2}}} & -A^{-\frac{1}{2}} \\ A^{-\frac{1}{2}} & 0 \end{pmatrix}.$$

Thus, we infer by the Lumer-Phillips theorem that $\overline{\mathcal{A}_B}$ generates a strongly continuous semigroup $\{T_B(t)\}_{t \geq 0}$ of contractions on $H \times H$.

How does the $\{T_B(t)\}_{t \geq 0}$ behave ? In this section, we will reveal various interesting behaviors of $\{T_B(t)\}_{t \geq 0}$: analyticity, exponential stability, differentiability and norm continuity.

Assume that

$$f : [\sigma_0, +\infty) \longrightarrow (\sigma_1, +\infty)$$

(for some $\sigma_1 > 0$) is measurable with

$$\lim_{s \to +\infty} f(s) = +\infty$$

such that

$$(s + 1)^{-1} f(s) \quad \text{is bounded on } [\sigma_0, \infty), \qquad (4.1)$$

$$|f^2(s) - 4s| \geq C_1 \max\left\{ f^2(s), \quad s \right\}, \quad \text{for any } s > L_1, \qquad (4.2)$$

for some constants C_1, $L_1 > 0$. We assume that B can be expressed as $B = B_1 + iB_2$, where B_1 is a self-adjoint and strictly positive operator and B_2 is a self-adjoint operator in H, satisfying

$$\langle f(A)u, \ u \rangle \leq \langle B_1 u, \ u \rangle \leq b\langle f(A)u, \ u \rangle, \quad u \in \mathcal{D}(A), \qquad (4.3)$$

$$|\langle B_2 u, \ u \rangle| \leq a\langle B_1 u, \ u \rangle, \quad u \in \mathcal{D}(A), \qquad (4.4)$$

for some constants $b \geq 1$, $a > 0$.

The following is a collection of some results in Sections 6.2, 6.1, 5.1 and 4.1, applied to $n = 1$.

Lemma 4.1. *Assume that A_0 generates a strongly continuous semigroup $\{T(t)\}_{t \geq 0}$ on H with $\|T(t)\| \leq Ce^{\omega t}$ $(t \geq 0)$ for some constants C, $\omega > 0$. Let σ, $\mu \in R$, $\mu_0 > 0$. Then*

(i) $\{T(t)\}_{t \geq 0}$ *is norm continuous for $t > 0$ provided* $\{\sigma + i\mu; \ |\mu| > \mu_0\} \subset \rho(A_0)$ *and*

$$\lim_{|\mu| \to \infty} \|R(\sigma + i\mu; \ A_0)\| = 0.$$

(ii) $\{T(t)\}_{t \geq 0}$ *is a differentiable semigroup provided* $\{\sigma + i\mu; \ |\mu| > \mu_0\} \subset \rho(A_0)$ *and*

$$\overline{\lim_{|\mu| \to \infty}} \ln|\mu| \|R(\sigma + i\mu; \ A_0)\| = 0.$$

(iii) $\{T(t)\}_{t \geq 0}$ *is an analytic semigroup provided* $\{\sigma + i\mu; \ |\mu| > \mu_0\} \subset \rho(A_0)$ *and*

$$\|R(\sigma + i\mu; \ A_0)\| \leq \frac{M_0}{|\mu|}, \quad |\mu| > \mu_0$$

for some constant $M_0 > 0$.

(iv) $\{T(t)\}_{t \geq 0}$ *is an analytic semigroup of angle $\frac{\pi}{2}$ if for each $\theta \in (0, \frac{\pi}{2})$ there exists C_θ, $\omega_\theta > 0$ such that*

$$\omega_\theta + \Sigma_{\frac{\pi}{2} + \theta} \subset \rho(A_0)$$

and

$$\|R(\lambda; \ A_0)\| \leq \frac{C_\theta}{|\lambda|}, \quad \text{for any } \lambda \in \omega_\theta + \Sigma_{\frac{\pi}{2} + \theta}.$$

(v) $\{T(t)\}_{t \geq 0}$ *is exponentially stable if*

$$\|R(\lambda; \ A_0)\| \leq \text{const} \quad (\text{Re}\lambda > 0).$$

Proof. Part (i) follows from Theorem 2.1, part (ii) from Corollary 1.2, and part (iv) from Theorem 4.1.3.

For part (iii), we let $\omega_0 > \omega$ and observe

$$R(\omega_0 + i\mu;\ A_0)$$
$$= \ R(\sigma + i\mu;\ A) + (\omega_0 - \sigma)R(\omega_0 + i\mu;\ A_0)R(\sigma + i\mu;\ A_0), \quad |\mu| > \mu_0.$$

It follows that

$$\|R(\omega_0 + i\mu;\ A_0)\| \leq \frac{M_1}{|\mu|}, \quad |\mu| > 0,$$

for some constant $M_1 > 0$. Put $\theta = \arctan(2M_1)^{-1}$. Then for any

$$\lambda \in \omega_0 + \left(\Sigma_{\frac{\pi}{2}+\theta} \setminus \Sigma_{\frac{\pi}{2}-\theta}\right),$$

$$\|[\lambda - (\omega_0 + i\,\mathrm{Im}\lambda)]R(\omega_0 + i\,\mathrm{Im}\lambda;\ A_0)\| \leq \frac{1}{2}$$

and therefore $\lambda \in \rho(A_0)$ with

$$R(\lambda;\ A_0) = R(\omega_0 + i\,\mathrm{Im}\lambda;\ A_0)\{I + [\lambda - (\omega_0 + i\,\mathrm{Im}\lambda)]R(\omega_0 + i\,\mathrm{Im}\lambda;\ A_0)\}^{-1}.$$

From this we get that for λ as above

$$\|R(\lambda;\ A_0)\| \leq \frac{2M_1}{|\mathrm{Im}\lambda|} \leq \frac{2M_1}{\cos\theta}|\lambda - \omega_0|^{-1}.$$

On the other hand, we have by hypothesis that for $\lambda \in \omega_0 + \Sigma_{\frac{\pi}{2}-\theta}$,

$$\|R(\lambda;\ A_0)\| \leq \frac{M}{\mathrm{Re}\lambda - \omega} \leq \frac{M}{\mathrm{Re}\lambda - \omega_0} \leq \frac{M}{\sin\theta}|\lambda - \omega_0|^{-1},$$

with some $M > 0$. Now an application of Theorem 4.1.3 gives part (iii).

Finally, we look at part (v). It is plain to see that $iR \subset \rho(A_0)$ with $\|R(\lambda;\ A_0)\| \leq \mathrm{const}\ (\mathrm{Re}\lambda \geq 0)$ by a similar argument as above. Thus Corollary 5.1.6 applies and verifies part (v).

Theorem 4.2. (i) $\{\mathcal{T}_{f(A)}(t)\}_{t\geq 0}$ *is exponentially stable and norm continuous for* $t > 0$.

(ii) $\{\mathcal{T}_{f(A)}(t)\}_{t\geq 0}$ *is differentiable for* $t > 0$, *if*

$$\lim_{s\to+\infty} \frac{\ln s}{f(s)} = 0. \tag{4.5}$$

(iii) $\{\mathcal{T}_{f(A)}(t)\}_{t\geq 0}$ *is an analytic semigroup, if*

$$\sup\left\{f^{-1}(s)s^{\frac{1}{2}};\ s \geq \sigma_0\right\} < +\infty. \tag{4.6}$$

(iv)$\{T_{f(A)}(t)\}_{t\geq 0}$ *is an analytic semigroup of angle* $\frac{\pi}{2}$, *if*

$$\lim_{s\to+\infty}\frac{s^{\frac{1}{2}}}{f(s)}=0. \tag{4.7}$$

Proof. Put

$$\tau_{\pm}(s)=-\frac{1}{2}f(s)\pm\frac{1}{2}\sqrt{f^2(s)-4s},\quad s\in[\sigma_0,\ +\infty).$$

We have immediately that for $s\in[\sigma_0,\ +\infty)$,

$$\lambda^2+f(s)\lambda+s=(\lambda-\tau_+(s))(\lambda-\tau_-(s)),\quad \lambda\in\mathbf{C} \tag{4.8}$$

and for some $\varepsilon>0$

$$\tau_{\pm}(s)\in\{\lambda\in\mathbf{C};\ \ \mathrm{Re}\lambda<-\varepsilon\}, \tag{4.9}$$

noting that if $f^2(s)>4s$,

$$-\tau_+(s)=\frac{2s}{f(s)+\sqrt{f^2(s)-4s}}\geq\frac{s}{f(s)},$$

and using (4.1).

Choose L_2 such that $L_2>L_1$ and, in the case of (4.7), $f^2(s)>4s$ for each $s>L_2$. It is easy to see by the above observations that there exist $C_2>0,\ \delta>1$ such that for $s\in[\sigma_0,\ L_2],\ \lambda\in\mathbf{C}$,

$$\left|s^{\frac{1}{2}}\left(\lambda^2+f(s)\lambda+s\right)^{-1}\right|,\ \ \left|f(s)\lambda^{-1}\left(\lambda^2+f(s)\lambda+s\right)^{-1}\right|,$$

$$\left|\lambda\left(\lambda^2+f(s)\lambda+s\right)^{-1}\right|,\ \ \left|f^{\frac{1}{2}}(s)s^{\frac{1}{2}}\left(\lambda^2+f(s)\lambda+s\right)^{-1}\right|,$$

$$\left|\lambda f^{\frac{1}{2}}(s)\left(\lambda^2+f(s)\lambda+s\right)^{-1}\right|, \tag{4.10}$$

$$\leq\begin{cases}C_2|\lambda|^{-1}, & \text{if } |\lambda|>\delta,\\[2mm] C_2, & \text{if } \mathrm{Re}\lambda>0.\end{cases}$$

In the sequel, we let $s>L_2$. Then $f^2(s)\neq 4s$ by (4.2). An easy computation yields that

$$\left(\lambda^2+f(s)\lambda+s\right)^{-1} \tag{4.11}$$

$$=\frac{1}{\sqrt{f^2(s)-4s}}\left[(\lambda-\tau_+(s))^{-1}-(\lambda-\tau_-(s))^{-1}\right],$$

$$\lambda\left(\lambda^2+f(s)\lambda+s\right)^{-1} \tag{4.12}$$

$$=\frac{\tau_+(s)}{\sqrt{f^2(s)-4s}}(\lambda-\tau_+(s))^{-1}-\frac{\tau_-(s)}{\sqrt{f^2(s)-4s}}(\lambda-\tau_-(s))^{-1}.$$

It is also clear that

$$\sup_{s>L_2}\left[\left(s^{\frac{1}{2}}+f(s)+|\tau_\pm(s)|\right)\left|f^2(s)-4s\right|^{-\frac{1}{2}}\right]<+\infty. \tag{4.13}$$

Next, we distinguish three cases to carry on our discussion.

Case 1: $f^2(s)>4s$.

Observe by (4.9) that for any $\theta\in(0,\ \frac{\pi}{2})$,

$$|(\lambda-\tau_\pm(s))^{-1}|\leq\frac{1}{\cos\theta|\lambda+\varepsilon|},\quad\lambda\in\Sigma_{\frac{\pi}{2}+\theta}. \tag{4.14}$$

On the other hand

$$\begin{aligned}|\lambda-\tau_+(s)|&\geq&|\tau_+(s)|=\frac{2s}{f(s)+\sqrt{f^2(s)-4s}}\\[2mm]&\geq&\frac{s}{f(s)},\quad\operatorname{Re}\lambda>0,\end{aligned} \tag{4.15}$$

$$\begin{aligned}|\lambda-\tau_-(s)|&\geq&|\tau_-(s)|=\frac{1}{2}(f(s)+\sqrt{f^2(s)-4s})\\[2mm]&\geq&\frac{1}{2}\left(C_1^{\frac{1}{2}}+1\right)f(s),\quad\operatorname{Re}\lambda>0\end{aligned} \tag{4.16}$$

by (4.2). Now, let $\operatorname{Re}\lambda>0$. From (4.2), (4.14) and the estimate $|\tau_\pm(s)|\leq f(s)$, we obtain that

$$\begin{aligned}\frac{|\tau_\pm(s)|f^{\frac{1}{2}}(s)}{\sqrt{f^2(s)-4s}}|(\lambda-\tau_\pm(s))^{-1}|&\leq&C_1^{-\frac{1}{2}}f^{\frac{1}{2}}(s)|\lambda+\varepsilon|^{-1}\\[4mm]&\leq&C_1^{-\frac{1}{2}}|\lambda+\varepsilon|^{-\frac{1}{2}},\end{aligned}$$

whenever $f(s)\leq|\lambda+\varepsilon|$. Furthermore, we have by (4.2), (4.15) and (4.16) that

$$\begin{aligned}\frac{|\tau_\pm(s)|f^{\frac{1}{2}}(s)}{\sqrt{f^2(s)-4s}}|(\lambda-\tau_\pm(s))^{-1}|&\leq&\frac{f^{\frac{1}{2}}(s)}{\sqrt{f^2(s)-4s}}\\[4mm]&\leq&C_1^{-\frac{1}{2}}\frac{f^{\frac{1}{2}}(s)}{f(s)}\\[4mm]&\leq&C_1^{-\frac{1}{2}}\frac{1}{f^{\frac{1}{2}}(s)}\\[4mm]&\leq&C_1^{-\frac{1}{2}}|\lambda+\varepsilon|^{-\frac{1}{2}},\end{aligned}$$

whenever $f(s) > |\lambda + \varepsilon|$. Likewise, using (4.2), (4.14) – (4.16) we obtain

$$\frac{s^{\frac{1}{2}} f^{\frac{1}{2}}(s)}{\sqrt{f^2(s) - 4s}} |(\lambda - \tau_{\pm}(s))^{-1}|$$

$$\leq \begin{cases} C_1^{-\frac{1}{2}} \left(\dfrac{s}{f(s)}\right)^{\frac{1}{2}} |\lambda + \varepsilon|^{-1} \leq C_1^{-\frac{1}{2}} |\lambda + \varepsilon|^{-\frac{1}{2}}, & \text{if } \dfrac{f(s)}{s} > \dfrac{1}{|\lambda + \varepsilon|}, \\[3mm] C_3 \left(\dfrac{s}{f(s)}\right)^{\frac{1}{2}} \dfrac{f(s)}{s} = C_3 \left(\dfrac{f(s)}{s}\right)^{\frac{1}{2}}, & \text{if } \dfrac{f(s)}{s} \leq \dfrac{1}{|\lambda + \varepsilon|}, \end{cases}$$

for some constant $C_3 > 0$. Therefore, for $\lambda \in \mathbf{C}$ with $\mathrm{Re}\,\lambda > 0$,

$$\left.\begin{array}{l} \dfrac{|\tau_{\pm}(s)| f^{\frac{1}{2}}(s)}{\sqrt{f^2(s) - 4s}} |(\lambda - \tau_{\pm}(s))^{-1}| \\[5mm] \dfrac{s^{\frac{1}{2}} f^{\frac{1}{2}}(s)}{\sqrt{f^2(s) - 4s}} |(\lambda - \tau_{\pm}(s))^{-1}| \end{array}\right\} \leq \mathrm{const}\,|\lambda + \varepsilon|^{-\frac{1}{2}}. \qquad (4.17)$$

Case 2: $f^2(s) < 4s$, $s \geq \frac{1}{4} |\mathrm{Im}\,\lambda|^2$.

Observe that for $\lambda \in \mathbf{C}$ with $\mathrm{Re}\,\lambda > 0$,

$$|\lambda - \tau_{\pm}(s)| \geq \frac{1}{2} f(s),$$

$$\left| f^{-\frac{1}{2}}(s)(\lambda - \tau_{\pm}(s)) \right| \geq \frac{1}{2} f^{\frac{1}{2}}(s).$$

We thus see that for λ as above,

$$|(\lambda - \tau_{\pm}(s))^{-1}|, \quad \left| f^{\frac{1}{2}}(s)(\lambda - \tau_{\pm}(s))^{-1} \right| \leq 2\max\{\sigma_1^{-1},\ \sigma_1^{-\frac{1}{2}}\},$$

and

$$\longrightarrow 0 \quad (\text{uniformly for } s \text{ as above}), \quad \text{as } |\mathrm{Im}\,\lambda| \to +\infty; \qquad (4.18)$$

moreover, there exists a constant $C_4 > 0$ such that for λ as above

$$|(\lambda - \tau_{\pm}(s))^{-1}|$$

$$\begin{cases} = o\left((\ln|\mathrm{Im}\,\lambda|)^{-1}\right) & \text{as } |\mathrm{Im}\,\lambda| \to \infty, \quad \text{if (4.5) holds,} \qquad (4.19) \\[3mm] \leq C_4 (|\mathrm{Im}\,\lambda|)^{-1} & \text{for } |\mathrm{Im}\,\lambda| > \delta, \quad \text{if (4.6) holds,} \end{cases}$$

$$\left| f^{\frac{1}{2}}(s)(\lambda - \tau_\pm(s))^{-1} \right|$$

$$
\begin{cases}
= o\left((\ln|\mathrm{Im}\lambda|)^{-\frac{1}{2}}\right) & \text{as } |\mathrm{Im}\lambda| \to \infty, \quad \text{if (4.5) holds,} \\[2mm]
\leq C_4(|\mathrm{Im}\lambda|)^{-\frac{1}{2}} & \text{for } |\mathrm{Im}\lambda| > \delta, \quad \text{if (4.6) holds.}
\end{cases}
\tag{4.20}
$$

Case 3: $f^2(s) < 4s$, $s < \frac{1}{4}|\mathrm{Im}\lambda|^2$.

We have that for $\lambda \in \mathbf{C}$ with $\mathrm{Re}\lambda > 0$,

$$
\begin{aligned}
|\lambda - \tau_\pm(s)| &\geq \left|\frac{1}{2}f(s)\right| + |\mathrm{Im}\lambda| - \frac{1}{2}\left(4s - f^2(s)\right)^{\frac{1}{2}} \\[2mm]
&\geq \frac{1}{2}\sigma_1 + |\mathrm{Im}\lambda| - s^{\frac{1}{2}} \\[2mm]
&\geq \frac{1}{2}\sigma_1 + |\mathrm{Im}\lambda| - \frac{1}{2}|\mathrm{Im}\lambda| \\[2mm]
&= \frac{1}{2}(\sigma_1 + |\mathrm{Im}\lambda|)
\end{aligned}
$$

and therefore

$$|(\lambda - \tau_\pm(s))^{-1}| \leq 2(\sigma_1 + |\mathrm{Im}\lambda|)^{-1}, \tag{4.21}$$

$$
\begin{aligned}
\left| f^{\frac{1}{2}}(s)(\lambda - \tau_\pm(s))^{-1} \right| &\leq 2f^{\frac{1}{2}}(s)(\sigma_1 + |\mathrm{Im}\lambda|)^{-1} \\[2mm]
&\leq 2|\mathrm{Im}\lambda|^{\frac{1}{2}}(\sigma_1 + |\mathrm{Im}\lambda|)^{-1}.
\end{aligned}
\tag{4.22}
$$

Now, collect the results in the three cases (especially (4.14), (4.17) – (4.22)), recall (4.10) – (4.13) and put

$$R_{f(A)}(\lambda) = \left(\lambda^2 + f(A)\lambda + A\right)^{-1}, \quad \lambda \in \rho(A, \, f(A)).$$

We obtain by applying the self-adjoint calculus that for all $\lambda \in \mathbf{C}$ with $\mathrm{Re}\lambda > 0$,

$$\left\| f^{\frac{1}{2}}(A)A^{\frac{1}{2}}R_{f(A)}(\lambda) \right\|, \quad \left\| \lambda f^{\frac{1}{2}}(A)R_{f(A)}(\lambda) \right\|$$

$$
\begin{cases}
\leq \text{const and} \longrightarrow 0 \text{ as } |\mathrm{Im}\lambda| \to +\infty, \\[2mm]
= o\left((\ln|\mathrm{Im}\lambda|)^{-\frac{1}{2}}\right) \text{ as } |\mathrm{Im}\lambda| \to +\infty, \quad \text{if (4.5) holds,} \\[2mm]
\leq \text{const}\,(|\mathrm{Im}\lambda|)^{-\frac{1}{2}} \text{ for } |\mathrm{Im}\lambda| > \delta, \quad \text{if (4.6) holds,}
\end{cases}
\tag{4.23}
$$

$$\left\| A^{\frac{1}{2}} R_{f(A)}(\lambda) \right\|, \quad \left\| f(A) R_{f(A)}(\lambda) \right\|, \quad \left\| \lambda R_{f(A)}(\lambda) \right\|,$$

$$\left\| \lambda^{-1} A R_{f(A)}(\lambda) \right\|$$

$$
\begin{cases}
\leq \text{ const and } \longrightarrow 0 \text{ as } |\text{Im}\lambda| \to +\infty, & \\
= o\left((\ln |\text{Im}\lambda|)^{-1} \right) \text{ as } |\text{Im}\lambda| \to +\infty, & \text{if (4.5) holds,} \\
\leq \text{ const } (|\text{Im}\lambda|)^{-1} \text{ for } |\text{Im}\lambda| > \delta, & \text{if (4.6) holds;}
\end{cases}
\tag{4.24}
$$

moreover, in the case of (4.7) we have that for each $\theta \in (0, \frac{\pi}{2})$,

$$\left\| A^{\frac{1}{2}} R_{f(A)}(\lambda) \right\|, \quad \left\| f(A) R_{f(A)}(\lambda) \right\|, \quad \left\| \lambda R_{f(A)}(\lambda) \right\|,$$

$$\left\| \lambda^{-1} A R_{f(A)}(\lambda) \right\|$$

$$\tag{4.25}$$

$$\leq \quad C_\theta |\lambda|^{-1}, \quad \text{for any } \lambda \in \delta + \Sigma_{\frac{\pi}{2}+\theta}.$$

for some $C_\theta > 0$. Note

$$R\left(\lambda; \overline{f(A)}\right) = \begin{pmatrix} \lambda^{-1} - \lambda^{-1} A R_{f(A)} & A^{\frac{1}{2}} R_{f(A)} \\ A^{\frac{1}{2}} R_{f(A)} & \lambda R_{f(A)} \end{pmatrix}.$$

In view of (4.24) and (4.25), we can apply Lemma 4.1 to obtain the desired results. The proof is then complete.

More generally, we have the following three theorems.

Theorem 4.3. *Let* $\mathcal{D}(B) \supset \mathcal{D}\left(A^{\frac{1}{2}}\right)$. *Then* $\overline{\mathcal{A}_B} = \mathcal{A}_B$, *and*
    (i) $\{\mathcal{T}_B(t)\}_{t \geq 0}$ *is exponentially stable and norm continuous for* $t > 0$.
    (ii) $\{\mathcal{T}_B(t)\}_{t \geq 0}$ *is differentiable for* $t > 0$ *if* (4.5) *holds.*

Proof. A standard verification shows that $\overline{\mathcal{A}_B} = \mathcal{A}_B$ and for $\text{Re}\lambda > 0$,

$$R(\lambda; \mathcal{A}_B) = \begin{pmatrix} \lambda^{-1} - \lambda^{-1}\overline{A^{\frac{1}{2}} R_B(\lambda) A^{\frac{1}{2}}} & A^{\frac{1}{2}} R_B(\lambda) \\ -\overline{R_B(\lambda) A^{\frac{1}{2}}} & \lambda R_B(\lambda) \end{pmatrix}, \tag{4.26}$$

where

$$R_B(\lambda) := (\lambda^2 + \lambda B + A)^{-1} \in \mathbf{L}(H).$$

Set $f_1 = \dfrac{1}{2} f$. We define

$$G_{01} = f_1^{-\frac{1}{2}}(A) B_1 f_1^{-\frac{1}{2}}(A) - I, \quad \mathcal{D}(G_{01}) = \mathcal{D}\left(A f_1^{-\frac{1}{2}}(A)\right),$$

$$G_{02} = f_1^{-\frac{1}{2}}(A) B_2 f_1^{-\frac{1}{2}}(A), \quad \mathcal{D}(G_{02}) = \mathcal{D}\left(A f_1^{-\frac{1}{2}}(A)\right).$$

Obviously, $\mathcal{D}\left(Af_1^{-\frac{1}{2}}(A)\right)$ is dense in H. Therefore we deduce from (4.3) and (4.4) that

$$G_1 := \overline{G}_{01}, \quad G_2 := \overline{G}_{02} \in \mathbf{L}(H), \tag{4.27}$$

$$I \leq G_1 \leq (2b-1)I, \quad -aG_1 \leq G_2 \leq aG_1.$$

Now, observing that for each $u \in H$,

$$\|(G_1 + iG_2)u\|\|u\| \geq |\langle(G_1 + iG_2)u, \, u\rangle|$$

$$\geq |\langle G_1 u, \, u\rangle| \geq \|u\|^2,$$

we obtain that $(G_1 + iG_2)$ is boundedly invertible and

$$\|(G_1 + iG_2)^{-1}\| \leq 1. \tag{4.28}$$

On the other hand, for $u \in E$, $\lambda \in \mathbf{C}$ with $\mathrm{Re}\,\lambda > 0$, we see

$$\|((G_1 + iG_2)^{-1} + \lambda f_1(A)R_{f_1(A)}(\lambda))u\| \, \|u\|$$

$$\geq |\mathrm{Re}\,\langle(G_1 + iG_2)^{-1}u, \, u\rangle + \mathrm{Re}\,\langle \lambda f_1(A)R_{f_1(A)}(\lambda)u, \, u\rangle|$$

$$\geq \mathrm{Re}\,\langle(G_1 + iG_2)^{-1}u, \, u\rangle$$

$$= \langle(G_1 + iG_2)^{-1}u, \, G_1(G_1 + iG_2)^{-1}u\rangle$$

$$\geq \|(G_1 + iG_2)^{-1}u\|^2$$

$$\geq \|G_1 + iG_2\|^{-2}\|u\|^2,$$

and therefore $((G_1 + iG_2)^{-1} + \lambda f_1(A)R_{f_1(A)}(\lambda))$ is boundedly invertible and

$$\left\|[(G_1 + iG_2)^{-1} + \lambda f_1(A)R_{f_1(A)}(\lambda)]^{-1}\right\| \leq \|G_1 + iG_2\|^2.$$

Combining this with (4.28) shows that $[I + \lambda f_1(A)R_{f_1(A)}(\lambda)(G_1+iG_2)]$ is boundedly invertible and

$$\left\|[I + \lambda f_1(A)R_{f_1(A)}(\lambda)(G_1 + iG_2)]^{-1}\right\| \leq \|G_1 + iG_2\|^2$$

for all $\lambda \in \mathbf{C}$ with $\mathrm{Re}\,\lambda > 0$.

It is easy to see that for any $\lambda \in \mathbf{C}$ with $\mathrm{Re}\,\lambda > 0$,

$$\begin{aligned} R_B(\lambda) &= R_{f_1(A)}(\lambda) - R_{f_1(A)}(\lambda)(\lambda B - \lambda f_1(A))R_B(\lambda) \\[4pt] &= R_{f_1(A)}(\lambda) - \lambda f_1^{\frac{1}{2}}(A)R_{f_1(A)}(\lambda)(G_1 + iG_2)f_1^{\frac{1}{2}}(A)R_B(\lambda), \end{aligned} \tag{4.29}$$

$$f_1^{\frac{1}{2}}(A)R_B(\lambda)$$

$$= [I + \lambda f_1(A)R_{f_1(A)}(\lambda)(G_1 + iG_2)]^{-1}f_1^{\frac{1}{2}}(A)R_{f_1(A)}(\lambda). \tag{4.30}$$

By virtue of (4.23) with f replaced by f_1, we obtain from (4.30) that for all $\lambda \in \mathbf{C}$ with $\mathrm{Re}\lambda > 0$,

$$\left\|\overline{f_1^{\frac{1}{2}}(A)R_B(\lambda)A^{\frac{1}{2}}}\right\|, \quad \|\lambda f_1^{\frac{1}{2}}(A)R_B(\lambda)\|,$$

$$\begin{cases} \leq \text{ const and } \longrightarrow 0 \text{ as } |\mathrm{Im}\lambda| \to +\infty, \\ = o\left((\ln|\mathrm{Im}\lambda|)^{-\frac{1}{2}}\right) \text{ as } |\mathrm{Im}\lambda| \to +\infty, \quad \text{if (4.5) holds.} \end{cases}$$

This combined with (4.23), (4.24) and (4.29) yields that for any $\lambda \in \mathbf{C}$ with $\mathrm{Re}\lambda > 0$,

$$\left\|\lambda^{-1}\overline{A^{\frac{1}{2}}R_B(\lambda)A^{\frac{1}{2}}}\right\|, \quad \|A^{\frac{1}{2}}R_B(\lambda)\|, \quad \left\|\overline{R_B(\lambda)A^{\frac{1}{2}}}\right\|, \quad \|\lambda R_B(\lambda)\|$$

$$\begin{cases} \leq \text{ const and } \longrightarrow 0 \text{ as } |\mathrm{Im}\lambda| \to +\infty, \\ = o\left((\ln|\mathrm{Im}\lambda|)^{-1}\right) \text{ as } |\mathrm{Im}\lambda| \to +\infty, \quad \text{if (4.5) holds.} \end{cases} \tag{4.31}$$

According to this and (4.26), Lemma 4.1 applies and yields the results as required.

Theorem 4.4. *If (4.6) holds, then $\{T_B(t)\}_{t\geq 0}$ is an analytic semigroup, which is exponentially stable.*

Proof. Write

$$\Xi := \left\{(\mu;\ u,\ v);\ \mu \in R \setminus \{0\},\ u \in \mathcal{D}\left(A^{\frac{1}{2}}\right) \text{ and } v \in \mathcal{D}(A)\right.$$

$$\left. \text{with } \frac{1}{2} \leq \|u\| + \|v\| \leq \frac{3}{2}\right\}.$$

We claim

$$\inf\left\{|\mu|^{-1}\left\|(i\mu - A_B)\binom{u}{v}\right\|;\ (\mu;\ u,\ v) \in \Xi\right\} \neq 0. \tag{4.32}$$

In fact, if this is not true then there exists a sequence

$$\{(\mu_k;\ u_k,\ v_k)\}_{k \in N} \subset \Xi$$

such that

$$\lim_{k \to \infty} |\mu_k|^{-1}\left(i\mu_k u_k - A^{\frac{1}{2}}v_k\right) = 0, \tag{4.33}$$

$$\lim_{k \to \infty} |\mu_k|^{-1} \left(A^{\frac{1}{2}} u_k + i\mu_k v_k + Bv_k \right) = 0. \tag{4.34}$$

Therefore

$$\lim_{k \to \infty} |\mu_k|^{-1} \left(i\mu_k \|u_k\|^2 - \left\langle v_k, \ A^{\frac{1}{2}} u_k \right\rangle \right) = 0, \tag{4.35}$$

$$\lim_{k \to \infty} |\mu_k|^{-1} \left(\left\langle A^{\frac{1}{2}} u_k, \ v_k \right\rangle + i\mu_k \|v_k\|^2 + \langle Bv_k, \ v_k \rangle \right) = 0. \tag{4.36}$$

Accordingly,

$$\lim_{k \to \infty} |\mu_k|^{-1} \mathrm{Re} \left\langle A^{\frac{1}{2}} u_k, \ v_k \right\rangle = 0,$$

$$\lim_{k \to \infty} |\mu_k|^{-1} \left(\mathrm{Re} \left\langle A^{\frac{1}{2}} u_k, \ v_k \right\rangle + \langle B_1 v_k, \ v_k \rangle \right) = 0.$$

Hence,

$$\lim_{k \to \infty} |\mu_k|^{-1} \langle B_1 v_k, \ v_k \rangle = 0. \tag{4.37}$$

Consequently,

$$\lim_{k \to \infty} |\mu_k|^{-1} \langle Bv_k, \ v_k \rangle = 0 \tag{4.38}$$

by (4.4). Moreover, according to (4.3) we have

$$|\mu_k|^{-1} \left\| f^{\frac{1}{2}}(A) v_k \right\|^2 = |\mu_k|^{-1} \left\langle f^{\frac{1}{2}}(A) v_k, \ f^{\frac{1}{2}}(A) v_k \right\rangle$$

$$\leq |\mu_k|^{-1} \langle B_1 v_k, \ v_k \rangle, \quad k \in N.$$

Combining this with (4.37), we know

$$\lim_{k \to \infty} |\mu_k|^{-\frac{1}{2}} \left\| f^{\frac{1}{2}}(A) v_k \right\| = 0. \tag{4.39}$$

In view of (4.27), we obtain that $f^{-\frac{1}{2}}(A) B f^{-\frac{1}{2}}(A)$ is bounded on $\mathcal{D}\left(A f^{-\frac{1}{2}}(A) \right)$. This implies that

$$\left\| f^{-\frac{1}{2}}(A) Bv \right\| \leq \text{const} \left\| f^{\frac{1}{2}}(A) v \right\|, \quad v \in \mathcal{D}(A). \tag{4.40}$$

It follows from (4.39) and (4.40) that

$$\lim_{k \to \infty} |\mu_k|^{-\frac{1}{2}} \left\| f^{-\frac{1}{2}}(A) Bv_k \right\| = 0. \tag{4.41}$$

Observing by (4.34) that

$$\lim_{k \to \infty} |\mu_k|^{-1} \left(\left\| A^{\frac{1}{2}} f^{-\frac{1}{2}}(A) u_k \right\|^2 + i\mu_k \left\langle v_k, \ A^{\frac{1}{2}} f^{-1}(A) u_k \right\rangle \right.$$

$$\left. + \left\langle f^{-\frac{1}{2}}(A) Bv_k, \ A^{\frac{1}{2}} f^{-\frac{1}{2}}(A) u_k \right\rangle \right) = 0, \tag{4.42}$$

we conclude by (4.6) and (4.41) that

$$|\mu_k|^{-1} \left\| A^{\frac{1}{2}} f^{-\frac{1}{2}}(A) u_k \right\|^2 \leq \text{const} \left(1 + |\mu_k|^{-\frac{1}{2}} \left\| A^{\frac{1}{2}} f^{-\frac{1}{2}}(A) u_k \right\| \right), \quad k \in N.$$

Therefore

$$|\mu_k|^{-\frac{1}{2}} \left\| A^{\frac{1}{2}} f^{-\frac{1}{2}}(A) u_k \right\| \leq \text{const}, \quad k \in N. \tag{4.43}$$

A combination of (4.39) and (4.43) shows

$$\lim_{k \to \infty} |\mu_k|^{-1} \left\langle A^{\frac{1}{2}} u_k, \ v_k \right\rangle = 0.$$

Taking (4.35), (4.36) and (4.38) into account, we get

$$\lim_{k \to \infty} \|u_k\| = \lim_{k \to \infty} \|v_k\| = 0.$$

This is in contradiction with the definition of Ξ. In other words, (4.32) holds.
It is easy to see that

$$\mathcal{A}_B \left(\mathcal{D} \left(A^{\frac{1}{2}} \right) \times \mathcal{D}(A) \right) = \mathcal{D} \left(A^{\frac{1}{2}} \right) \times H.$$

This together with $0 \in \rho \left(\overline{\mathcal{A}_B} \right)$ suggests

$$\overline{\mathcal{A}_B} \Big|_{\mathcal{D}\left(A^{\frac{1}{2}} \right) \times \mathcal{D}(A)} = \overline{\mathcal{A}_B}.$$

Thus, by (4.32) we have

$$\inf \left\{ |\mu|^{-1} \left\| (i\mu - \overline{\mathcal{A}_B}) \begin{pmatrix} u \\ v \end{pmatrix} \right\|; \ \mu \in R \setminus \{0\}, \ \begin{pmatrix} u \\ v \end{pmatrix} \in \mathcal{D}\left(\overline{\mathcal{A}_B} \right) \right.$$

$$\left. \text{with } \|u\| + \|v\| = 1 \right\} \neq 0.$$

From this we deduce that for each $\mu \in R \setminus \{0\}$, $i\mu - \overline{\mathcal{A}_B}$ is injective and its inverse is bounded on $\mathcal{R}\left(i\mu - \overline{\mathcal{A}_B} \right)$. Since $\tau + i\mu \in \rho\left(\overline{\mathcal{A}_B} \right)$ for each $\tau > 0$, it follows that $i\mu \in \rho\left(\overline{\mathcal{A}_B} \right)$. As a consequence,

$$\left\| R\left(i\mu; \overline{\mathcal{A}_B} \right) \right\| \leq \frac{\text{const}}{|\mu|}, \quad \mu \in R \setminus \{0\},$$

which implies

$$\left\| R\left(\lambda; \overline{\mathcal{A}_B} \right) \right\| \leq \text{const}, \quad \text{Re}\lambda \geq 0.$$

Thus, an application of Lemma 4.1 ends the proof.

Corollary 4.5. *Let $B_0 \in L(H)$. Then the closure of the operator matrix*

$$\mathcal{A}_{B+B_0} = \begin{pmatrix} 0 & A^{\frac{1}{2}} \\ -A^{\frac{1}{2}} & -B - B_0 \end{pmatrix}$$

with domain $\mathcal{D}(\mathcal{A}_{B+B_0}) = \mathcal{D}(\mathcal{A}_B)$, generates a strongly continuous semigroup $\{T_{B+B_0}(t)\}_{t\geq 0}$ on $H \times H$, and the conclusions of Theorem 4.3 and 4.4 (except the exponential stability) hold with $\{T_{B+B_0}(t)\}_{t\geq 0}$ instead of $\{T_B(t)\}_{t\geq 0}$.

6.5 Notes

With respect to the differentiability of a strongly continuous semigroup, the earlier results were obtained by Hille [3] and Yosida [2]. The full characterization of a differentiable semigroup, which depends only on the properties of its generator, is due to Pazy [2, Theorem 2.4.8]. This theorem also characterizes a semigroup of operators which is infinitely differentiable in the uniform operator topology (i.e., in $L(E)$), since if $T(t)$ is differentiable for $t > 0$, then $T(t)$ is infinitely differentiable in $L(E)$ for $t > 0$ (Pazy [2, Corollary 2.4.4]). Some generalizations of Pazy's characterization were given by Barbu [1] and Watanabe [1].

Section 6.1 is from Xiao-Liang [24].

The first necessary condition, for an operator A to generate a strongly continuous semigroup which is norm continuous for $t > 0$, was given by Hille-Phillips [1]. In 1991, You [1] showed that this condition is also sufficient in the case of Hilbert spaces; that is, the norm continuity for $t > 0$ of a strongly continuous semigroup in a Hilbert space is equivalent to the decay to zero of the resolvent of its generator along a vertical line. A much simpler and more straightforward proof of this result was given by El-Mennaoui-Engel [1]. The analogous characterization (Theorem 2.1) of the norm continuity for $t > 0$ of the propagators of an arbitrary order strongly wellposed (ACP_n) was established by Liang-Xiao [8]. The extensions (Theorems 2.2 and 2.3, Corollary 2.4) to general operator families are due to Liang-Xiao [11].

In Kurepa [1] it is showed that for a cosine operator function $S_0(t)$, the norm continuity for $t \in R$ implies the boundedness of the generator A_0. This result was also proved by Lutz [1] using a different method. Both the proofs depend heavily on the norm continuity at $t = 0$ of $S_0(t)$. Theorem 3.2 is taken from Liang-Xiao [13], which (even Corollary 3.4) is a nontrivial generalization of Kurepa's result; and the approach used, which is different from that in Kurepa [1] or Lutz [1] and reveals the boundedness of A_0 without the assumption of norm continuity at $t = 0$ (of $S_0(t)$), is nontrivial as well. Lemma 3.1 is an improved version of Lemma 5.2 in Goldstein [2].

Section 6.4 comes from Xiao-Liang [20]. Stimulated by the two conjectures posed by Chen-Russell [1], Chen-Triggiani [1-3] and independently F. L. Huang [6-8] investigated the behaviors of $\{T_B(t)\}_{t\geq 0}$ for the case when B is 'comparable' to A^α ($0 < \alpha \leq 1$). They showed (among others) the differentiability as well as the analyticity (if $\frac{1}{2} \leq \alpha \leq 1$) of $\{T_B(t)\}_{t \geq 0}$. Theorems 4.2 and 4.3 generalize Theorem 1.1 of Chen-Triggiani [2, 3] .

Chapter 7

Almost periodicity

Summary

This chapter investigates almost periodic (in short a.p.) solutions of the strongly wellposed (ACP_2) in a Banach space E, such solutions corresponding to "almost standing waves" in applications.

In Section 7.1, we look at the Cauchy problem for the incomplete second order equation $u''(t) = Au(t)$ $(t \in R)$. It is known that in this case the first propagator $C(t)$ is a cosine operator function, and the second propagator $S(t)$ a sine operator function. We characterize (Theorem 1.2) a.p. cosine (or sine) operator functions in terms of the spectral properties of A. In the case when E is a Hilbert space, we clarify the relation between a.p. cosine operator functions and sine operator functions and show their structures in terms of the mean value $\mathcal{P}_\lambda$ (see Theorem 1.3 and the statement above it). For general Banach spaces, we show (Theorem 1.7) that, if $S(t)$ is a.p., so are all the solutions of the incomplete (ACP_2).

In Section 7.2, we consider complete (ACP_2) and characterize those (ACP_2) which have a.p. generalized solutions. Also, the relation between almost periodicity of $S_0(t)$ and $S_1(t)$ is discussed. The final part is devoted to the case when the equation takes the form

$$u''(t) + (aB + bI)u'(t) + (cB + dI)u(t) = 0, \quad t \geq 0;$$

Theorems 2.6 and 2.8 together can be viewed as a generalization of Theorem 1.2, and Theorem 2.9 a generalization of Theorem 1.3.

7.1 Incomplete second order equations

Let E denote a Banach space. Consider the incomplete second order equation

$$u''(t) = Au(t) \quad (t \in R), \tag{1.1}$$

where A is a closed densely defined linear operator in E. It is well known that the Cauchy problem for (1.1) is wellposed (or equivalently by Theorem 1.1.5, strongly wellposed), which will be assumed throughout this section, if and only if A is the generator of a strongly continuous cosine function $\{C(t)\}_{t\in R}$; that any solution $u(t)$ $(t \in R)$ of (1.1) admits a representation

$$u(t) = C(t)u(0) + S(t)u'(0) \quad (t \in R),$$

where $\{S(t)\}_{t\in R}$ is the strongly continuous sine function, defined by

$$S(t)u := \int_0^t C(s)u\,ds \quad (t \in R, \ u \in E);$$

that for each $u \in \mathcal{D}(A)$, $t \in R$,

$$\begin{cases} C''(t)u = AC(t)u = C(t)Au, \quad C(-t)u = C(t)u, \\[2mm] S''(t)u = AS(t)u = S(t)Au, \quad -S(-t)u = S(t)u. \end{cases} \tag{1.2}$$

Let f map R into E. Given $\varepsilon > 0$, we call $\tau \in R$ an ε-period for f if

$$\|f(t+\tau) - f(t)\| \leq \varepsilon, \quad t \in R.$$

The set of all ε-periods for f is denoted by $\vartheta(f, \varepsilon)$. We say that f is almost periodic, written a.p., if for every $\varepsilon > 0$, the set $\vartheta(f, \varepsilon)$ is relatively dense in R. A subset V of R is called relatively dense (in R) if there exists an $l > 0$ such that every subinterval of R of length l meets V. We say that f is weakly almost periodic, written w.a.p., if for each $u^* \in E^*$, the scalar function $u \circ f$ is almost periodic.

Definition 1.1. $\{S(t)\}_{t\in R}$ (resp. $\{C(t)\}_{t\in R}$) is almost periodic, if for each $u \in E$ the function $t \mapsto S(t)u$ $(t \in R)$ (resp. $t \mapsto C(t)u$ $(t \in R)$) is almost periodic.

$\{S(t)\}_{t\in R}$ (resp. $\{C(t)\}_{t\in R}$) is weakly almost periodic, if for each $u \in E$ the function $t \mapsto S(t)u$ $(t \in R)$ (resp. $t \mapsto C(t)u$ $(t \in R)$) is weakly almost periodic.

Theorem 1.2. $\{S(t)\}_{t\in R}$ *(resp. $\{C(t)\}_{t\in R}$) is a.p. if and only if $\{S(t)\}_{t\in R}$ (resp. $\{C(t)\}_{t\in R}$) is uniformly bounded and the set D of eigenvectors of the A in equation (1.1) is total in E.*

Proof. We only consider the case of $\{S(t)\}_{t\in R}$. $\{C(t)\}_{t\in R}$ can be similarly treated.

Necessity. Immediately, the assertion that $\{S(t)\}_{t\in R}$ is uniformly bounded follows from the fact that an a.p. function is bounded and from the uniform boundedness principle. For each $u \in E$, $\lambda \in R$, set

$$\mathcal{P}_\lambda u = \lim_{t\to\infty} \frac{1}{2t} \int_{-t}^{t} e^{-i\lambda s} S(s)u\,ds. \tag{1.3}$$

Since $\{S(t)\}_{t\in R}$ is a.p., the limit in (1.3) exists. Clearly, $\mathcal{P}_\lambda \in L(E)$. For each $u \in \mathcal{D}(A)$, $\lambda \in R$, we have

$$\lambda^2 \mathcal{P}_\lambda u = \lim_{t\to\infty} \frac{1}{2t} \left[i\lambda e^{-i\lambda s} S(s)u \Big|_{-t}^{t} - \int_{-t}^{t} i\lambda e^{-i\lambda s} S'(s)u\,ds \right]$$

$$= -\lim_{t\to\infty} \frac{1}{2t} \int_{-t}^{t} i\lambda e^{-i\lambda s} S'(s)u\,ds$$

$$= \lim_{t\to\infty} \frac{1}{2t} \left[e^{-i\lambda s} S'(s)u \Big|_{-t}^{t} - \int_{-t}^{t} e^{-i\lambda s} S''(s)u\,ds \right]$$

$$= \lim_{t\to\infty} \frac{1}{2t} \left[-2i(\sin \lambda t)S'(t)u - \int_{-t}^{t} e^{-i\lambda s} S(s)Au\,ds \right].$$

But

$$\lim_{t\to\infty} \frac{1}{2t} \int_{-t}^{t} e^{-i\lambda s} S(s)Au\,ds = \mathcal{P}_\lambda(Au)$$

exists, so does $\lim_{t\to\infty} \frac{1}{t}(\sin \lambda t)S'(t)u$.

Denote

$$W := \lim_{t\to\infty} \frac{1}{t}(\sin \lambda t)S'(t)u,$$

then

$$W = \lim_{t\to\infty} \frac{1}{t}\left\{ \frac{d}{dt}[(\sin \lambda t)S(t)u] - \lambda(\cos \lambda t)S(t)u \right\}$$

$$= \lim_{t\to\infty} \frac{1}{t}\frac{d}{dt}\{(\sin \lambda t)S(t)u\}.$$

We now claim $W = 0$. In fact, assume $W \neq 0$ then there exists a real bounded linear functional g on E such that $g(W) > 0$. Obviously,

$$g(W) = \lim_{t\to\infty} \frac{1}{t}\frac{d}{dt}\{(\sin \lambda t)g[S(t)u]\}.$$

Therefore, there exists an ε with $0 < \varepsilon < g(W)$ and a $k > 0$ such that

$$\frac{1}{t}\frac{d}{dt}\{(\sin \lambda t)g[S(t)u]\} > g(W) - \varepsilon, \quad \text{for } |t| \geq k.$$

Thus

$$\int_k^t \frac{d}{ds}\{(\sin \lambda s)g[S(s)u]\}ds > \int_k^t [g(W) - \varepsilon]s\,ds, \quad \text{for } |t| \geq k.$$

Hence

$$g[S(t)u] > \frac{1}{2}(t^2 - k^2)[g(W) - \varepsilon] + (\sin \lambda k)g[S(k)u]$$

for

$$t = \frac{\pi}{2\lambda}(4n+1) \quad (n \in N \text{ large enough}),$$

which is in contradiction with the boundedness of the function $t \rightarrow g[S(t)u]$ $(t \in R)$. So

$$
\begin{aligned}
\lambda^2 \mathcal{P}_\lambda u &= -\lim_{t\to\infty} \frac{1}{2t} \int_{-t}^{t} e^{-i\lambda s} S(s) Au\, ds \\[2mm]
&= -\mathcal{P}_\lambda Au \\[2mm]
&= -A \lim_{t\to\infty} \frac{1}{2t} \int_{-t}^{t} e^{-i\lambda s} S(s) u\, ds \\[2mm]
&= -A\mathcal{P}_\lambda u,
\end{aligned}
\tag{1.4}
$$

i.e.,

$$(\lambda^2 + A)\mathcal{P}_\lambda u = 0, \quad \mathcal{P}_\lambda(\lambda^2 + A)u = 0, \quad u \in \mathcal{D}(A), \ \lambda \in R. \tag{1.5}$$

Suppose now that D is not total in E. Then there exists a $u^* \in E^*$, $u^* \neq 0$ such that $u^*(u) = 0$ for each $u \in D$. From (1.5), we deduce that $u^*(\mathcal{P}_\lambda u) = 0$ for each $u \in \mathcal{D}(A)$, $\lambda \in R$, and this implies that the a.p. function $t \rightarrow u^*[S(t)u]$ vanishes due to all its "Fourier coefficients" being zero. Hence the function

$$t \rightarrow \{u^*[S(t)u]\}' = u^*[S'(t)u]$$

vanishes. In particular,

$$u^*[S'(0)u] = u^*(u) = 0, \quad \text{for each } u \in \mathcal{D}(A).$$

This implies $u^* = 0$ due to the denseness of $\mathcal{D}(A)$, which is a contradiction.

Sufficiency. We know that there exists an $\omega > 0$ such that

$$R(\lambda^2; A)u = \int_0^\infty e^{-\lambda t} S(t)u\, dt \quad (u \in E, \ \mathrm{Re}\,\lambda > \omega). \tag{1.6}$$

In view of the uniform boundedness of $\{S(t)\}_{t\in R}$, we have that the right-hand side of (1.6) defines an E-valued analytic function in $\mathrm{Re}\,\lambda > 0$. Since the norm of the resolvent $R(\mu; B)$ of an arbitrary operator B must tend to infinity as μ approaches the boundary of the resolvent set $\rho(B)$, it follows that every λ^2 with $\mathrm{Re}\,\lambda > 0$ belongs to $\rho(A)$. Thus the spectrum

$$\sigma(A) \subset (-\infty, \ 0].$$

Moreover, there exists an $a > 0$ such that the point spectrum

$$\sigma_p(A) \subset (-\infty, \ -a^2]. \tag{1.7}$$

Indeed, let the sequence $\{-\lambda_n^2; \ \lambda_n > 0\}_{n\in N}$ satisfy $-\lambda_n^2 \in \sigma_p(A)$ for each $n \in N$ and $\lambda_n \to 0$ as $n \to \infty$. Then

$$S(t)u_n = \frac{1}{\lambda_n}(\sin \lambda_n t)u_n, \quad \text{for } t \in R,$$

where for each $n \in N$, u_n is an eigenvector of A corresponding to $-\lambda_n^2$ with $\|u_n\| = 1$. Accordingly,

$$\frac{1}{\lambda_n}|\sin \lambda_n t| = \|S(t)u_n\| \le \sup_{t \in R} \|S(t)\| \qquad (1.8)$$

for each $n \in N$, $t \in R$. The left-hand side of (1.8) approaches $|t|$ as n approaches ∞, which yields a contradiction. On the other hand, let u be such that $Au = 0$, then $S(t)u = tu$ for $t \in R$, which implies $u = 0$.

For each $u \in E$, there exist by assumption, $u_1, u_2, \cdots, u_n, \cdots$, eigenvectors of A, corresponding to eigenvalues $-\lambda_1^2, -\lambda_2^2, \cdots, -\lambda_n^2, \cdots$, $(\lambda_j \in R, \lambda_j \ne 0, j \in N$) such that

$$u = \lim_{i \to \infty} \sum_{j=1}^{n_i} \alpha_j u_j$$

for some $n_i \in N$ $(n_i \to \infty$ as $i \to \infty)$, $\alpha_j \in C$, $i, j \in N$. Observe that the function

$$t \longmapsto S(t) \left(\sum_{j=1}^{n_i} \alpha_j u_j \right) = \sum_{j=1}^{n_i} \left(\frac{\alpha_j}{\lambda_j} \sin \lambda_j t \right) u_j,$$

for each n_i, is a.p. and $t \mapsto S(t)u$ is the uniform limit on R of

$$t \longmapsto S(t) \left(\sum_{j=1}^{n_i} \alpha_j u_j \right), \quad \text{as } i \to \infty.$$

This ends the proof by the uniform boundedness of $\{S(t)\}_{t \in R}$.

Suppose now E is a Hilbert space. Fattorini in [7, §V.7] has shown that when $\{C(t)\}_{t \in R}$ is a.p., $\{S(t)\}_{t \in R}$ may not be so.

Theorem 1.3. *Let E be a Hilbert space. If $\{S(t)\}_{t \in R}$ is a.p., then so is $\{C(t)\}_{t \in R}$. In this case, we obtain, for each $u \in E, t \in R$,*

$$S(t)u = \sum_{\lambda \ge a} 2i(\sin \lambda t)\mathcal{P}_\lambda u, \quad C(t)u = \sum_{\lambda \ge a} 2i\lambda(\cos \lambda t)\mathcal{P}_\lambda u,$$

with a as in (1.7).

Proof. For each $u \in E, \lambda \geq a, \mu \in R$,

$$
\begin{aligned}
\mathcal{P}_\mu \mathcal{P}_\lambda u &= \lim_{t \to \infty} \frac{1}{2t} \int_{-t}^t e^{-i\mu s} S(s) \mathcal{P}_\lambda u \, ds \\[2mm]
&= \lim_{t \to \infty} \frac{1}{2t} \int_{-t}^t e^{-i\mu s} \frac{1}{\lambda} (\sin \lambda s) \mathcal{P}_\lambda u \, ds \\[2mm]
&= \begin{cases} 0, & \text{if } \mu \neq \pm\lambda, \\[2mm] \dfrac{1}{2\lambda} \mathcal{P}_\lambda u, & \text{if } \mu = -\lambda, \\[2mm] -\dfrac{i}{2\lambda} \mathcal{P}_\lambda u, & \text{if } \mu = \lambda. \end{cases}
\end{aligned}
\tag{1.9}
$$

Here we used the fact that $\mathcal{P}_\lambda u \in \mathcal{N}\left(\lambda^2 + A\right)$, derived from (1.5) and the closedness of A. Accordingly, $\mathcal{Q}_\lambda := 2i\lambda \mathcal{P}_\lambda \ (\lambda \geq a)$ is a family of mutually orthogonal projections. Set, for each $u \in E$, $t \in R$,

$$
\tilde{S}(t)u = \sum_{\lambda \geq a} \frac{\sin \lambda t}{\lambda} \mathcal{Q}_\lambda u;
$$

then it is clear that $t \mapsto \tilde{S}(t)u$ is a.p. Combining (1.5), (1.7) and (1.9), and noting $-\mathcal{P}_\mu = \mathcal{P}_{-\mu}$ for each $\mu \in R$, we deduce that for each $\mu \in R$,

$$
\begin{aligned}
&\lim_{t \to \infty} \frac{1}{2t} \int_{-t}^t e^{-i\mu s} \tilde{S}(s) u \, ds \\[2mm]
&\quad = \begin{cases} \lim\limits_{t \to \infty} \dfrac{2i}{2t} \displaystyle\int_{-t}^t e^{-i\mu s} (\sin|\mu|s) \mathcal{P}_{|\mu|} u \, ds, & \text{if } \mu \notin (-a, a) \\[4mm] 0, & \text{if } \mu \in (-a, a) \end{cases} \\[4mm]
&\quad = \mathcal{P}_\mu u \\[2mm]
&\quad = \lim_{t \to \infty} \frac{1}{2t} \int_{-t}^t e^{-i\mu s} S(s) u \, ds,
\end{aligned}
$$

so that both of the a.p. functions $t \mapsto \tilde{S}(t)u$ and $t \mapsto S(t)u$ have the same Fourier series, which implies $S(t) = \tilde{S}(t)$ $(t \in R)$. Hence

$$
C(t)u = (S(t)u)' = \sum_{\lambda \geq a} (\cos \lambda t) \mathcal{Q}_\lambda u
$$

for each $u \in E$, $t \in R$. It follows immediately that $\{C(t)\}_{t \in R}$ is a.p.

Theorem 1.4. Let $\{S(t)\}_{t \in R}$ (resp. $\{C(t)\}_{t \in R}$) be a.p., and suppose that for some $\lambda \in R$, $-\lambda^2$ is an isolated point of $\sigma(A)$. Then $-\lambda^2$ is a simple pole of the resolvent $R(\cdot; A)$ with the residue $2i\lambda P_\lambda$ (resp. $2P_\lambda$ if $\lambda \neq 0$, or P_0 if $\lambda = 0$).

Proof. We present a detailed proof only for the case of $\{S(t)\}_{t \in R}$. $\{C(t)\}_{t \in R}$ can be dealt with in the same way.

Let $u \in \mathcal{N}\left(\lambda^2 + A\right)$ ($\lambda \in R$, $\lambda \neq 0$). Then

$$S(t)u = \frac{1}{\lambda}(\sin \lambda t)u, \quad t \in R.$$

Therefore

$$\mathcal{P}_\lambda u = \lim_{t \to \infty} \frac{1}{2t} \int_{-t}^{t} e^{-i\lambda s} \frac{1}{\lambda}(\sin \lambda s)u \, ds = -\frac{i}{2\lambda}u. \tag{1.10}$$

Accordingly, $\mathcal{N}\left(\lambda^2 + A\right) \subset \mathcal{R}(\mathcal{P}_\lambda)$. The converse inclusion follows from (1.5). Hence

$$\mathcal{N}\left(\lambda^2 + A\right) = \mathcal{R}(\mathcal{P}_\lambda) \quad (\lambda \in R, \, \lambda \neq 0). \tag{1.11}$$

Moreover, by (1.5) again, we obtain

$$\mathcal{R}\left(\lambda^2 + A\right) \subset \mathcal{N}(\mathcal{P}_\lambda) \quad (\lambda \in R, \lambda \neq 0). \tag{1.12}$$

A combination of (1.10) and (1.11) shows

$$\mathcal{P}_\lambda^2 u = -\frac{i}{2\lambda}\mathcal{P}_\lambda u \quad (u \in E, \, \lambda \in R, \, \lambda \neq 0).$$

Thus it is clear that

$$2i\lambda\mathcal{P}_\lambda \quad (\lambda \in R, \, \lambda \neq 0) \quad \text{are projections.} \tag{1.13}$$

Let P be the spectral projection associated with A and $-\lambda^2$. Then A is completely reduced by the decomposition $E = \mathcal{N}(P) \oplus \mathcal{R}(P)$, that is $A = A_{\mathcal{N}} \oplus A_P$. So

$$-\lambda^2 \notin \sigma(A_{\mathcal{N}}) \quad \text{and} \quad \sigma(A_P) = \{-\lambda^2\}.$$

Since $S(t)$ and A commute, we know that $S(t)$ and P commute. Hence $S(t)$ is also completely reduced by $E = \mathcal{N}(P) \oplus \mathcal{R}(P)$. Write

$$S(t) = S_{\mathcal{N}}(t) \oplus S_P(t).$$

As $\{S_P(t)\}_{t \in R}$ is a.p. with generator A_P in $\mathcal{R}(P)$, by Theorem 1.2 the set of eigenvectors of A_P is total in $\mathcal{R}(P)$, that is

$$\mathcal{N}\left(\lambda^2 + A_P\right) = \mathcal{R}(P),$$

i.e.,

$$A_P = \lambda^2 I \Big|_{\mathcal{R}(P)}.$$

Thus $R(\cdot;\ A_{\mathcal{P}})$ has a simple pole at $-\lambda^2$ and this implies that also

$$R(\cdot;\ A) = R(\cdot;\ A_{\mathcal{N}}) \oplus R(\cdot;\ A_{\mathcal{P}})$$

has a simple pole at $-\lambda^2$.

To prove that $P = 2i\lambda\mathcal{P}_\lambda$, we observe that by (1.11)

$$\begin{aligned}
\mathcal{R}(\mathcal{P}_\lambda) &= \mathcal{N}\left(\lambda^2 + A_{\mathcal{N}}\right) \oplus \mathcal{N}\left(\lambda^2 + A_{\mathcal{P}}\right) \\[2mm]
&= \{0\} \oplus \mathcal{R}(P) \\[2mm]
&= \mathcal{R}(P).
\end{aligned}$$

Therefore $\mathcal{R}(2i\lambda\mathcal{P}_\lambda) = \mathcal{R}(P)$. On the other hand, we have

$$\begin{aligned}
\mathcal{R}\left(\lambda^2 + A\right) &= \mathcal{R}\left(\lambda^2 + A_{\mathcal{N}}\right) \oplus \mathcal{R}\left(\lambda^2 + A_{\mathcal{P}}\right) \\[2mm]
&= \mathcal{N}(P) \oplus \{0\} \\[2mm]
&= \mathcal{N}(P)
\end{aligned}$$

so that (1.12) gives

$$\mathcal{N}(P) \subset \mathcal{N}(\mathcal{P}_\lambda) = \mathcal{N}(2i\lambda\mathcal{P}_\lambda).$$

According to the above observations we obtain $P = 2i\lambda\mathcal{P}_\lambda$, noting (1.13) and the fact that $0 \notin \sigma_p(A)$.

Theorem 1.5. *Suppose E is weakly sequentially complete. Then each w.a.p. sine (or cosine) function is a.p.*

Proof. Again, we only take care of $\{S(t)\}_{t\in R}$.

Clearly, $\{S(t)\}_{t\in R}$ is uniformly bounded. Since E is weakly sequentially complete, we have that for each $\lambda \in R, u \in \mathcal{D}(A)$ and $u^* \in E^*$, there is a $V_{u,\lambda} \in E$ such that

$$\lim_{t\to\infty} u^*\left[\frac{1}{2t}\int_{-t}^{t} e^{-i\lambda s} S(s)u\,ds\right] = u^*(V_{u,\lambda}).$$

Therefore,

$$u^*\left[\lambda^2(I - A)^{-1}V_{u,\lambda}\right]$$

$$= \lim_{t\to\infty} u^*\left[\frac{\lambda^2}{2t}(I - A)^{-1}\int_{-t}^{t} e^{-i\lambda s} S(s)u\,ds\right]$$

$$= -\lim_{t\to\infty}\frac{1}{t}u^*\left[i(I - A)^{-1}(\sin\lambda t)S'(t)u\right] + u^*\left[V_{u,\lambda} - (I - A)^{-1}V_{u,\lambda}\right].$$

Proceeding now as in the proof of (1.4), we conclude

$$u^* \left[\lambda^2 (I - A)^{-1} V_{u,\lambda} \right] = u^* \left[V_{u,\lambda} - (I - A)^{-1} V_{u,\lambda} \right].$$

Consequently,

$$\lambda^2 (I - A)^{-1} V_{u,\lambda} = V_{u,\lambda} - (I - A)^{-1} V_{u,\lambda},$$

i.e.

$$(\lambda^2 + A) V_{u,\lambda} = 0.$$

The fact that D is total in E is then derived as in the proof of Theorem 1.2 with $V_{u,\lambda}$ instead of $\mathcal{P}_\lambda u$. The result now follows by Theorem 1.2.

Theorem 1.6. *Assume that $\{S(t)\}_{t \in R}$ is w.a.p. Then all the solutions of* (1.1) *are w.a.p.*

Proof. First, we prove $t \mapsto C(t)u$ is w.a.p. for every $u \in \mathcal{D}(A)$.
  It is clear that there exist $M, \omega > 0$ such that

$$\|S'(t)\| = \|C(t)\| \le M e^{\omega |t|}, \quad t \in R. \tag{1.14}$$

Moreover, the almost periodicity of $\{S(t)\}_{t \in R}$ implies that

$$\sup_{t \in R} \|S(t)\| < \infty. \tag{1.15}$$

For any $u \in \mathcal{D}(A)$, $t \in R$, set

$$g(t, u) := S(t)Au - \omega_0^2 S(t)u, \tag{1.16}$$

where $\omega_0 > \omega$ is a constant. It is clear that $g(t, u)$ is continuous in t, and

$$\|g(t, u)\| \le \|S(t)\|\|Au\| + \omega_0^2 \|S(t)\|\|u\|$$

$$\le \text{const} \left(\|Au\| + \|u\| \right), \quad t \in R.$$

Therefore, the integral $\displaystyle\int_{-\infty}^{\infty} e^{-\omega_0 |t-s|} g(s, u) ds$ exists. By virtue of (1.2), (1.14), (1.15) and (1.16), we obtain that for any $u \in \mathcal{D}(A)$,

$$-\frac{1}{2\omega_0} \int_{-\infty}^{\infty} e^{-\omega_0 |t-s|} g(s, u) ds$$

$$= -\frac{1}{2\omega_0} \int_{-\infty}^{t} e^{-\omega_0 (t-s)} g(s, u) ds - \frac{1}{2\omega_0} \int_{t}^{\infty} e^{\omega_0 (t-s)} g(s, u) ds$$

$$= -\frac{1}{2\omega_0} \int_{-\infty}^{t} e^{-\omega_0 (t-s)} S''(s) u ds - \frac{1}{2\omega_0} \int_{t}^{\infty} e^{\omega_0 (t-s)} S''(s) u ds$$

$$+\frac{\omega_0}{2}\int_{-\infty}^{t}e^{-\omega_0(t-s)}S(s)u\,ds+\frac{\omega_0}{2}\int_{t}^{\infty}e^{\omega_0(t-s)}S(s)u\,ds$$

$$=\ -\frac{1}{2\omega_0}e^{-\omega_0(t-s)}S'(s)u\Big|_{-\infty}^{t}+\frac{1}{2}\int_{-\infty}^{t}e^{-\omega_0(t-s)}S'(s)u\,ds$$

$$-\frac{1}{2\omega_0}e^{\omega_0(t-s)}S'(s)u\Big|_{t}^{\infty}-\frac{1}{2}\int_{t}^{\infty}e^{\omega_0(t-s)}S'(s)u\,ds$$

$$+\frac{\omega_0}{2}\int_{-\infty}^{t}e^{-\omega_0(t-s)}S(s)u\,ds+\frac{\omega_0}{2}\int_{t}^{\infty}e^{\omega_0(t-s)}S(s)u\,ds$$

$$=\ \frac{1}{2}e^{-\omega_0(t-s)}S(s)u\Big|_{-\infty}^{t}-\frac{1}{2}e^{\omega_0(t-s)}S(s)u\Big|_{t}^{\infty}$$

$$=\ S(t)u,\quad t\in R.$$

Thus, we see that for any $u\in\mathcal{D}(A),\ t\in R$,

$$C(t)u\ =\ [S(t)u]'$$

$$=\ \frac{1}{2}\int_{-\infty}^{t}e^{-\omega_0(t-s)}g(s,\ u)ds-\frac{1}{2}\int_{t}^{\infty}e^{\omega_0(t-s)}g(s,\ u)ds \qquad (1.17)$$

$$=\ \frac{1}{2}\int_{-\infty}^{0}e^{\omega_0\mu}[g(t+\mu,\ u)-g(t-\mu,\ u)]d\mu.$$

Observe that $g(t,\ u)$ is w.a.p. for all $u\in\mathcal{D}(A)$ since $\{S(t)\}_{t\in R}$ is so. Fix $u^*\in E^*$. We then have that for every sequence of real numbers $\{\alpha_n\}$, there exists a subsequence $\{\alpha_{n_k}\}$ satisfying that to each $\varepsilon>0$ corresponds a positive integer K such that for any $k,\ l\geq K$,

$$\sup_{t\in R}|u^*(g(t+\alpha_{n_k},\ u))-u^*(g(t+\alpha_{n_l},\ u))|\leq\omega_0\varepsilon.$$

So for any $k,\ l\geq K$,

$$\sup_{t\in R}|u^*(C(t+\alpha_{n_k})u)-u^*(C(t+\alpha_{n_l})u)|$$

$$\leq\ \frac{1}{2}\int_{-\infty}^{0}e^{\omega_0\mu}\Bigg[\sup_{t\in R}|u^*(g(t+\mu+\alpha_{n_k},\ u))-u^*(g(t+\mu+\alpha_{n_l},\ u))|$$

$$+\sup_{t\in R}|u^*(g(t-\mu+\alpha_{n_k},\ u))-u^*(g(t-\mu+\alpha_{n_l},\ u))|\Bigg]d\mu$$

$$\leq\ \varepsilon,$$

by (1.17). Hence, the function $t \mapsto C(t)u$ is w.a.p. for every $u \in \mathcal{D}(A)$.

On the other hand, it is known that if $u(t)$ is an arbitrary solution of (1.1), then $u(0) \in \mathcal{D}(A)$ and

$$u(t) = C(t)u(0) + S(t)u'(0).$$

Combining this fact with the conclusion above, we see that all the solutions of (1.1) are w.a.p. The proof is then complete.

From the proof of Theorem 1.6, we also find that the following result is true.

Theorem 1.7. *Assume that* $\{S(t)\}_{t \in R}$ *is a.p. Then all the solutions of* (1.1) *are a.p.*

Combining Theorem 1.5 and Theorem 1.7 together gives immediately

Theorem 1.8. *Assume that* E *is weakly sequentially complete and* $\{S(t)\}_{t \in R}$ *is w.a.p. Then all the solutions of* (1.1) *are a.p.*

7.2 Complete second order equations

Of concern is the complete second order equation

$$u''(t) + A_1 u'(t) + A_0 u(t) = 0 \quad (t \geq 0), \tag{2.1}$$

where A_0, A_1 are densely defined closed linear operators in a Banach space E. We assume throughout this section that the Cauchy problem for (2.1) is strongly wellposed. Recall (see Sections 2.1 and 2.2) that $\mathcal{D}(A_0) \bigcap \mathcal{D}(A_1)$ is dense in E; that if $u(\cdot)$ is a solution of (2.1), then

$$u(t) = S_0(t)u(0) + S_1(t)u'(0), \quad t \geq 0;$$

that for $u \in \mathcal{D}(A_0)$, $v \in \mathcal{D}(A_0) \bigcap \mathcal{D}(A_1)$, (2.1) has a solution $u(\cdot)$ satisfying $u(0) = u$, $u'(0) = v$;

$$S_0(t)u = u - \int_0^t S_1(s)A_0 u \, ds \quad (u \in \mathcal{D}(A_0)), \tag{2.2}$$

$$S_1(t)u = \int_0^t [S_0(s)u - S_1(s)A_1 u] ds \quad (u \in \mathcal{D}(A_0) \bigcap \mathcal{D}(A_1)); \tag{2.3}$$

that there exist $C, \omega > 0$ such that for $\mathrm{Re}\lambda > \omega$, $\lambda \in \rho_0(A_0, A_1)$ and

$$\frac{1}{\lambda}u - \frac{1}{\lambda}R_\lambda A_0 u = R_\lambda(\lambda + A_1)u$$

$$= \int_0^\infty e^{-\lambda s} S_0(s)u \, ds \quad (u \in \mathcal{D}(A_0) \bigcap \mathcal{D}(A_1)), \tag{2.4}$$

$$R_\lambda u = \int_0^\infty e^{-\lambda s} S_1(s) u \, ds \quad (u \in E); \qquad (2.5)$$

moreover,

$$S_0(t)u = \lim_{m \to \infty} \frac{(-1)^m}{m!} \left(\frac{m}{t}\right)^{m+1} \left(\frac{1}{\lambda}I - \frac{1}{\lambda}\overline{R_\lambda A_0}\right)^{(m)} u \Bigg|_{\lambda = m/t} \qquad (2.6)$$

$$(t > 0, \ u \in E),$$

$$S_1(t)u = \lim_{m \to \infty} \frac{(-1)^m}{m!} \left(\frac{m}{t}\right)^{m+1} (R_\lambda)^{(m)} u \Bigg|_{\lambda = m/t} \qquad (2.7)$$

$$(t > 0, \ u \in E).$$

These facts will be used in the following discussion.

Almost periodic $\{S_0(t)\}_{t \geq 0}$ (resp. $\{S_1(t)\}_{t \geq 0}$) is defined exactly in the same way as in Definition 1.1 (see also the definitions above it) with R^+ replacing R everywhere.

We will investigate a.p. solutions of the complete second order equation (2.1). First of all we characterize, in terms of A_0 and A_1, (2.1) whose generalized solutions are bounded (for each $u, v \in E$, $S_0(t)u + S_1(t)v$ is called a generalized solution of (2.1)).

Proposition 2.1. $\{S_0(t)\}_{t \geq 0}$ *(resp. $\{S_1(t)\}_{t \geq 0}$) is uniformly bounded if and only if there exist C, $\omega \geq 0$ such that for $\mathrm{Re}\,\lambda > \omega$, $m \in N_0$*

$$\left\|\left(\frac{1}{\lambda}\overline{R_\lambda A_0}\right)^{(m)}\right\| \leq C m! (\mathrm{Re}\,\lambda)^{-m-1}$$

$$\left(resp. \ \left\|(R_\lambda)^{(m)}\right\| \leq C m! (\mathrm{Re}\,\lambda)^{-m-1}\right).$$

Proof. *Necessity.* Let $M = \sup_{t \geq 0} \|S_0(t)\|$. From (2.4), we have

$$\left\|\left(\frac{1}{\lambda}u - \frac{1}{\lambda}R_\lambda A_0 u\right)^{(m)}\right\| \leq M m! (\mathrm{Re}\,\lambda)^{-m-1} \|u\|$$

$$(u \in \mathcal{D}(A_0) \bigcap \mathcal{D}(A_1), \ \mathrm{Re}\,\lambda > \omega, \ m \in N_0).$$

This justifies our result using the plain equality

$$\left(\left(\frac{1}{\lambda}\right)u\right)^{(m)} = \left(\frac{(-1)^m m!}{\lambda^{m+1}}\right)u.$$

The case of $\{S_1(t)\}_{t \geq 0}$ is treated similarly using (2.5).

Sufficiency. It follows immediately from (2.6) (resp. (2.7)).

Theorem 2.2. *Almost periodicity of $\{S_1(t)\}_{t\geq0}$ implies almost periodicity of $\{S_0(t)\}_{t\geq0}$, provided $\{S_0(t)\}_{t\geq0}$ is uniformly bounded.*

Proof. As $\{S_1(t)\}_{t\geq0}$ is a.p., for each $u \in E$, $\lambda \in R$, the Fourier coefficient

$$\mathbf{P}_\lambda u := \lim_{t\to\infty} \frac{1}{t} \int_0^t e^{-i\lambda s} S_1(s)u\,ds$$

exists. Clearly $\mathbf{P}_\lambda \in \mathbf{L}(E)(\lambda \in R)$. For each $u \in \mathcal{D}(A_0)\bigcap\mathcal{D}(A_1)$, $\lambda \in R$, we have

$$\lambda^2 \mathbf{P}_\lambda u = -\lim_{t\to\infty}\frac{1}{t}\int_0^t i\lambda e^{-i\lambda s} S_1'(s)u\,ds$$

$$= \lim_{t\to\infty}\left[\frac{1}{t}e^{-i\lambda s}(S_0(s)u - S_1(s)A_1 u)\right]\Big|_0^t - \lim_{t\to\infty}\frac{1}{t}\int_0^t e^{-i\lambda s} S_1''(s)u\,ds$$

$$= \lim_{t\to\infty}\frac{1}{t}\int_0^t e^{-i\lambda s}\left[A_1 S_1'(s)u + A_0 S_1(s)u\right]ds.$$

Here we used (2.3). Take now λ_0 with $\mathrm{Re}\lambda_0 > \omega$. Then

$$R_{\lambda_0}\lambda^2\mathbf{P}_\lambda u$$

$$= \lim_{t\to\infty}\frac{1}{t}\left\{\left[e^{-i\lambda s}\overline{R_{\lambda_0}A_1}S_1(s)u\right]\Big|_0^t\right.$$

$$\left. + (i\lambda\overline{R_{\lambda_0}A_1} + \overline{R_{\lambda_0}A_0})\int_0^t e^{-i\lambda s} S_1(s)u\,ds\right\}$$

$$= (i\lambda\overline{R_{\lambda_0}A_1} + \overline{R_{\lambda_0}A_0})\,\mathbf{P}_\lambda u$$

and therefore

$$\overline{R_{\lambda_0}\left[(i\lambda)^2 + i\lambda A_1 + A_0\right]}\mathbf{P}_\lambda u = 0. \tag{2.8}$$

Due to the denseness of $\mathcal{D}(A_0)\bigcap\mathcal{D}(A_1)$, (2.8) holds also for $u \in E$. Next, we prove that for each $u \in E$,

$$u \in D_u := \overline{\mathrm{span}\{\mathbf{P}_\lambda u; \lambda \in R\}}.$$

Indeed, if $u \notin D_u$ then there exists $u^* \in E^*$ such that $u^*(u) \neq 0$ but $u^*(v) = 0$ for each $v \in D_u$. Thus the a.p. function $t \mapsto u^*(S_1(t)u)$ vanishes since all its Fourier coefficients are zero, and for this so does the function

$$t \mapsto [u^*(S_1(t)u)]' = u^*(S_1'(t)u).$$

In particular, $u^*(S_1'(0)u) = u^*(u) = 0$, a contradiction. So for each $u \in E$,

$$u = \lim_{n \to \infty} \sum_j \alpha_{nj} \mathbf{P}_{\lambda_{nj}} u \qquad (2.9)$$

for some $\lambda_{nj} \in R$, $\alpha_{nj} \in C$ (finite in j). It is easily verified from (2.8) that for $u \in E$, $\text{Re}\mu > \omega$, $\lambda \in R$,

$$\overline{R_{\lambda_0} P_\mu} \int_0^\infty \left(e^{-\mu t} e^{i\lambda t} \mathbf{P}_\lambda u \right) dt$$

$$= \mu R_{\lambda_0} \mathbf{P}_\lambda u + \overline{R_{\lambda_0} A_1} \mathbf{P}_\lambda u + i\lambda R_{\lambda_0} \mathbf{P}_\lambda u. \qquad (2.10)$$

On the other hand, according to (2.4) and (2.5), we obtain that for $\text{Re}\mu > \omega$, $\lambda \in R$, $u \in \mathcal{D}(A_0) \bigcap \mathcal{D}(A_1)$,

$$\overline{R_{\lambda_0} P_\mu} \int_0^\infty e^{-\mu t} (S_0(t)u + i\lambda S_1(t)u) dt$$

$$= \mu R_{\lambda_0} u + \overline{R_{\lambda_0} A_1} u + i\lambda R_{\lambda_0} u. \qquad (2.11)$$

Again by the denseness of $\mathcal{D}(A_0) \bigcap \mathcal{D}(A_1)$, (2.11) is also satisfied for $u \in E$. Since $\overline{R_{\lambda_0} P_\mu}$ ($\text{Re}\mu > \omega$) is one-to-one, combining (2.10) with (2.11) implies that for $u \in E$, $\text{Re}\mu > \omega$, $\lambda \in R$,

$$\int_0^\infty e^{-\mu t} e^{i\lambda t} \mathbf{P}_\lambda u \, dt = \int_0^\infty e^{-\mu t} \left(S_0(t) \mathbf{P}_\lambda u + i\lambda S_1(t) \mathbf{P}_\lambda u \right) dt.$$

From this, it follows that for each $u \in E$, $\lambda \in R$,

$$e^{i\lambda t} \mathbf{P}_\lambda u = S_0(t) \mathbf{P}_\lambda u + i\lambda S_1(t) \mathbf{P}_\lambda u \quad (t \geq 0),$$

and therefore the function $t \mapsto S_0(t) \mathbf{P}_\lambda u$ ($t \geq 0$) is a.p. Making use of (2.9), we deduce that $\{S_0(t)\}_{t \geq 0}$ is a.p. because of its uniform boundedness, and the proof is complete.

Remark. When E is a uniformly convex Banach space, the above result is immediate from (2.2) and Corduneanu [1, Theorem 6.20].

We say that the almost periodicity of $\{S_0(t)\}_{t \geq 0}$ does not imply the almost periodicity of $\{S_1(t)\}_{t \geq 0}$ even if $\{S_1(t)\}_{t \geq 0}$ is assumed to be uniformly bounded. For example, let

$$E = C \times C,$$

$$A_1(u, \ v) = (3u, \ -3iv), \quad (u, v) \in E,$$

$$A_0(u, \ v) = (0, \ -2v), \quad (u, v) \in E.$$

Then it is not hard to see that for each $(u, v) \in E$,

$$t \mapsto S_0(t)(u, \ v) = \left(u, \ \left(2e^{it} - e^{2it}\right) v\right)$$

is a.p., while the bounded function

$$t \mapsto S_1(t)(u, \ v) = \left(3^{-1}\left(1 - e^{-3t}\right) u, \ i\left(e^{it} - e^{2it}\right) v\right)$$

is not.

Theorem 2.3. *Let $\lambda_0 \in \mathbf{C}$ with $\mathrm{Re}\lambda_0 > \omega$. Both $\{S_0(t)\}_{t\geq 0}$ and $\{S_1(t)\}_{t\geq 0}$ are a.p. if and only if $\{S_0(t)\}_{t\geq 0}$ and $\{S_1(t)\}_{t\geq 0}$ are uniformly bounded, and for each $u \in \mathcal{D}(A_0)\bigcap\mathcal{D}(A_1)$, there exist double sequences $\{u_{nj}\}, \{\lambda_{nj}\}$ (finite in $j, u_{nj} \in E, \lambda_{nj} \in R$) satisfying*

$$\overline{R_{\lambda_0}P_{i\lambda_{nj}}}u_{nj} = 0 \tag{2.12}$$

such that

$$\lim_{n\to\infty}\sum_j(i\lambda_{nj}u_{nj}) = u \quad and \quad \lim_{n\to\infty}\sum_j u_{nj} = 0.$$

In this case, $\left\{\begin{pmatrix} u \\ i\lambda u \end{pmatrix}; \ \overline{R_{\lambda_0}P_\lambda}u = 0, \ \lambda \in R, \ u \in E\right\}$ is total in $E \times E$.

Proof. *Sufficiency.* Let $v \in E$ with $\overline{R_{\lambda_0}P_{i\lambda}}v = 0$ for some $\lambda \in R$. Proceeding as in the proof of Theorem 2.2, we obtain that the function

$$t \mapsto S_0(t)v + i\lambda S_1(t)v = e^{i\lambda t}v,$$

for $t \geq 0$, is a.p. Now for each $u \in \mathcal{D}(A_0)\bigcap\mathcal{D}(A_1)$, there exist, by hypothesis, double sequences $\{u_{nj}\}, \{\lambda_{nj}\}$ (finite in $j, \lambda_{nj} \in R$) with (2.12) such that

$$\lim_{n\to\infty}\sum_j(i\lambda_{nj}u_{nj}) = u, \quad \lim_{n\to\infty}\sum_j u_{nj} = 0.$$

But

$$t \mapsto S_0(t)u_{nj} + i\lambda_{nj}S_1(t)u_{nj}$$

is a.p. by the preceding comment, so is

$$S_1(t)u = \lim_{n\to\infty}\sum_j\left(S_0(t)u_{nj} + i\lambda_{nj}S_1(t)u_{nj}\right) \quad (t \geq 0),$$

since the limit is uniform on $t \geq 0$. This establishes our claim noting Theorem 2.2.

Necessity. The uniform boundedness of either $\{S_0(t)\}_{t\geq 0}$ or $\{S_1(t)\}_{t\geq 0}$ is obvious. Observe that for each $u \in \mathcal{D}(A_0)\bigcap\mathcal{D}(A_1)$,

$$S_1'(t)u = S_0(t)u - S_1(t)A_1u$$

by (2.3). Then the function $t \mapsto \begin{pmatrix} S_1(t)u \\ S_1'(t)u \end{pmatrix}$ from R^+ to $E \times E$ is a.p., whose Fourier coefficients are

$$\lim_{t \to \infty} \frac{1}{t} \int_0^t e^{-i\lambda s} \begin{pmatrix} S_1(t)u \\ S_1'(t)u \end{pmatrix} ds = \begin{pmatrix} P_\lambda u \\ i\lambda P_\lambda u \end{pmatrix}$$

$$(u \in \mathcal{D}(A_0) \bigcap \mathcal{D}(A_1), \ \lambda \in R).$$

Arguing as in the proof of (2.9), we conclude that for each $u \in \mathcal{D}(A_0) \bigcap \mathcal{D}(A_1)$,

$$\begin{pmatrix} 0 \\ u \end{pmatrix} = \lim_{n \to \infty} \sum_j \alpha_{nj} \begin{pmatrix} P_{\lambda_{nj}} u \\ i\lambda_{nj} P_{\lambda_{nj}} u \end{pmatrix}$$

for some $\lambda_{nj} \in R$, $\alpha_{nj} \in \mathbf{C}$ (finite in j), i.e.,

$$\lim_{n \to \infty} \sum_j i\lambda_{nj} \left(\alpha_{nj} P_{\lambda_{nj}} u \right) = u,$$

$$\lim_{n \to \infty} \sum_j \alpha_{nj} P_{\lambda_{nj}} u = 0.$$

Here $\alpha_{nj} P_{\lambda_{nj}} u$ satisfies

$$\overline{R_{\lambda_0} P_{i\lambda_{nj}}} \alpha_{nj} P_{\lambda_{nj}} u = 0$$

by (2.8).

Finally, we set, for $u \in E$, $\lambda \in R$,

$$M_\lambda u = \lim_{t \to \infty} \frac{1}{t} \int_0^t e^{-i\lambda s} S_0(s) u \, ds. \tag{2.13}$$

Using arguments similar to those in the treatment of $\{S_1(t)\}_{t \geq 0}$ and noting (2.2), we obtain

$$\overline{R_{\lambda_0} P_{i\lambda}} M_\lambda u = 0;$$

moreover $t \mapsto \begin{pmatrix} S_0(t)u \\ S_0'(t)u \end{pmatrix}$ $(u \in \mathcal{D}(A_0))$ is a.p., whose Fourier coefficients are $\begin{pmatrix} M_\lambda u \\ i\lambda M_\lambda u \end{pmatrix}$ $(\lambda \in R)$, so that for each $u \in \mathcal{D}(A_0)$,

$$\begin{pmatrix} u \\ 0 \end{pmatrix} = \lim_{n \to \infty} \sum_j \alpha_{nj} \begin{pmatrix} M_{\lambda_{nj}} u \\ i\lambda_{nj} M_{\lambda_{nj}} u \end{pmatrix}$$

for some sequence $\lambda_{nj} \in R$, $\alpha_{nj} \in \mathbf{C}$. This together with (2.13) completes the proof of Theorem 2.3.

Next, we study more explicitly the case when (2.1) appears in the form

$$u''(t) + (aB + bI)u'(t) + (cB + dI) = 0 \quad (t \geq 0) \tag{2.14}$$

with $a, b, c, d \in C$.

According to Corollaries 2.4.8 and 2.4.9, the Cauchy problem for (2.14) is strongly wellposed if and only if $-(aB + bI)$ is the generator of a strongly continuous semigroup when $a \neq 0$, or $-(cB + dI)$ is the generator of a strongly continuous cosine function when $a = 0$. Moreover, making use of (2.4) and (2.5), we obtain immediately

Lemma 2.4. $S_1(t)Bu = BS_1(t)u, S_0(t)Bu = BS_0(t)u, t \geq 0, u \in \mathcal{D}(B)$.

From now on, the discussion is carried out under the following assumption.

Assumption 2.5. $a^2d + c^2 \neq abc$ if $a \neq 0$ and $ia^{-1}c \in R; c \neq 0$ if $a = 0$.

Theorem 2.6. $\{S_0(t)\}_{t \geq 0}$ *is a.p. if and only if* $\{S_0(t)\}_{t \geq 0}$ *is uniformly bounded, the set of eigenvectors of B is total in E, and for each $\Lambda \in \sigma_p(B) - T_0$ (T_0 indicates the set $\{-c^{-1}d\}$ if $c \neq 0$, or the empty set if $c = 0$), $i(a\Lambda + b)$, $c\Lambda + d \in R$.*

Proof. Let $\Lambda \in \sigma_p(B)$. Then there exists $v \neq 0$ such that $Bv = \Lambda v$. Set λ, λ' satisfying

$$i\lambda + i\lambda' = -(a\Lambda + b), \quad (i\lambda)(i\lambda') = c\Lambda + d. \quad (2.15)$$

Then it is not difficult to verify that for $t \geq 0$,

$$S_0(t)v = \begin{cases} (\lambda - \lambda')^{-1} \left(\lambda e^{i\lambda' t} - \lambda' e^{i\lambda t} \right) v, & \text{if } \lambda \neq \lambda', \\ (1 - i\lambda t)e^{i\lambda t} v, & \text{if } \lambda = \lambda'. \end{cases} \quad (2.16)$$

Sufficiency. Since for $\Lambda \in \sigma_p(B) - T_0$, $i(a\Lambda + b)$, $c\Lambda + d \in R$ and $\{S_0(t)\}_{t \geq 0}$ is uniformly bounded, we conclude from (2.16) that the corresponding $\lambda, \lambda' \in R - \{0\}$ and $\lambda \neq \lambda'$. Let $\lambda < \lambda'$. Define a map $g : \sigma_p(B) \to R$ by

$$g(\Lambda) = \begin{cases} \lambda, & \text{if } \Lambda \notin T_0, \\ 0, & \text{otherwise.} \end{cases}$$

By (2.15), for $\Lambda \in \sigma_p(B)$, $\lambda = g(\Lambda)$,

$$(ia\lambda + c)\Lambda + (-\lambda^2 + ib\lambda + d) = 0.$$

Combining this with Assumption 2.5, we deduce that $ia\lambda + c \neq 0$, and therefore

$$\Lambda = (ia\lambda + c)^{-1} \left(\lambda^2 - ib\lambda - d \right)$$

(similarly,

$$\Lambda = (ia\lambda' + c)^{-1} \left((\lambda')^2 - ib\lambda' - d \right),$$

which implies $\lambda' \notin \mathcal{R}(g) - \{0\}$). Accordingly, g is one-to-one and

$$g^{-1}(\lambda) = (ia\lambda + c)^{-1} \left(\lambda^2 - ib\lambda - d \right), \quad \lambda \in \mathcal{R}(g). \quad (2.17)$$

Now, for each $u \in E$, there exists a scalar sequence $\{\Lambda_{nj}\}$ (finite in j) with

$$u_{\Lambda_{nj}} \in \mathcal{N}(\Lambda_{nj} - B)$$

such that

$$u = \lim_{n \to \infty} \sum_j u_{\Lambda_{nj}}.$$

From the previous observations, we know that for each $u_\Lambda \in \mathcal{N}(\Lambda - B)$ with $\Lambda \in \sigma_p(B)$, $t \geq 0$,

$$S_0(t)u_\Lambda = \begin{cases} (\lambda - \lambda')^{-1}\left(\lambda e^{i\lambda' t} - \lambda' e^{i\lambda t}\right)u_\Lambda \\ \qquad (\lambda = g(\Lambda),\ \lambda' = i(a\Lambda + b) - \lambda),\ \text{if}\ \ \Lambda \notin T_0, \qquad (2.18) \\ u_\Lambda,\ \ \text{if}\ \ \Lambda \in T_0. \end{cases}$$

So $t \mapsto S_0(t)u_\Lambda$ is a.p. The function

$$t \mapsto S_0(t)u = \lim_{n \to \infty} \sum_j S_0(t)u_{\Lambda_{nj}}$$

is then a.p. because of the uniform boundedness of $\{S_0(t)\}_{t \geq 0}$, and this gives the desired result.

Necessity. For each $\Lambda \in \sigma_p(B) - T_0$, applying (2.16) again we deduce that the corresponding λ, $\lambda' \in R$ and $\lambda \neq \lambda'$ due to the almost periodicity of $\{S_0(t)\}_{t \geq 0}$. From this, it follows immediately that $i(a\Lambda + b)$, $c\Lambda + d \in R$. Next, noting (2.13) and integrating by parts, we obtain that for each $u \in \mathcal{D}(B)$, $\lambda \in R$,

$$\lambda^2 M_\lambda u = \lim_{t \to \infty} \left\{ \frac{1}{t}\frac{d}{dt}\left[e^{-i\lambda t}S_0(t)u\right] - \frac{1}{t}\int_0^t e^{-i\lambda s}S_0''(s)u\,ds \right\}. \qquad (2.19)$$

But

$$-\frac{1}{t}\int_0^t e^{-i\lambda s}S_0''(s)u\,ds$$

$$= \frac{1}{t}\int_0^t e^{-i\lambda s}\left[(aB + bI)S_0'(s)u + (cB + dI)S_0(s)u\right]ds$$

$$= \frac{1}{t}\left[e^{-i\lambda t}S_0(t)(aB + bI)u - (aB + bI)u\right]$$

$$+ \frac{1}{t}[i\lambda(aB + bI) + cB + dI]\int_0^t e^{-i\lambda s}S_0(s)u\,ds.$$

Hence

$$\lim_{t \to \infty}\left[-\frac{1}{t}\int_0^t e^{-i\lambda s}S_0''(s)u\,ds\right] = [i\lambda(aB + bI) + cB + dI]M_\lambda u.$$

This together with (2.19) yields

$$\lim_{t \to \infty} \frac{1}{t} \frac{d}{dt} \left[e^{-i\lambda t} S_0(t)u \right] = -P_{i\lambda} M_\lambda u.$$

Observing the function $t \mapsto e^{-i\lambda t} S_0(t)u$ is bounded, we assert that $P_{i\lambda} M_\lambda u = 0$ by the same reasoning as for the claim $W = 0$ in the proof of Theorem 1.2. Therefore

$$(ia\lambda + c)BM_\lambda u + \left(-\lambda^2 + ib\lambda + d\right) M_\lambda u = 0 \quad (u \in \mathcal{D}(B), \ \lambda \in R).$$

From this and Assumption 2.5, we know

$$
\begin{cases}
M_\lambda u = 0, & \text{if } ia\lambda + c = 0, \\[2mm]
BM_\lambda u = (ia\lambda + c)^{-1} \left(\lambda^2 - ib\lambda - d\right) M_\lambda u, & \text{if } ia\lambda + c \neq 0.
\end{cases}
\tag{2.20}
$$

Clearly, (2.20) holds for each $u \in E$. Accordingly, each non-zero $M_\lambda u$ ($u \in E$, $\lambda \in R$) belongs to the set of eigenvectors of B. But $u \in \mathrm{span}\{M_\lambda u; \ \lambda \in R\}$ for each $u \in E$, using arguments similar to those in the proof of Theorem 1.2. This ends the proof.

Our next theorem provides an explicit expression for an a.p. $\{S_0(t)\}_{t \geq 0}$ when E is a Hilbert space.

Theorem 2.7. *Let E be a Hilbert space and let $\{S_0(t)\}_{t \geq 0}$ be a.p. Then*
(i) $\{N_\lambda\}$ ($\lambda \in \mathcal{R}(g)$) *is a family of mutually orthogonal projections onto* $\mathcal{N}\left(g^{-1}(\lambda) - B\right)$, *where*

$$
N_\lambda :=
\begin{cases}
(\lambda')^{-1}(\lambda' - \lambda)M_\lambda, & \text{if } \lambda \in \mathcal{R}(g) - \{0\}, \\[2mm]
M_0, & \text{if } \lambda = 0.
\end{cases}
$$

(ii) *For each $u \in E, t \geq 0$,*

$$S_0(t)u = M_0 u + \sum_{\lambda \neq 0} (\lambda - \lambda')^{-1} \left(\lambda e^{i\lambda' t} - \lambda' e^{i\lambda t}\right) N_\lambda u.$$

Proof. Combining (2.17) with (2.20), we get

$$M_\lambda u \in \mathcal{N}(g^{-1}(\lambda) - B) \quad (u \in E, \ \lambda \in \mathcal{R}(g)). \tag{2.21}$$

Then for each $u \in E$, $\lambda \in \mathcal{R}(g)$, $t \geq 0$,

$$
S_0(t)M_\lambda u =
\begin{cases}
(\lambda - \lambda')^{-1} \left(\lambda e^{i\lambda' t} - \lambda' e^{i\lambda t}\right) M_\lambda u, & \text{if } \lambda \neq 0, \\[2mm]
M_\lambda u, & \text{if } \lambda = 0,
\end{cases}
\tag{2.22}
$$

according to (2.18). Therefore for $u \in E$, λ, $\mu \in \mathcal{R}(g)$,

$$M_\mu M_\lambda u = \lim_{t\to\infty} \frac{1}{t} \int_0^t e^{-i\mu s} S_0(s) M_\lambda u \, ds$$

$$= \begin{cases} (\lambda' - \lambda)^{-1}\lambda' M_\lambda u, & \text{if } \mu = \lambda \neq 0, \\ 0, & \text{if } \mu \neq \lambda, \\ M_\lambda u, & \text{if } \mu = \lambda = 0, \end{cases} \tag{2.23}$$

by recalling that for $\lambda \in \mathcal{R}(g) - \{0\}$, the corresponding λ' does not belong to $\mathcal{R}(g)$. Now observe that for each $v \in \mathcal{N}(\Lambda - B)$, $\lambda = g(\Lambda)$,

$$M_\lambda v = \begin{cases} \displaystyle\lim_{t\to\infty} \frac{1}{t} \int_0^t e^{-i\lambda s}(\lambda - \lambda')^{-1}\left(\lambda e^{i\lambda' s} - \lambda' e^{i\lambda s}\right) v \, ds, & \text{if } \lambda \neq 0 \\ \displaystyle\lim_{t\to\infty} \frac{1}{t} \int_0^t v \, ds, & \text{if } \lambda = 0 \end{cases}$$

$$= \begin{cases} (\lambda' - \lambda)^{-1}\lambda' v, & \text{if } \lambda \neq 0, \\ v, & \text{if } \lambda = 0, \end{cases}$$

in view of (2.18). So $\mathcal{N}(\Lambda - B) \subset \mathcal{R}(M_\lambda)$ for any $\Lambda \in \sigma_p(B)$. This, together with (2.21) and (2.23), suggests (i) (if $0 \notin \mathcal{R}(g)$, then $c = 0$ and therefore $M_0 = 0$ by (2.20)).

(ii) follows immediately from (i) and (2.22) by applying Theorem 2.6. Thus Theorem 2.7 is proved.

Now, we look at the case of $\{S_1(t)\}_{t\geq 0}$. Proceeding analogously as in the proofs of Theorem 2.6 and Theorem 2.7 and noting that for $\{S_1(t)\}_{t\geq 0}$ the equalities corresponding to (2.16), (2.22) and (2.23) are respectively

$$S_1(t)v = \begin{cases} i(\lambda - \lambda')^{-1}\left(e^{i\lambda' t} - e^{i\lambda t}\right) v, & \text{if } \lambda \neq \lambda', \\ te^{i\lambda t} v, & \text{if } \lambda = \lambda', \end{cases} \tag{2.24}$$

(from (2.24), we know that if $t \mapsto S_1(t)v$ is bounded then $\lambda \neq \lambda'$ for each $\Lambda \in \sigma_p(B)$),

$$S_1(t)\mathbf{P}_\lambda u = i(\lambda - \lambda')^{-1}\left(e^{i\lambda' t} - e^{i\lambda t}\right) \mathbf{P}_\lambda u,$$

$$\mathbf{P}_\mu \mathbf{P}_\lambda u = \begin{cases} i(\lambda' - \lambda)^{-1}\mathbf{P}_\lambda u, & \text{if } \mu = \lambda, \\ 0, & \text{if } \mu \neq \lambda, \end{cases}$$

we obtain

Theorem 2.8. $\{S_1(t)\}_{t\geq 0}$ *is a.p. if and only if* $\{S_1(t)\}_{t\geq 0}$ *is uniformly bounded, the set of eigenvectors of* B *is total in* E, *and for each* $\Lambda \in \sigma_p(B)$, $i(a\Lambda + b)$, $c\Lambda + d \in R$.

Moreover, if we assume that E is a Hilbert space and $\{S_1(t)\}_{t\geq 0}$ is a.p., then $\{\mathbf{Q}_\lambda\}(\lambda \in \mathcal{R}(g))$ is a family of mutually orthogonal projections onto $\mathcal{N}(g^{-1}(\lambda) - B)$ and

$$S_1(t)u = \sum_\lambda i(\lambda - \lambda')^{-1}\left(e^{i\lambda' t} - e^{i\lambda t}\right)\mathbf{Q}_\lambda u \quad (u \in E, \ t \geq 0),$$

where $\mathbf{Q}_\lambda := i(\lambda - \lambda')\mathbf{P}_\lambda$.

About the relation between almost periodicity of the two propagators in the special case, the following result is interesting. Here we need not assume that $\{S_0(t)\}_{t\geq 0}$ is uniformly bounded.

Theorem 2.9. *Almost periodicity of* $\{S_1(t)\}_{t\geq 0}$ *implies almost periodicity of* $\{S_0(t)\}_{t\geq 0}$, *provided* E *is a Hilbert space.*

Proof. According to Theorem 2.8, $\mathcal{R}(\mathbf{Q}_\lambda) = \mathcal{N}\left(g^{-1}(\lambda) - B\right)$ for $\lambda \in \mathcal{R}(g)$ and

$$u = S_1'(t)u\Big|_{t=0} = \sum_{\lambda \in \mathcal{R}(g)} \mathbf{Q}_\lambda u \quad (u \in E).$$

Making use of (2.16), we get

$$S_0(t)u = \sum_{\lambda \in \mathcal{R}(g)} (\lambda - \lambda')^{-1}\left(\lambda e^{i\lambda' t} - \lambda' e^{i\lambda t}\right)\mathbf{Q}_\lambda u \quad (u \in E, \ t \geq 0). \qquad (2.25)$$

Observe that for any $\lambda \in \mathcal{R}(g)$, if $|\lambda| + |\lambda'| \neq |\lambda + \lambda'|$ then

$$|\lambda| + |\lambda'| = |\lambda - \lambda'|;$$

otherwise,

$$|\lambda| + |\lambda'| = \left|ag^{-1}(\lambda) + b\right|.$$

But

$$|\lambda - \lambda'| = \left[\left|ag^{-1}(\lambda) + b\right|^2 + 4(cg^{-1}(\lambda) + d)\right]^{1/2}$$

by (2.15), so

$$\sup_{\lambda \in \mathcal{R}(g), |\lambda| > M} \left[|\lambda - \lambda'|^{-1}\left(|\lambda| + |\lambda'|\right)\right] < +\infty$$

for some $M > 0$ noting (2.17). When $|\lambda| \leq M$ $(\lambda \in \mathcal{R}(g))$,

$$\left|(\lambda - \lambda')^{-1}\left(\lambda e^{i\lambda' t} - \lambda' e^{i\lambda t}\right)\right|$$

$$= \left| e^{i\lambda t} + \lambda(\lambda - \lambda')^{-1} \left(e^{i\lambda' t} - e^{i\lambda t} \right) \right|$$

$$\leq 1 + M \sup_{t \geq 0, \lambda \in \mathcal{R}(g)} \left| (\lambda - \lambda')^{-1} \left(e^{i\lambda t} - e^{i\lambda' t} \right) \right|$$

$$< +\infty$$

due to the boundedness of $\{S_1(t)\}_{t \geq 0}$. From these observations and (2.25), we deduce

$$\sup_{t \geq 0} \|S_0(t)\| \leq \sup_{t \geq 0, \lambda \in \mathcal{R}(g)} \left| (\lambda - \lambda')^{-1} \left(\lambda e^{i\lambda' t} - \lambda' e^{i\lambda t} \right) \right|$$

$$< +\infty.$$

$\{S_0(t)\}_{t \geq 0}$ is therefore a.p.

Example 2.10. Let $E = L^2(0,1)$, $B = \frac{\partial^2}{\partial x^2}$ with

$$\mathcal{D}(B) = \left\{ u \in H^2(0,1); \ u(x) \Big|_{x=0,1} = 0 \right\}.$$

Clearly, $\sin(2n\pi x)$ $(n \in N)$ are eigenvectors of B corresponding to eigenvalues $\Lambda_n = -(2n\pi)^2$. By the Fourier expansion method, $\{\sin(2n\pi x); \ n \in N\}$ is total in $C[0,1]$, as well as in E, since $C[0,1]$ is dense in E. Given a, b, c, d with $ia, ib, c, d \in R$ and

$$\inf_{n \in N} \left[(ia\Lambda_n + ib)^2 + 4(c\Lambda_n + d) \right] > 0$$

(e.g., $ia, \ c < 0, \ ib, \ d > 0$), then the equation

$$x^2 - (ia\Lambda_n + ib)x - (c\Lambda_n + d) = 0$$

has two unequal real roots λ_n, λ'_n, and

$$\sup_{t \geq 0, n \in N} \left| i(\lambda_n - \lambda'_n)^{-1} (e^{i\lambda'_n t} - e^{i\lambda_n t}) \right|$$

$$\leq \sup_{n \in N} \left[(ia\Lambda_n + ib)^2 + 4(c\Lambda_n + d) \right]^{-1/2}$$

$$< +\infty.$$

It is easy to verify that $i(aB + b)$ is self-adjoint, and therefore $-(aB + b)$ is the generator of a strongly continuous group. It follows that the Cauchy problem for (2.14) is strongly wellposed. Denote by $\mathbf{Q}_{\Lambda_n}$ $(n \in N)$ the projection onto $\mathcal{N}(\Lambda_n - B)$. Then for each $u \in E$,

$$u = \sum_{n=1}^{\infty} \mathbf{Q}_{\Lambda_n} u,$$

any two terms in the expansion being orthogonal (in fact, eigenvectors corresponding to different eigenvalues are always orthogonal). Hence

$$S_1(t)u = \sum_{n=1}^{\infty} i(\lambda_n - \lambda'_n)^{-1} \left(e^{i\lambda'_n t} - e^{i\lambda_n t} \right) \mathbf{Q}_{\Lambda_n} u \quad (t \geq 0),$$

and therefore

$$\sup_{t \geq 0} \|S_1(t)\| \leq \sup_{t \geq 0, n \in N} \left| i(\lambda_n - \lambda'_n)^{-1} \left(e^{i\lambda'_n t} - e^{i\lambda_n t} \right) \right| < +\infty.$$

Accordingly we conclude that $\{S_1(t)\}_{t \geq 0}$ is a.p., and so is $\{S_0(t)\}_{t \geq 0}$ by Theorem 2.9.

7.3   Notes

Almost periodic scalar functions were treated firstly by Bohr [1]. The extension to the vector-valued case is due to Bochner [1].

In 1970, Fattorini [4] set up the theory of almost periodic groups and almost periodic cosine functions in Hilbert spaces; see also Fattorini [7, Chapter 5] . The theory of almost periodic strongly continuous groups and semigroups in Banach spaces was investigated by Bart-Goldberg [1], and a characterization in terms of their generators was given there. Following this work, Cioranescu [2] and Piskarev [1] studied further the almost periodic cosine functions in Banach spaces. Theorems 1.2, 1.4 and 1.5, for the case of $\{C(t)\}_{t \in R}$, are due to Cioranescu [2] with proofs depending on the characteristic equality

$$C(t+s) + C(t-s) = 2C(t)C(s) \quad (t, \ s \in R).$$

For the case of sine operator function $\{S(t)\}_{t \in R}$, Theorems 1.2, 1.4 and 1.5 are taken from Xiao-Liang [7]. Theorem 1.3 is also from Xiao-Liang [7]. Theorems 1.6, 1.7 and 1.8 are due to Xiao-Liang [8].

Section 7.2 comes from Xiao-Liang [8].

Appendix

A1 Fractional powers of nonnegative operators

Let A be a nonnegative operator in a Banach space E, and let $\alpha \in \mathbf{C}$ with $\operatorname{Re}\alpha > 0$.

Definition A1.1. (i) If $A \in \mathbf{L}(E)$ and $0 \in \rho(A)$,

$$A^\alpha := \frac{1}{2\pi i} \int_C \lambda^\alpha R(\lambda;\, A) d\lambda,$$

where the contour C surrounds $\sigma(A)$, avoiding the negative real axis and the origin, and λ^α is taken to be positive for $\lambda > 0$.

(ii) If $A \in \mathbf{L}(E)$ and $0 \in \sigma(A)$,

$$A^\alpha := \lim_{\varepsilon \to 0+} (A + \varepsilon)^\alpha.$$

(iii) If $A \notin \mathbf{L}(E)$ and $0 \in \rho(A)$,

$$A^\alpha := \left[\left(A^{-1} \right)^\alpha \right]^{-1}.$$

(iv) If $A \notin \mathbf{L}(E)$ and $0 \in \sigma(A)$, A^α is defined by

$$A^\alpha u := \lim_{\varepsilon \to 0+} (A + \varepsilon)^\alpha u, \quad u \in \mathcal{D}(A^\alpha),$$

where $\mathcal{D}(A^\alpha)$ is the set of all $u \in E$ for which the above limit exists.

Theorem A1.2. $\mathcal{D}[(A + \varepsilon)^\alpha] = \mathcal{D}(A^\alpha)$ *for each* $\varepsilon > 0$.

Theorem A1.3. A^α *is a closed linear operator in* E.

Theorem A1.4. *If* $\alpha \in N$, *then* A^α *is the usual power of* A.

Theorem A1.5. (i) *For* $0 < \operatorname{Re}\alpha < 1$, $u \in \mathcal{D}(A)$,

$$A^\alpha u = \frac{\sin \alpha \pi}{\pi} \int_0^\infty \lambda^{\alpha-1}(\lambda + A)^{-1} A u \, d\lambda.$$

(ii) *For* $0 < \operatorname{Re}\alpha < 2$, $u \in \mathcal{D}(A^2)$,

$$A^\alpha u = \frac{\sin \alpha \pi}{\pi} \int_0^\infty \lambda^{\alpha-1} \left[(\lambda + A)^{-1} - \lambda(1 + \lambda^2)^{-1} \right] Au d\lambda + \sin\left(\frac{\alpha\pi}{2}\right) Au.$$

Theorem A1.6. *Let* $\operatorname{Re}\alpha$, $\operatorname{Re}\beta > 0$. *Then*

$$A^{\alpha+\beta} = A^\beta A^\alpha = A^\alpha A^\beta.$$

Theorem A1.7. *Let* $0 < \alpha < 1$. *Then* A^α *is a nonnegative operator and for* $\beta \in \mathbf{C}$ *with* $\operatorname{Re}\beta > 0$ *we have*

$$(A^\alpha)^\beta = A^{\alpha\beta}.$$

Theorem A1.8. $\sigma(A^\alpha) = \{|z|^\alpha e^{i\alpha \arg z}; \; z \in \sigma(A), \; |\arg z| < \pi\}.$

Theorem A1.9. *Let* $0 < \alpha < 1$. *Then*

$$\|A^\alpha u\| \leq \text{const} \left(a^\alpha \|u\| + a^{\alpha-1} \|Au\| \right)$$

for all $u \in \mathcal{D}(A)$, $a > 0$.

Definition A1.10. If $0 \in \rho(A)$,

$$A^{-\alpha} := (A^\alpha)^{-1}.$$

For details on fractional powers of nonnegative operators, please see, e.g., Balakrishnan [1], Martínez-Sanz-Marco [1], Fattorini [6, 7] and Pazy [2].

A2 Strongly continuous semigroups and cosine functions

Definition A2.1. Let E be a SCLCS. A family $\{T(t)\}_{t \geq 0}$ of continuous linear operators on E is a (exponentially equicontinuous) strongly continuous (operator) semigroup if
   (i) $T(0) = I$, $T(s + t) = T(t)T(s)$ $(t, s \geq 0)$,
   (ii) $\lim_{t \to 0+} T(t)u = u$ $(u \in E)$,
   (iii) there exists $\omega > 0$ such that $\{e^{-\omega t}T(t); \; t \geq 0\}$ is equicontinuous.
The generator A of a strongly continuous semigroup $\{T(t)\}_{t \geq 0}$ is defined by

$$Au = \lim_{h \to 0+} \frac{1}{h}[T(h)u - u], \quad u \in \mathcal{D}(A).$$

where $\mathcal{D}(A)$ is the set of all $u \in E$ for which the above limit exists.

A strongly continuous (operator) group (as well as its generator) is defined analogously, with the parameter t running over R instead of R^+ (with $h \to 0$ instead of $h \to 0^+$).

Condition (iii) is implicitly contained in conditions (i) and (ii) in the case when E is a Banach space.

Definition A2.2. Let E be a SCLCS. A family $\{C(t)\}_{t \geq 0}$ of continuous linear operators on E is a strongly continuous cosine (operator) function if
(i) $C(0) = I$, $\quad 2C(t)C(s) = C(s+t) + C(|s-t|) \quad (t,\ s \geq 0)$,
(ii) $\lim_{t \to 0^+} C(t)u = u \quad (u \in E)$,
(iii) there exists $\omega > 0$ such that $\{e^{-\omega t}C(t);\ t \geq 0\}$ is equicontinuous.
The generator A of a strongly continuous cosine function $\{C(t)\}_{t \geq 0}$ is defined by

$$Au = \lim_{t \to 0^+} \frac{2}{t^2}[C(t)u - u], \quad u \in \mathcal{D}(A),$$

where $\mathcal{D}(A)$ is the set of all $u \in E$ for which the above limit exists. Clearly, condition (iii) is implicitly contained in (i) and (ii) in the case when E is a Banach space; a strongly continuous cosine function $\{C(t)\}_{t \geq 0}$ can be extended to the real axis R by defining

$$C(t) = C(-t), \quad t < 0$$

such that $\{C(t)\}_{t \in R}$ satisfies

$$2C(t)C(s) = C(s+t) + C(s-t), \quad t,\ s \in R.$$

Theorem A2.3. *Let A be a closed linear operator in a SCLCS. Then the following statements are equivalent.*
(i) *A generates a strongly continuous semigroup $T(t)$.*
(ii) *The Cauchy problem for $u'(t) = Au(t)$ $(t \geq 0)$ is wellposed (with the propagator $T(t)$).*
(iii) *$\{\lambda \in \mathbf{C};\ \mathrm{Re}\lambda > \omega\} \subset \rho(A)$ for some $\omega > 0$ and*

$$\lambda \longmapsto R(\lambda;\ A) \in LT - \mathbf{L}(E)$$

(with the determining function $T(t)$).

Theorem A2.4. *Let A be a closed linear operator in a SCLCS. Then the following statements are equivalent.*
(i) *A generates a strongly continuous cosine function $C(t)$.*
(ii) *The Cauchy problem for $u''(t) = Au(t)$ $(t \geq 0)$ is wellposed (with the two propagators $C(t)$, $\int_0^t C(s)ds$).*
(iii) *$\{\lambda^2;\ \mathrm{Re}\lambda > \omega\} \subset \rho(A)$ for some $\omega > 0$ and*

$$\lambda \longmapsto \lambda R(\lambda^2;\ A) \in LT - \mathbf{L}(E)$$

(with the determining function $C(t)$).

In the sequel, we assume that E is a Banach space.

Definition A2.5. Let $\{T(t)\}_{t\geq 0}$ be a strongly continuous semigroup on E.
(i) $T(t)$ is a semigroup of contractions on E if

$$\|T(t)\| \leq 1, \quad \text{for all } t \geq 0.$$

(ii) $T(t)$ is an analytic semigroup of angle θ $(\theta \in (0, \frac{\pi}{2}])$ if it extends to a semigroup $\{T(z); z \in \Sigma_\theta\}$, analytic (in the norm of $\mathbf{L}(E)$) in Σ_θ and satisfying that

$$T(z)u \longrightarrow u, \quad \text{as } z \longrightarrow 0 \ (z \in \Sigma_{\theta'}),$$

for each $u \in E$ and each fixed $\theta' \in (0, \theta)$.
(iii) $T(t)$ is a differentiable semigroup (or differentiable for $t > 0$) if for every $u \in E$, $t \longmapsto T(t)u$ is differentiable for $t > 0$.

Theorem A2.6 (Phillips' perturbation theorem). *Let A be the generator of a strongly continuous semigroup on E. If $B \in \mathbf{L}(E)$, then $A + B$ is also the generator of a strongly continuous semigroup on E.*

Theorem A2.7 (Stone). *A is the generator of a strongly continuous group of unitary operators on a Hilbert space if and only if iA is self-adjoint.*

Theorem A2.8. *Let A generate a strongly continuous semigroup $T(t)$ on E. Then $T(t)$ is norm continuous at $t = 0$ if and only if $A \in \mathbf{L}(E)$.*

For the proofs, please see Davies [1, Section 1.3] and Pazy [2, Section 1.1].

Theorem A2.9. *Let A generate a strongly continuous semigroup on E. Then $-(-A)^{\frac{1}{2}}$ generates an analytic semigroup of angle $\frac{\pi}{2}$.*

Please see Fattorini [6, Section 6.4] for a proof.

Theorem A2.10. *Assume A is the generator of a strongly continuous cosine function $\{C(t)\}_{t\geq 0}$ on E. Then A generates an analytic semigroup $\{T(t)\}_{t\geq 0}$ of angle $\frac{\pi}{2}$, given by the abstract Weierstrass formula*

$$T(t)u = \frac{1}{(\pi t)^{\frac{1}{2}}} \int_0^\infty e^{-\frac{s^2}{4t}} C(s)u\,ds, \quad t > 0, \ u \in E.$$

Please see Fattorini [7, §VI.2] for a proof.

Definition A2.11. A linear operator A in E is dissipative if for every $u \in \mathcal{D}(A)$ there is a $u^* \in E^*$ with $\langle u^*, u \rangle = \|u\|^2 = \|u^*\|^2$ such that $\mathrm{Re}\,\langle u^*, Au \rangle \leq 0$.

Theorem A2.12. *A linear operator A in E is dissipative if and only if*

$$\|(\lambda - A)u\| \geq \lambda\|u\|, \quad u \in \mathcal{D}(A), \ \lambda > 0.$$

Theorem A2.13. *Let A be a dissipative operator in E.*
(i) *If A is closable, then $\overline{A}$ is also dissipative.*
(ii) *If $\overline{\mathcal{D}(A)} = E$, then A is closable.*

Please see Pazy [2, Section 1.4] for a proof.

Theorem A2.14 (Lumer-Phillips). *Let A be a densely defined linear operator in E. If A is dissipative and there exists a $\lambda_0 > 0$ such that $\mathcal{R}(\lambda_0 - A) = E$, then A is the generator of a strongly continuous semigroup of contractions on E.*

Theorem A2.15. *If A generates a strongly continuous semigroup of contractions on a Hilbert space H, then so does A^*.*

Bibliography

J. Alvarez and J. Hounie

[1] *Functions of pseudo-differential operators of non-positive order,* J. Funct. Anal. 141 (1996), 45-59.

R. A. Adams

[1] "Sobolev Spaces," Academic Press, New York, 1975.

S. Agmon and L. Nirenberg

[1] *Properties of solutions of ordinary differential equations in Banach spaces,* Comm. Pure Appl. Math. 16 (1963), 121-239.

H. Amann

[1] "Linear and Quasilinear Parabolic Problem," Vol. I, Birkhäuser, Basel, 1995.

L. Amerio and G. Prouse

[1] "Almost-Periodic Functions and Functional Equations," Van Nostrand-Reinhold, New York, 1971.

W. Arendt

[1] *Resolvent positive operators,* Proc. London Math. Soc. 54 (1987), 321-349.

[2] *Vector-valued Laplace transforms and Cauchy problems,* Israel J.Math. 59 (1987), 327-352.

W. Arendt and C. J. K. Batty

[1] *Tauberian theorems and stability of one-parameter semigroups,* Trans. Amer. Math. Soc. 306 (1988), 837-852.

[2] *Domination and ergodicity for positive semigroups,* Proc. Amer. Math. Soc. 114 (1992), 743-747.

[3] *A complex tauberian theorem and mean ergodic semigroups,* Semigroup Forum 50 (1995), 351-366.

W. Arendt and H. Kellermann

[1] *Integrated solutions of Volterra integro-differential equations and applications*, Volterra Integro-differential Equations in Banach Spaces and Applications, Proc. Conf. Trento (1987), G. Da Prato and M. Iannelli (eds.), Pitman 1989, 21-51.

W. Arendt, F. Neubrander and U. Schlotterbeck

[1] *Interpolation of semigroups and integrated semigroups*, Semigroup Forum 45 (1992), 26-37.

P. Aviles and J. T. Sanderfur

[1] *A new approach to nonlinear second order equations with applications to partial differential equations*, J. Diff. Equations 58 (1985), 404-427.

V. A. Babalola

[1] *Semigroups of operators on locally convex spaces*, Trans. Amer. Math. Soc. 199 (1974), 163-179.

A. V. Balakrishnan

[1] *Fractional powers of closed operators and the semi-groups generated by them*, Pacific J. Math. 10 (1960), 419-437.

[2] "Applied Functional Analysis," Springer-Verlag, New York, 1976.

V. Barbu

[1] *Differentiable distribution semi-groups*, Anal. Scuola Norm. Sup. Pisa 23 (1969), 413-429.

[2] "Nonlinear Semigroups and Differential Equations in Banach Spaces," Noordhoff Int. Publ. Leyden, the Netherlands, 1976.

H. Bart and S. Goldberg

[1] *Characterizations of almost periodic strongly continuous groups and semigroups*, Math. Ann. 236 (1978), 105-116.

C. J. K. Batty

[1] *Tauberian theorems for the Laplace-Stieltjes transform*, Trans. Amer. Math. Soc. 322 (1990), 783-804.

[2] *Asymptotic stability of Schrödinger semigroups*, Math. Ann. 292 (1992), 457-492.

[3] *Asymptotic behaviour of semigroups of operators*, Funct. Anal. Operator Theory, 30 (1994), 35-52.

[4] *Spectral conditions for stability of one-parameter semigroups*, J. Diff. Equations 127 (1996), 87-96.

C. J. K. Batty and D. A. Greenfield

[1] *On the invertibility of isometric semigroup representations*, Studia Math. 110 (3) (1994), 235-250.

C. J. K. Batty and Vũ Quôc Phóng

[1] *Stability of individual elements under one-parameter semigroups*, Trans. Amer. Math. Soc. 322 (1990), 805-818.

[2] *Stability of strongly continuous representations of abelian semigroups*, Math. Z. 209 (1992), 75-88

R. Beals

[1] *On the abstract Cauchy problem*, J. Funct. Anal. 10 (1972), 281-299.

[2] *Semigroups and abstract Gevrey spaces*, J. Funct. Anal. 10 (1972), 300-308.

M. Becker

[1] *Linear approximation processes in locally convex spaces*, (I), *Semigroups of operators and saturation*, Aeq. Math. 14 (1976), 73-81.

J. Bergh and J. Löfström

[1] "Interpolation Spaces," Springer-Verlag, Berlin, 1976.

A. Beurling

[1] *On analytic extension of semigroups of operators*, J. Funct. Anal. 6 (1970), 387-400.

S. Bochner

[1] *Abstrakte fastperiodische Funktionen*, Acta Math. 61 (1933), 149-184.

S. Bochner and J. von Neumann

[1] *On compact solutions of operational-differential equations I*, Ann. of Math. 36 (1935), 255-291.

H. Bohr

[1] *Zur Theorie der fastperiodischen Funktionen*, I Teil, Acta Math. 45 (1925), 29-127.

H. Brézis

[1] "Opérateurs maximaux monotones et semigroupes de contractions dans les espaces de Hilbert," North-Holland, Amsterdam, 1973.

H. Brézis, M. G. Crandall and E. Kappel

[1] "Semigroups, Theory and Applications," Vol II (eds.), Proc. Trieste 84, Pitman, 1986.

P. L. Butzer and H. Berens

[1] "Semig-roups of Operators and Approximation," Springer-Verlag, New York, 1967.

M. W. Certain

[1] *One-parameter semigroups holomorphic away from zero*, Trans. Amer. Math. Soc. 187 (1974), 377-389.

K. C. Chang

[1] *Solutions of asymptotically linear equations via Morse theory*, Comm. Pure Appl. Math. 34 (1981), 693-712.

[2] "Critical Point Theory and Its Applications," Shanghai Science and Technology Press, Shanghai, 1986. (In Chineses)

[3] "Infinite Dimensional Morse Theory and Multiple Solution Problems," Birkhäuser Basel, 1992.

K. C. Chang and M. Z. Guo

[1] "Teaching Materials on Functional Analysis," Part II, Beijing Univ. Press, 1990. (In Chineses)

J. Chazarain

[1] *Problémes de Cauchy au sens des distributions vectorielles et applications*, C. R. Acad. Sci. Paris. 266 (1968), 10-13.

[2] *Problèmes de Cauchy abstraits et applications a quelques problèmes mixtes*, J. Funct. Anal. 7 (1971), 386-446.

G. Chen, M. P. Coleman and H. H. West

[1] *Pointwise stabilization in the middle of the span for second order systems, nonuniform and uniform exponential decay of solutions*, SIAM J. Appl. Math. 47 (1987), 751-780.

G. Chen, S. G. Krantz, D. L. Russell, C. E. Wayne, H. H. West, and M. P. Coleman

[1] *Analysis, designs, and behavior of dissipative joints for coupled beams*, SIAM J. Appl. Math. 47 (1987), 1665-1693.

G. Chen and D. L. Russell

[1] *A mathematical model for linear elastic systems with structural damping,* Quart. Appl. Math. January (1982), 433-454.

S. Chen and R. Triggiani

[1] *Proof of two conjectures of G. Chen and D. L. Russell on structural damping for elastic systems,* Proc. Seminar in Approximation and optimization, Lecture Notes in Math. 1354, Springer-Verlag, Berlin, 1987, 234-256.

[2] *Proof of extensions of two conjectures on structural damping for elastic systems,* Pacific J. Math. 136 (1989), 15-55.

[3] *Gevrey class semigroups arising from elastic systems with gentle dissipation: the case $0 < \alpha < \frac{1}{2}$,* Proc. Amer. Math. Soc. 110 (1990), 401-415.

Y. H. Choe

[1] *C_0-Semigroups on a locally convex space,* J. Math. Anal. Appl. 106 (1985), 293-320.

I. Ciorănescu

[1] *On the abstract Cauchy problem for the operator $\frac{d^2}{dt^2} - A$,* Int. Eqns. Oper. Th. 7 (1984), 27-35.

[2] *Characterizations of almost periodic strongly continuous cosine operator functions,* J. Math. Anal. Appl. 116 (1986), 222-229.

[3] *On the second order Cauchy problem associated with a linear operator,* J. Math. Anal. Appl. 154 (1991), 238-243.

PH. Clément, S. Invernizzi, E. Mitidieri and I. I. Vrabie

[1] *"Semigroup Theory and Applications,"* (eds.) Lecture Notes in Pure and Applied Math. 116, Marcel Dekker, 1989.

PH. Clément and J. Prüss

[1] *On second order differential equations in Hilbert space,* Boll. Un. Mat. Ital. 44 (1989), 623-638.

C. Corduneanu

[1] *"Almost Periodic Functions,"* Interscience, New York, 1968.

M. G. Crandall and A. Pazy

[1] *On the differentiability of weak solutions of s differential equation in Banach space,* J. Math. Mech. 18 (1969), 1007-1016.

M. G. Crandall, A. Pazy and L. Tartar

[1] *Remarks on generators of analytic semigroups,* Israel J. Math 32 (1979), 363-374.

G. Da Prato

[1] *Semigruppi regolarizzabili*, Ricerche di Mat. 15 (1966), 223-248.

[2] *R-semigruppi analitici ed equazioni di evoluzione in L^p*, Ricerche di Mat. 16 (1967), 233-249.

[3] *Semigruppi periodici*, Ann. Mat. Pura Appl. 78 (1968), 55-67.

G. Da Prato and E. Giusti

[1] *Une charatterizzazioni dei generatori di funzioni coseno astratte*, Boll. Un. Mat. Ital. 22 (1967), 357-368.

G. Da Prato and M. Iannelli

[1] *On a method for studying abstract evolution equations in the hyperbolic case*, Comm. in Partial Diff. Eqs. 1 (1976), 585-608.

[2] "Volterra Integrodifferential Equations in Banach Spaces and Applications," (eds.) R. N. M. 190, Pitman, Boston, London, Melbourne, 1989.

E. B. Davies

[1] "One Parameter Semigroups," Academic Press, London, 1980.

[2] *The harmonic functions of mean ergodic Markov semigroups*, Math. Z. 181 (1982), 543-552.

[3] "Spectral Theory and Differential Operators," Cambridge Univ. Press, Cambridge, 1995.

E. B. Davies and M. M. Pang

[1] *The Cauchy problem and a generalization of the Hille-Yosida Theorem*, Proc. London Math. Soc. 55 (1987), 181-208.

R. deLaubenfels

[1] *Polynomials of generators of integrated semigroups*, Proc. Amer. Math. Soc. 107 (1989), 197-204.

[2] *Integrated semigroups, C-semigroups and the abstract Cauchy problem*, Semigroup Forum 41 (1990), 83-95.

[3] *Integrated semigroups and integrodifferential equations*, Math. Z. 204 (1990), 501-514.

[4] *Entire solutions of the abstract Cauchy problem*, Semigroup Forum 42 (1991), 83-195.

[5] *Existence and uniqueness families for the abstract Cauchy problem*, J. London Math. Soc. 44 (1991), 310-338.

[6] *Incomplete iterated Cauchy problems*, J. Math. Anal. Appl. 168 (1992), 552-578.

[7] *C-semigroups and the Cauchy problem*, J. Funct. Anal. 111 (1993), 44-61.

[8] *Matrices of operators and regularized semigroups*, Math. Z. 212 (1993), 619-629.

[9] "Existence Families, Functional Calculi and Evolution Equations," in: Lect. Notes in Math. 1570, Springer-Verlag, Berlin, 1994.

R. deLaubenfels and Y. Lei

[1] *Regularized functional calculi, semigroups, and cosine functions, for pseudodifferential operators*, preprint, 1997.

R. deLaubenfels and S. Kantorovitz

[1] *Laplace and Laplace-Stieltjes spaces*, J. Funct. Anal. 116 (1993), 1-61.

R. deLaubenfels and S. W. Wang

[1] *Spectral conditions guaranteeing a nontrivial solution of the abstract Cauchy problem*, Proc. Amer. Math. Soc., to appear.

R. deLaubenfels, G. Sun and S. W. Wang

[1] *Regularized semigroups, existence families and the abstract Cauchy problem*, J. Diff. and Int. Eqns. 8 (1995), 1477-1496.

K. Deimling

[1] "Nonlinear Functional Analysis," Springer-Verlag, Berlin, 1985.

G. Dore, A. Favini, E. Obrecht and A. Venni

[1] "Differential Equations in Banach Spaces," (eds.) Lecture Notes in Pure and Applied Math. 148, Marcel Dekker, 1993.

N. Dunford and J. T. Schwartz

[1] "Linear Operators," Part I: *General Theory*, Interscience, New York, 1958.

[2] "Linear Operators," Part II: *Spectral Theory*, Interscience, New York, 1963.

[3] "Linear Operators," Part III: *Spectral Operators*, Interscience, New York, 1971.

O. El-Mennaoui and K. -J. Engel

[1] *On the characterization of eventually norm continuous semigroups in Hilbert spaces*, Arch. Math. 63 (1994), 437-440.

K. -J. Engel

[1] *Polynomial operator matrices as semigroup generators: the* 2×2 *case*, Math. Ann. 284 (1989), 563-576.

[2] *Polynomial operator matrices as semigroup generators: the general case*, Int. Eqns. Oper. Th. 13 (1990), 175-192.

[3] *Growth estimates for semigroups generated by* 2×2 *operator matrices*, Results in Math. 20 (1991), 444-453.

[4] *On singular perturbations of second order Cauchy problems*, Pacific J. Math. 152 (1992), 79-91.

[5] *Systems of evolution equations*, Conf. Sem. Mat. Univ. Bari. 260 (1994), 61-109.

[6] *On dissipative wave equations in Hilbert spaces*, J. Math. Anal. Appl. 184 (1994), 302-316.

K. -J. Engel and G. Hengstberger

[1] *On the well-posedness of finite-dimensionally coupled systems*, Tübinger Berichte zur Funktionalanalysis 5 (1995/96), 117-127.

K. -J. Engel and R. Nagel

[1] *Cauchy problems for polynomial operator matrices on abstract energy spaces*, Forum Math. 2 (1990), 89-102.

H. O. Fattorini

[1] *Ordinary differential equations in linear topological spaces* I, J. Diff. Equations 5 (1968), 72-105.

[2] *Ordinary differential equations in linear topological spaces* II, J. Diff. Equations 6 (1969), 50-70.

[3] *Extension and behavior at infinity of solutions of certain linear operational differential equations*, Pacific J. Math. 33 (1970), 583-615.

[4] *Uniformly bounded cosine functions in Hilbert space*, Indiana Univ. Math. J. 20 (1970), 411-425.

[5] *Some remarks on second-order abstract Cauchy problems*, Funkcialaj Ekvacioj 24 (1981), 331-344.

[6] "The Cauchy Problem," Addison-Wesley, Reading, Mass. 1983.

[7] "Second Order Linear Differential Equations in Banach Spaces," Elsevier Science Publishers B. V., Amsterdam, 1985.

A. Favini

[1] *Laplace transform method for a class of degenerate evolution problems,* Rend. Mat. 3-4 (1979), 511-536.

[2] *Degenerate and singular evolution equations in Banach spaces,* Math. Ann. 273 (1985), 17-44.

[3] *An operational method for abstract degenerate evolution equations of hyperbolic type,* J. Funct. Anal. 76 (1988), 432-456.

A. Favini and E. Obrecht

[1] *Conditions for parabolicity of second order abstract differential equations,* Diff. Int. Equations 4 (1991), 1005-1020.

C. Fefferman and E. M. Stein

[1] H^p *spaces of several variables,* Acta Math. 129 (1972), 137-193.

W. Feller

[1] *On the generation of unbounded semi-groups of bounded linear operators,* Ann. of Math. 58 (1953), 166-174.

[2] *On second-order differential operators,* Ann. of Math. 61 (1955), 90-105.

[3] *On boundaries and lateral conditions for the Kolmogorov differential equations,* Ann. of Math. 65 (1957), 527-570.

W. E. Fitzgibbon

[1] *Strongly damped quasilinear evolution equations,* J. Math. Anal. Appl. 79 (1981), 536-550.

C. Foias

[1] *On strongly continuous semigroups of spectral operators in Hilbert spaces,* Acta Sci. Math. Szeged 19 (1958), 188-191.

[2] *On the Lax-Phillips nonconservative scattering theory,* J. Funct. Anal. 19 (1975), 273-301.

C. Foias and B. Sz.-Nagy

[1] "Harmonic Analysis of Operators in Hilbert Spaces," North-Holland, Amsterdam, 1970.

A. Friedman

[1] "Generalized Functions and Partial Differential Equations," Prentice Hall, 1963.

L. M. Gearhart

[1] *Spectral theory for contraction semigroups on Hilbert spaces*, Trans. Amer. Math. Soc. 236 (1978), 385-394.

I. M. Gelfand and G. E. Shilov

[1] "Generalized Functions," Vol. 3, Academic Press, New York, 1968.

J. A. Goldstein

[1] *Abstract evolution equations*, Trans. Amer. Math. Soc. 141 (1969), 159-186.

[2] *Semigroups and second order differential equations*, J. Funct. Anal. 4 (1969), 50-70.

[3] *Some remarks on infinitesimal generators of analytic semigroups*, Proc. Amer. Math. Soc. 22 (1969), 91-93.

[4] *On the growth of solutions of inhomogeneous abstract wave equations*, J. Math. Anal. Appl. 37 (1972), 650-654.

[5] *A perturbation theorem for evolution equations and some applications*, Illinois J. Math. 18 (1974), 196-207.

[6] *Some developments in semigroups of operators since Hille-Phillips*, Int. Eqns. Oper. Th. 4 (1981), 350-365.

[7] "Semigroups of linear operators and applications," Oxford, New York, 1985.

[8] *Extremal properties of contraction semigroups on Hilbert and Banach spaces*, Bull. London Math. Soc. 25 (1993), 369-376.

[9] *Applications of operator semigroups to Fourier analysis*, Semigroup Forum 52 (1996), 37-47.

J. A. Goldstein, R. deLaubenfels and J. T. Sandefur

[1] *Regularized semigroups, iterated Cauchy problems and equipartition of energy*, Monat. Math. 115 (1993), 47-66.

J. A. Goldstein and J. T. Sandefur

[1] *Equipartition of energy for higher order hyperbolic equations*, Comm. P. D. E. 7 (1982), 1217-1251.

[2] *An abstract D'Alembert formula*, SIAM J. Math. Anal. 18 (1987), 842-856.

J. A. Goldstein and R. Svirsky

[1] *On a domain characterization of Schrödinger operators with magnetic vector potentials and singular potentials*, Proc. Amer. Math. Soc. 105 (1989), 317-323.

G. Greiner

[1] *Spectral properties and asymptotic behavior of the linear transport equation*, Math. Z. 185 (1984), 167-177.

[2] *A spectral decomposition of strongly continuous groups of positive operators*, Quart. J. Oxford 35 (2) (1984), 37-47

[3] *A short proof of Gearhart's theorem*, Semesterbericht Funktionalanalysis Tübingen 16 (1989), 89-92.

G. Greiner and R. Nagel

[1] *On the stability of strongly continuous semigroups of positive operators on* $L^2(\mu)$, Ann. Scuola Norm Sup. Pisa 10 (1983), 257-262.

G. Greiner, J. Voigt and M. P. H. Wolff

[1] *On the spectral bound of the generator of semigroups of positive operators*, J. Operator Theory 5 (1981), 245-256.

G. Greiner and M. Schwarz

[1] *Weak spectral mapping theorems for functional differential equations*, J. Diff. Equations 94 (1991), 205-216.

R. Grimmer and H. Liu

[1] *Integrated semigroups and Integrodifferential equations*, Semigroup Forum 48 (1994), 79-95.

J. Hadamard

[1] "Lectures on Cauchy's Problem in Linear Partial Differential Equations," Yale Univ. Press, New Haven, 1923. Reprinted by Dover, New York, 1952.

J. K. Hale

[1] "Theory of Functional Differential Equations," Springer-Verlag, New York, 1977.

P. R. Halmos

[1] "A Hilbert Space Problem Book," (2nd ed.) Springer-Verlag, New York, 1982.

A. Haraux

[1] "Nonlinear Evolution Equations – Global Behavior of Solutions," Lecture Notes in Math. 841, Springer-Verlag, Berlin, 1981.

I. W. Herbst

[1] *The spectrum of Hilbert space semigroups*, J. Operator Theory 10 (1983), 87-94.

E. Hewitt

[1] "Real and Abstract Analysis," Springer-Verlag, Berlin, 1965.

H. Heyer

[1] *Transient Feller semigroups on certain Gelfand pairs*, Bull. Inst. Math. Acad. Sinica 11 (1983), 227-256.

M. Hieber

[1] *Integrated semigroups and differential operators on L^p*, Dissertation, Universität Tübingen, 1989.

[2] *Laplace transforms and α-times integrated semigroups*, Forum Math. 3 (1991), 595-612.

[3] *Integrated semigroups and differential operators on L^p spaces*, Math. Ann. 291 (1991), 1-16.

[4] *Spectral theory and Cauchy problems on L^p spaces*, Math. Z. 216 (1994), 613-628.

[5] *L^p spectra of pseudodifferential operators generating integrated semigroups*, Trans. Amer. Math. Soc. 347 (1995) 4023-4035.

M. Hieber, A. Holderrieth and F. Neubrander

[1] *Regularized semigroups and systems of linear partial differential equations*, Ann. Scuola Norm. di Pisa 19 (1992), 363-379.

E. Hille

[1] *Representation of one-parameter semi-groups of linear transformations*, Proc. Nat. Acad. Sc. U. S. A. 28 (1942), 175-178.

[2] "Functional Analysis and Semi-groups," Amer. Math. Soc. Coll. Publ. 31, New York, 1948.

[3] *On the differentiability of semig-roup of operators*, Acta Sci. Math. (Szeged) 12B (1950), 19-24.

[4] *Une généralisation du problème de Cauchy*, Ann. Inst. Fourier 4 (1952), 31-48.

[5] *A note on Cauchy's problem*, Ann. Soc. Polon. Math. 25 (1952), 56-68.

[6] *The abstract Cauchy problem and Cauchy's problem for parabolic differential equations*, J. Anal. Math. 3 (1954), 81-196.

E. Hille and R. S. Phillips

[1] "Functional Analysis and Semi-Groups, Amer. Math. Soc. Colloquium Publications, vol. 31, Providence, R. I., 1957.

J. S. Howland

[1] *On a theorem of Gearhart*, Integral Eqs. Operator Theory 7 (1984), 138-142.

L. Hörmander

[1] *Estimates for translation invariant operators in L^p spaces*, Acta Math. 104 (1960), 93-140.

[2] *Pseudo-differential operators and hypoelliptic equations*, Ann. of Math. 83 (1966), 129-209.

[3] *On the characteristic Cauchy problem*, Ann. of Math. 88 (1968), 341-370.

[4] *On the existence of real analytic solutions of partial differential equations with constant coefficients*, Invent. Math. 21 (1973), 151-182.

[5] *Propagation of singularities and semiglobal existence theorems for (pseudo-) differential operators of principal type*, Ann. of Math. 108 (1978), 569-609.

[6] *Uniqueness theorems for second order elliptic differential equations*, Comm. Partial Diff. Eqns. 8 (1983), 21-64.

[7] "The Analysis of Linear Partial Differential Operators *I*," (2nd ed.) Springer-Verlag, Berlin, New York, 1983.

[8] "The Analysis of Linear Partial Differential Operators *II*," Springer-Verlag, Berlin, New York, 1983.

F. L. Huang

[1] *Stability of linear semigroup and stabilizability problems*, J. Sichuan Univ. (Sichuan Daxue Xuebao) 17 (3) (1980), 17-35. (In Chinese)

[2] *Characteristic property of semigroups of isometric linear operators in Banach spaces*, J. Sichuan Univ. (Sichuan Daxue Xuebao) 20 (3) (1983), 1-8. (In Chinese)

[3] *Asymptotic stability theory for linear dynamical systems in Banach spaces*, Chin. Sci. Bull. (Kexue Tongbao) 10 (1983), 584-586. (In Chinese)

[4] *Characteristic conditions for exponential stability of linear dynamical systems in Hilbert spaces*, Ann. of Diff. Eqs. 1 (1985), 43-56.

[5] *On the holomorphic property of the semigroup associated with linear elastic systems with structural damping*, Acta Math. Sci. 5 (1985), 271-277.

[6] *A problem for linear elastic systems with structural damping,* Acta Math. Sci. 6 (1986), 101-107. (In Chinese)

[7] *On the mathematical model for linear elastic systems with analytic damping,* SIAM Control and Opti. 26 (1988), 714-724.

[8] *Some problems for linear elastic systems with damping,* Acta. Math. Sci. 10 (1990), 319-326.

[9] *Strong asymptotic stability of linear dynamical systems in Banach spaces,* J. Diff. Equations 104 (1993), 307-324.

[10] *Spectral properties and stability of one-parameter semigroups,* J. Diff. Equations 104 (1993), 182-195.

F. L. Huang and T. W. Huang

[1] *Local C-cosine family theory and applications,* Chin. Ann. of Math. 16B (1995), 213-232.

F. L. Huang, Y. Z. Huang and F. M. Guo

[1] *Analyticity and differentiability of C_0 semigroups associated with Euler-Bernoulli beam equation with damping,* Sci. in China 35A, 1992, 122-133. (In Chinese)

F. L. Huang, J. Liang and T. J. Xiao

[1] *On a generalization of Horn's fixed point theorem,* J. Math. Anal. Appl. 164 (1992), 34-39.

F. L. Huang and K. S. Liu

[1] *A problem of exponential stability for linear dynamical systems in Hilbert spaces,* Chin. Sci. Bull. (Kexue Tongbao) 33 (1988), 460-462.

F. L. Huang, K. S. Liu and G. Chen

[1] *Differentiability of the semigroup associated with a structural damping model,* Proceeding of the 28th IEEE-CDC, Tampa, Florida, 1989, 2034-2038.

S. Z. Huang

[1] *An equivalent description of non-quasianalyticity through spectral theory of C_0-groups,* Tübinger Berichte zur Funktionalanalysis 3 (1993/94), 81-90.

[2] *Stability properties characterizing the spectra of operators on Banach spaces,* J. Funct. Anal. 132 (1995), 361-382.

P. E. T. Jorgensen

[1] *Spectral theory for one-parameter groups of isometries*, J. Math. Anal. Appl. 168 (1992), 131-146.

S. Kantorovitz

[1] *Characterization of unbounded spectral operators with spectrum in a half-line*, Comment. Math. Helvetici 56 (1981), 163-178.

[2] *Spectrality criteria for unbounded operators with real spectrum*, Math. Ann. 256 (1981), 19-28.

[3] "Spectral Theory of Banach Space Operators," Lecture Notes in Math. 1012, Springer, New York, 1983.

[4] *The Hille-Yosida space of an arbitrary operator*, J. Math. Anal. Appl. 138 (1988), 107-111.

[5] "Semigroups of Operators and Spectral Theory," R. N. M. 330, Pitman, New York, 1995.

T. Kato

[1] "Perturbation Theory for Linear Operators," Springer-Verlag, New York, 1966.

[2] *A characterization of holomorphic semigroups*, Proc. Amer. Math. Soc. 25 (1970), 495-498.

[3] *Linear evolution equations of "hyperbolic" type* I, J. Fac. Sci. Univ. Tokyo, 17 (1970), 241-258.

[4] *Linear evolution equations of "hyperbolic" type* II, J. Math. Soc. Japan 25 (1973), 648-666.

H. Kellermann and M. Hieber

[1] *Integrated semigroups*, J. Funct. Anal. 84 (1989), 160-180.

V. Kéyantuo

[1] *The Laplace transform and the ascent method for abstract wave equations*, J. Diff. Equations 122 (1995), 27-47.

Y. Konishi

[1] *Cosine functions of operators in locally convex spaces*, J. Fac. Sci. Univ. Tokyo, Sect. IA Math. 18 (1971), 443-463.

M. A. Krasnosel'skii and P. E. Sobolevskii

[1] *Fractional power of operators acting in Banach spaces*, Dokl. Akad. Nauk. SSSR 129 (1959), 499-502.

S. G. Krein

[1] "Linear Differential Equations in Banach Spaces," Amer. Math. Soc., Providence, R. I. 1971.

C.-C. Kuo and S.-Y. Shaw

[1] *On α-times integrated C-semigroups and the abstract Cauchy problem,* preprint 1995.

[2] *On strong and weak solutions of the abstract Cauchy problem,* preprint 1995.

S. Kurepa

[1] *A cosine functional equation in Banach algebras,* Acta Sci. Math. Szeged 23 (1962), 255-267.

S. Lang

[1] "Real and Functional Analysis," (3rd Edi.) GTM 142, Springer-Verlag, 1993.

R. Lange and B. Nagy

[1] *Semigroups and scalar-type operators in Banach spaces,* J. Funct. Anal. 119 (1994), 468-480.

P. D. Lax

[1] *A stability theorem for solutions of abstract differential equations, and its application to the study of local behavior of solutions of elliptic equations,* Comm. Pure. Appl. Math. 9 (1956), 747-766.

P. D. Lax and R. S. Phillips

[1] "Scattering Theory," Academic Press, New York, 1967.

[2] *On the scattering frequencies of the Laplace operator for exterior domains,* Comm. Pure. Appl. Math. 25 (1972), 85-101.

[3] *Scattering theory for dissipative hyperbolic systems,* J. Funct. Anal. 14 (1973), 172-235.

P. D. Lax and R. Richtmyer

[1] *Survey of the stability of linear finite difference equations,* Comm. Pure. Appl. Math. 9 (1956), 267-293.

Y. Lei and Q. Zheng

[1] *The application of C-semigroups to differential operators in $L^p(R^n)$,* J. Math. Anal. Appl. 188 (1994), 809-818.

B. R. Li

[1] "Introduction to Operator Algebras," World Sci., Singapore, 1992.

B. R. Li, S. W. Wang, S. Z. Yan and C. -C. Yang

[1] "Functional Analysis in China," (eds.) Kluwer Academic Publishers, Dordrecht, Boston, London, 1996.

S. J. Li and A. Szulkin

[1] *Periodic solutions for an asymptotically linear wave equation*, Topological Methods in Nonlin. Anal. 1 (1993), 211-230.

X. J. Li and Yao Yunlong

[1] *Time optimal control of distributed parameter systems*, Scientia Sinica, 24 (1981), 455-465.

Y.-C. Li

[1] *Integrated C-semigroups and C-cosine functions of operators on locally convex spaces*, Ph. D. dissertation, National Central University, 1991.

Y.-C. Li and S.-Y. Shaw

[1] *On generators of integrated C-semigroups and C-cosine functions*, Semigroup Forum 47 (1993), 29-35.

[2] *N-times integrated C-semigroups and the abstract Cauchy problem*, Chinese J. Math., to appear.

J. Liang

[1] *Studies on operator semigroups, differential operators and abstract differential equations*, Ph. D. dissertation, Sichuan Union University, 1997.

J. Liang, F. L. Huang and T. J. Xiao

[1] *Exponential stability for abstract linear autonomous functional differential equations with infinite delay*, Inter. J. Math. Math. Sci., 21 (1998), 255-260.

J. Liang and T. J. Xiao

[1] *On exponential stability of linear autonomous functional differential equations with infinite delay in Banach spaces*, Chin. Sci. Bull. (Kexue Tongbao) 34 (1989), 633-634. (In Chinese)

[2] *C-wellposedness of the Cauchy problem for complete second order equations*, Selected Works of Chinese Youth on ODE, National Science Press of China, 1991, 134-140. (In Chinese)

[3] *Solutions of abstract functional differential equations with infinite delay,* Acta Math. Sinica 34 (1991), 631-644. (In Chinese)

[4] *Functional differential equations with infinite delay in Banach spaces,* Inter. J. Math. Math. Sci. 14 (1991), 497-508.

[5] *A class of operator matrices and applications,* J. Kunming Inst. Tech. 17 (5) (1992), 88-97. (In Chinese)

[6] *Fundamental operators of functional differential equations with infinite delay in Banach spaces,* Appli. Funct. Anal. 2 (1995), 164-168.

[7] *Higher order abstract Cauchy problems in locally convex spaces,* Lecture Notes in Pure and Applied Math. 176, Marcel Dekker, 1996, 177-181,

[8] *Norm continuity (for $t > 0$) of propagators of arbitrary order abstract differential equations in Hilbert spaces,* J. Math. Anal. Appl. 204 (1996), 124-137.

[9] *Almost periodicity of the solutions of second order differential equations in Banach spaces,* Appli. Funct. Anal. 3 (1997), 87-91.

[10] *Wellposedness results for certain classes of higher order abstract Cauchy problems connected with integrated semigroups,* Semigroup Forum, 56 (1998), 84-103.

[11] *Norm continuity (for $t > 0$) of linear operator families,* Chin. Sci. Bull. (Kexue Tongbao) 43(1998), 719-722.

[12] *Integrated semigroups and higher order abstract equations,* J. Math. Anal. Appl. 222 (1998), 110-125.

[13] *A characterization of norm continuity of propagators for second order abstract differential equations,* Computers Math. Applic. 36 (1998), 87-94.

[14] *Evolution equations with Schrödinger-type operator coefficients in $L^p(\Omega)$,* submitted.

[15] *Differential operators and time-dependent second order equations,* preprint.

J. Liang, T. J. Xiao and F. L. Huang

[1] *Solvability and stability for abstract functional differential equations with infinite delay,* J. Sichuan Univ. (Sichuan Daxue Xuebao) 31 (1994), 8-14. (In Chinese)

J. L. Lions

[1] *Un remarque sur les applications du théorème de Hille-Yosida,* J. Math. Soc. Japan 9 (1957), 62-70.

[2] *Equations différentielles à coéfficients opérateurs non bornés*, Bull. Soc. Math. France 86 (1958), 321-330.

[3] *Les semi-groupes distributions*, Portugaliae Math. 19 (3) (1960), 141-164.

[4] "Équations Différentielles Opérationnelles et Problémes aux Limites," Springer-Verlag, Berlin, 1961.

[5] *Some linear and non-linear boundary value problems for evolution equations*, Lectures in Differential Equations, Vol. I, Van Nostrand, New York, 1969, 97-121.

[6] "Contrôlabilité Exacte, Perturbations et Stabilization de Systèmes Distribués," 1, Masson, Paris, 1989.

J. L. Lions and E. Magenes

[1] "Non-homogeneous Boundary Value Problems, and Applications, (I)," Springer-Verlag, New York, 1970.

K. S. Liu, F. L. Huang and G. Chen

[1] *Exponential stability analysis of a long chain of coupled vibrating strings with dissipative linkage*, SIAM J. Applied Math. 49 (1989), 1694-1707.

Ju. I. Ljubic

[1] *Conditions for the uniqueness of the solution of Cauchy's abstract problem*, Dokl. Akad. Nauk. SSSR 130 (1960), 969-972.

G. Lumer and R. S. Phillips

[1] *Dissipative operators in a Banach space*, Pacific J. Math. 11 (1961), 679-698.

D. Lutz

[1] *Strongly continuous operator cosine functions*, Lecture Notes in Math. 948, Springer-Verlag, New York, 1982, 73-97.

Z. M. Ma and M. Röckner

[1] "Introduction to the theory of (non-symmetric) Dirichlet Forms," Springer-Verlag, New York, 1992.

C. Martínez, M. Sanz and L. Marco

[1] *Fractional powers of operators*, J. Math. Soc. Japan 40 (1988), 331-347.

P. Massatt

[1] *Limiting behavior for strongly damped nonlinear wave equations*, J. Diff. Equations 48 (1983), 334-349.

I. V. Mel'nikova and A. I. Filinkov

[1] *Classification and well-posedness of the Cauchy problem for second-order equations in a Banach space*, Soviet Math. Dokl. 29 (1984), 646-651.

[2] *The connection between well-posedness of the Cauchy problem for an equation and for a system in a Banach space*, Soviet Math. Dokl. 37 (1988), 647-651.

A. Miyachi

[1] *On some Fourier multipliers for $H^p(R^n)$*, J. Fac. Sci. Univ. Tokyo 27 (1980), 157-179.

I. Miyadera

[1] *Generation of a strongly continuous semi-group of operators*, Tôhoku Math. J. 4 (1952), 109-114.

[2] *On the representation theorem by the Laplace transformation of vector-valued functions*, Tôhoku Math. J. 8 (1956), 170-180.

[3] *Semi-groups of operators in Fréchet space and applications to partial differential equations*, Tôhoku Math. J. 11 (1959), 162-183.

[4] *On the generators of exponentially bounded C-semigroups*, Proc. Japan Acad. 62 Ser. A (1986), 239-242.

[5] *A generalization of the Hille-Yosida theorem*, Proc. Japan Acad. Ser. A 64 (1988), 223-226.

I. Miyadera and N. Tanaka

[1] *Exponentially bounded C-semigroups and generation of semigroups*, J. Math. Anal. Appl. 143 (1989), 358-378.

R. T. Moore

[1] *Banach algebras of operators in locally convex spaces*, Bull. Amer. Math. Soc. 75 (1969), 68-73.

R. Nagel

[1] *Well-posedness of higher order abstract Cauchy problems*, Conf. Sem. Mat. Bari 203 (1985), 1-29.

[2] " One-parameter Semigroups of Positive Operators," (ed.) Lecture Notes Math. 1184, Springer-Verlag, Berlin, 1986.

[3] *Towards a "matrix theory" for unbounded operator matrices*, Math. Z. 201 (1989), 57-68.

[4] *The spectrum of unbounded operator matrices with non-diagonal domain*, J. Funct. Anal. 89 (1990), 291-302.

[5] *Stability criteria through characteristic equations of linear operators*, Tübinger Berichte zur Funktionalanalysis 5 (1995/96), 299-305.

B. Nagy

[1] *On cosine operator functions in Banach spaces*, Acta Sci. Math. 36 (1974), 281-290.

F. Neubrander

[1] *Well-posedness of abstract Cauchy problems*, Semigroup Forum 29 (1984), 75-85.

[2] *Well-posedness of higher order abstract Cauchy problems*, Trans. Amer. Math. Soc. 295 (1986), 257-290.

[3] *On the relation between the semigroup and its infinitesimal generator*, Proc. Amer. Math. Soc. 100 (1987), 104-107.

[4] *Integrated semigroups and their applications to the abstract Cauchy problem*, Pacific J. Math. 135 (1988), 111-155.

[5] *Integrated semigroups and their application to complete second order Cauchy problems*, Semigroup Forum 38 (1989), 233-251.

[6] *Abstract elliptic operators, analytic interpolation semigroups, and Laplace transforms of analytic functions*, Semesterbericht Funktionalanalysis Tübingen 15 (1988/1989), 163-186.

S. Nicaise

[1] *The Hille-Yosida and Trotter-Kato theorems for integrated semigroups*, J. Math. Anal. Appl. 180 (1993), 303-316.

L. Nirenberg

[1] *Remarks on strongly elliptic partial differential equations*, Comm. Pure. Appl. Math. 8 (1955), 648-675.

[2] *Uniqueness in Cauchy problems for differential equations with constant leading coefficients*, Comm. Pure. Appl. Math. 10 (1957), 89-105.

E. Obrécht

[1] *Sul problema di Cauchy per le equazioni paraboliche astratte di ordine n*, Rend. Sem. Mat. Univ. Padova 53 (1975), 231-256.

[2] *The Cauchy problem for time-dependent abstract parabolic equations of higher order*, J. Math. Anal. Appl. 125 (1987), 508-530.

S. Oharu

[1] *Semigroups of linear operators in a Banach space*, Publ. RIMS, Kyoto Univ. 7 (1971), 205-260.

A. Pazy

[1] *On the differentiability and compactness of semi-groups of linear operators*, J. Math. Mech. 17 (1968), 1131-1141.

[2] "Semigroups of Linear Operators and Applications to Partial Differential Equations," Springer-Verlag, New York, 1983

R. S. Phillips

[1] *Spectral theory for semi-groups of linear operators*, Trans. Amer. Math. Soc. 71 (1951), 393-415.

[2] *On the generation of semi-groups of linear transformations*, Proc. Amer. Math. Soc. 2 (1951), 234-237.

[3] *On the generation of semi-groups of linear operators*, Pacific J. Math. 2 (1952), 343-369.

[4] *Perturbation theory for semi-groups of linear operators*, Trans. Amer. Math. Soc. 74 (1953), 199-221.

[5] *A note on the abstract Cauchy problem*, Proc. Nat. Acad. Sci. U. S. A. 40 (1954), 244-248.

[6] *Dissipative hyperbolic systems*, Trans. Amer. Math. Soc. 86 (1957), 109-173.

[7] *Dissipative operators and parabolic differential equations*, Comm. Pure Appl. Math. 12 (1959), 249-276.

S. I. Piskarev

[1] *Periodic and almost periodic cosine operator functions*, Mat. Sb. 118 (1982), 386-398.

S. I. Piskarev and S. Y. Shaw

[1] *Perturbation and comparison of cosine operator functions*, Semigroup Forum 51 (1995), 225-246.

[2] *Multiplicative perturbations of C_0-semigroups and applications to step responses and cumulative outputs*, J. Funct. Anal. 128 (1995), 315-340.

A. J. Pritchard and J. Zabczyk

[1] *Stability and stabilizability of infinite dimensional systems*, SIAM Review 23 (1981), 25-52.

J. Prüss,

[1] *On the spectrum of C_0-semigroups*, Trans. Amer. Math. Soc. 284 (1984), 847-857.

[2] "Linear Volterra Equations and Applications," Birkhäuser, Dasel, 1993.

W. Rudin

[1] "Functional Analysis," McGraw-Hill, New York, 1973.

D. L. Russell

[1] "Mathematics of Finite-Dimensional Control Systems: Theory and Design," Lecture Notes in Pure and Appli. Math. 43, Marcel Dekker, New York, 1979.

[2] *A comparison of certain elastic dissipation mechanisms via decoupling and projection techniques*, Quart. Appl. Math. 49 (1991), 373-396.

[3] *A general framework for the study of indirect damping mechanisms in elastic systems*, J. Math. Anal. Appl. 173 (1993), 339-358.

J. T. Sandefur

[1] *Higher order abstract Cauchy problems*, J. Math. Anal. Appl. 60 (1977), 728-742.

[2] *Existence and uniqueness of solutions of second order nonlinear differential equations*, SIAM J. Math. Anal. 14 (1983), 477-487.

L. Schwartz

[1] "Lectures on Mixed Problems in Partial Differential Equations and the Representation of Semi-Groups," Tata Inst. Fund. Research, Bombay, 1958.

R. E. Showalter

[1] "Hilbert Space Methods for Partial Differential Equations," Pitman, London, 1977.

A. M. Sinclair

[1] *Continuous semigroups in Banach algebras*, Cambridge Univ. Press, Cambridge, England, 1981.

B. Simon

[1] *Schrödinger semigroups*, Bull. Amer. Math. Soc. 7 (1982), 447-526.

M. Slemrod

[1] *Asymptotic behavior of C_0 semi-groups as determined by the spectrum of the generator*, Indiana Univ. Math. J. 25 (1976), 783-792.

P. E. Sobolevskii

[1] *A certain type of differential equations in a Banach space*, Differencial'nye Uravnenija 4 (1968), 2278-2280.

J. Song, J. Y. Yu, S. J. Hu, Y. Z. Xiao and G. T. Zhu

[1] *Asymptotic property of the solution of a freely elastic beam with structural damping*, Scientia Sinica 27 (1984), 1307-1316.

M . Sova

[1] *Cosine operator functions*, Rozprawy Mat. 49 (1966), 1-47.

[2] *Problème de Cauchy pour equations hyperboliques opérationnelles a coefficients constants non-bornés*, Ann. Scuola Norm. Sup. Pisa, 22 (1968), 67-100.

[3] *Problèmes de Cauchy paraboliques abstraites de classes supérieures et les semi-groupes distributions*, Ricnerche Mat. 18 (1969), 215-238.

E. M. Stein

[1] "Singular Integrals and Differentiability Properties of Functions," Princeton University Press, New Jersey, 1970.

[2] "Harmonic Analysis, Real Variables, Orthogonality and Oscillatory Integrals," Princeton University Press, New Jersey, 1993.

H. B. Stewart

[1] *Generation of analytic semigroups by strongly elliptic operators*, Trans. Amer. Math. Soc. 199 (1974), 141-162.

B. Sz. -Nagy and C. Foias

[1] "Harmonic Analysis of Operators in Hilbert Space," North-Holland, Amsterdam American Elsevier, New York, Budapest, 1969.

N. Tanaka

[1] *Holomorphic C-semigroups and holomorphic semigroups*, Semigroup Forum 38 (1989), 253-261.

N. Tanaka and N. Okazawa

[1] *Local C-semigroups and local integrated semigroups*, Proc. London Math. Soc. 61 (3) (1990), 63-90.

N. Tanaka and I. Miyadera

[1] *Some remarks on C-semigroups and integrated semigroups*, Proc. Japan Acad. 63 (1987), 139-142.

[2] *Exponentially bounded C-semigroups and integrated semigroups*, Tokyo J. Math. 12 (1989), 99-115.

[3] *C-semigroups and the abstract Cauchy problem*, J. Math. Anal. Appl. 170 (1992), 196-206.

H. R. Thieme

[1] *"integrated semigroups" and integrated solutions to abstract Cauchy problems*, J. Math. Anal. Appl. 152 (1990), 416-447.

C. C. Travis and G. F. Webb

[1] *Existence and stability for partial differential equations*, Trans. Amer. Math. Soc. 200 (1974), 395-418.

[2] *Compactness, regularity and uniform continuity properties of strongly continuous cosine families*, Houston J. Math. 3 (1977), 555-567.

[3] *Cosine families and abstract nonlinear second order differential equations*, Acta Math. Sci. Hung. 32 (1978), 75-96.

R. Triggiani

[1] *On the stabilizability problem in Banach space*, J. Math. Anal. Appl. 52 (1975), 383-403.

[2] *Lack of uniform stabilization for noncontractive semigroups under compact perturbation*, Proc. Amer. Math. Soc. 105 (1989), 375-383.

[3] *Counterexamples to some stability questions for dissipative generators*, J. Math. Anal. Appl. 170 (1992), 49-64.

[4] *A sharp result on the exponential operator-norm decay of a family of strongly continuous semigroups*, Semigroup Forum 49 (1994), 387-395.

T. Ushijima

[1] *On the abstract Cauchy problem and semi-groups of linear operators in locally convex spaces*, Scientific Papers of the College of General Education Univ. Tokyo 21 (1971), 93-122.

J. A. van Casteren

[1] "Generators of Strongly Continuous Semigroups," Pitman, London, 1985.

J. M. A. M. van Neerven

[1] "The adjoint of a semigroup of linear operators," Lecture Notes in Math. 1529, Springer-Verlag, Berlin, 1992.

[2] *Exponential stability of operators and operator semigroups*, J. Funct. Anal. 130 (1995), 293-309.

[3] *Individual stability of C_0-semigroups with uniformly bounded local resolvent*, Semigroup Forum 53 (1996), 155-161.

J. M. A. M. van Neerven and B. Straub

[1] *On the existence and growth of mild solutions of the abstract Cauchy problem for operators with polynomially bounded resolvents*, Tübinger Berichte zur Funktionalanalysis 4 (1994/95), 182-206.

V. V. Vasilev, S. G. Krein and S. I. Piskarev

[1] *Operator semigroups, cosine operator functions, and linear differential equations*, translated in J. Sov. Math., Collection: Mathematical analysis 54 (1991), 1042-1129.

J. Voigt

[1] *On the perturbation theory for strongly continuous semigroups*, Math. Ann. 229 (1977), 163-171.

[2] *Interpolation for (positive) C_0-semigroups on L^p-spaces*, Math. Z. 188 (1985), 283-286.

[3] *Absorption semigroups, their generators, and Schrödinger semigroups*, J. Funct. Anal. 67 (1986), 167-205.

S. W. Wang

[1] *Properties of subgenerators of C-regularized semigroups*, Proc. Amer. Math. Soc., to appear.

[2] *Quasi-distribution semigroups and integrated semigroups*, J. Funct. Anal., to appear.

M. Watanabe

[1] *On the differentiability of semigroups of linear operators in locally convex spaces*, Sci. Rep. Niigata Univ. 9 Ser.A (1972), 23-34.

[2] *A new proof of the generation theorem of cosine families in Banach spaces*, Houston J. Math. 10 (1984), 285-290.

[3] *Weak conditions for generation of cosine families in linear topological spaces*, Proc. Amer. Math. Soc. 105 (1989), 151-158.

G. F. Webb

[1] *Continuous nonlinear perturbations of linear accretive operators in Banach spaces*, J. Funct. Anal. 10 (1972), 191-203.

[2] *Exponential representation of solutions to an abstract semi-linear differential equation*, Pacific J. Math. 70 (1977), 269-279. 241-255.

J. Weidmann

[1] "Linear Operators in Hilbert Spaces," Springer-Verlag, New York, 1980.

B. Weiss

[1] *Abstract vibrating systems*, J. Math. Mech. 17 (1967), 241-255.

G. Weiss

[1] *Weak L^p-stability of a linear semi-group on a Hilbert space implies exponential stability*, J. Diff. Equations 76 (1988), 269-285.

[2] *The resolvent growth assumption for semigroups on Hilbert spaces*, J. Math. Anal. Appl. 145 (1990), 154-171.

D. V. Widder

[1] "The Laplace Transform," Princeton University Press, New Jersey, 1946.

T. J. Xiao

[1] *On the well-posedness of the Cauchy problem for a kind of complete second order differential equations*, J. Sichuan Univ. (Sichuan Daxue Xuebao) 25 (1988), 421-428. (In Chinese)

[2] *The Cauchy problem for higher order operator differential equations*, Ph. D. dissertation, Sichuan Union University, 1994

T. J. Xiao and J. Liang,

[1] *Complete second order linear differential equations in Banach spaces*, Chin. Sci. Bull. (Kexue Tongbao) 33 (1988), 1274-1275. (In Chinese)

[2] *On complete second order linear differential equations in Banach spaces*, Pacific J. Math. 142 (1990), 175-195.

[3] *Well-posedness and exponential growth property of a class of complete second-order linear differential equations*, J. Sichuan Univ. (Sichuan Daxue Xuebao) 27 (1990), 396-401. (In Chinese)

[4] *On some problems in abstract differential equations*, J. Yunnan Teach. Univ. 11(4) (1991), 9-15. (In Chinese)

[5] *The Cauchy problem for higher order abstract differential equations in Banach spaces* (Abstract), J. Yunnan Teach. Univ. 11(4) (1991), 98-100. (In Chinese)

[6] *A note on the propagators of second order linear differential equations in Hilbert spaces*, Proc. Amer. Math. Soc. 113 (1991), 663-667.

[7] *Second order linear differential equations with almost periodic solutions*, Acta Math. Sinica, New Series 7 (1991), 354-359.

[8] *Complete second order linear differential equations with almost periodic solutions*, J. Math. Anal. Appl. 163 (1992), 136-146.

[9] *Well-posedness of the Cauchy problem for second order linear differential equations in Fréchet spaces*, Acta Math. Sinica 35 (1992), 354-363. (In Chinese)

[10] *Almost periodicity of solutions of incomplete second differential equations*, J. Yunnan Teach. Univ. 12(4) (1992), 10-13. (In Chinese)

[11] *Analyticity of the propagators of second order linear differential equations in Banach spaces*, Semigroup Forum 44 (1992), 356-363.

[12] *The Cauchy problem for higher order abstract differential equations in Banach spaces*, Chin. J. Contemporary Math. 14 (1993), 305-321.

[13] *Analyticity of solutions of a class of abstract differential equations*, J. Yunnan Teach. Univ. 13(1) (1993), 13-15. (In Chinese)

[14] *Parabolicity of a class of higher order abstract differential equations*, Proc. Amer. Math. Soc. 120 (1994), 173-181.

[15] *Integrated semigroups and cosine families and higher order abstract Cauchy problems*, Functional Analysis in China, B. R. Li, S. W. Wang, S. Z. Yan and C. C. Yang (eds), Kluwer Academic Publishers, 1996, 351-365.

[16] *Widder-Arendt theorem and integrated semigroups in locally convex spaces*, Sci. in China (Series A) 39 (1996), 1121-1130.

[17] *On the Cauchy problems of a class of higher order abstract equations*, Chin. Ann. of Math. 18A (1997), 135-144. (In Chinese)

[18] *Entire solutions of higher order abstract Cauchy problems*, J. Math. Anal. Appl. 208 (1997), 298-310.

[19] *Pseudodifferential operators and regularized semigroups*, Appli. Funct. Anal. 3 (1997), 213-218.

[20] *Semigroups arising from elastic systems with dissipation*, Computers Math. Applic. 33 (10) (1997), 1-9.

[21] *Laplace transforms and integrated, regularized semigroups in locally convex spaces*, J. Funct. Anal. 148 (1997), 448-479.

[22] *Exponential stability of solutions for higher order abstract Cauchy problems*, J. Math. Anal. Appl. 215 (1997), 485-498.

[23] *Differential operators and C-wellposedness of complete second order abstract Cauchy problems*, Pacific J. Math., to appear.

[24] *Pazy-type characterization for differentiability of propagators of higher order Cauchy problems in Banach spaces*, submitted.

[25] *Exponential growth bound of propagators for higher order differential equations in Hilbert spaces*, preprint.

[26] *Existence families and differential operators*, submitted.

A. Yagi

[1] *On the abstract linear evolution equation in Banach spaces*, J. Math. Soc. Japan 28 (1976), 290-303.

[2] *Applications of the purely imaginary powers of operators in Hilbert spaces*, J. Funct. Anal. 73 (1987), 216-231.

K. Yosida

[1] *On the differentiability and the representation of one parameter semi-groups of linear operators*, J. Math. Soc. Japan 1 (1948), 15-21.

[2] *On the differentiability of semi-groups of linear operators*, Proc. Japan Acad. 34 (1958), 337-340.

[3] *Time dependent evolution equations in a locally convex space*, Math. Ann. 162 (1965), 83-86.

[4] "Functional Analysis (6th edition)," Springer Verlag, New York, 1980.

P. H. You

[1] *Characteristic conditions for a C_0-semigroup with continuity in the uniform operator tolopogy for $t > 0$ in Hilbert space*, Proc. Amer. Math. Soc. 116 (1992), 991-997.

S. Zaidman

[1] *Sur un théorème de I. Miyadera concernant la représentation des fonctions vectorielles par des intégrales de Laplace*, Tôhoku Math. J. 12 (1960), 47-51.

[2] "Abstract Differential Equations," R. N. M. 36, Pitman, San Francisco, London, Melbourne, 1979.

[3] "Almost-periodic Functions in Abstarct Spaces," R. N. M. 126, Pitman, Boston, London, Melbourne, 1985.

[4] "Topics in Abstract Differential Equations," R. N. M. 304, Pitman, Boston, London, Melbourne, 1994.

Index

The numbers that follow the items indicate the sections where their meanings are explained.

Symbols

Springer
und
Umwelt

Als internationaler wissenschaftlicher
Verlag sind wir uns unserer besonderen
Verpflichtung der Umwelt gegenüber
bewußt und beziehen umweltorientierte
Grundsätze in Unternehmens-
entscheidungen mit ein. Von unseren
Geschäftspartnern (Druckereien,
Papierfabriken, Verpackungsherstellern
usw.) verlangen wir, daß sie sowohl
beim Herstellungsprozess selbst als
auch beim Einsatz der zur Verwendung
kommenden Materialien ökologische
Gesichtspunkte berücksichtigen.
Das für dieses Buch verwendete Papier
ist aus chlorfrei bzw. chlorarm
hergestelltem Zellstoff gefertigt und im
pH-Wert neutral.

 Springer

Lecture Notes in Mathematics

For information about Vols. 1–1504
please contact your bookseller or Springer-Verlag

Vol. 1647: D. Dias, P. Le Barz, Configuration Spaces over Hilbert Schemes and Applications. VII. 143 pages. 1996.

Vol. 1648: R. Dobrushin, P. Groeneboom, M. Ledoux, Lectures on Probability Theory and Statistics. Editor: P. Bernard. VIII, 300 pages. 1996.

Vol. 1649: S. Kumar, G. Laumon, U. Stuhler, Vector Bundles on Curves – New Directions. Cetraro, 1995. Editor: M. S. Narasimhan. VII, 193 pages. 1997.

Vol. 1650: J. Wildeshaus, Realizations of Polylogarithms. XI, 343 pages. 1997.

Vol. 1651: M. Drmota, R. F. Tichy, Sequences, Discrepancies and Applications. XIII, 503 pages. 1997.

Vol. 1652: S. Todorcevic, Topics in Topology. VIII, 153 pages. 1997.

Vol. 1653: R. Benedetti, C. Petronio, Branched Standard Spines of 3-manifolds. VIII, 132 pages. 1997.

Vol. 1654: R. W. Ghrist, P. J. Holmes, M. C. Sullivan, Knots and Links in Three-Dimensional Flows. X, 208 pages. 1997.

Vol. 1655: J. Azéma, M. Emery, M. Yor (Eds.), Séminaire de Probabilités XXXI. VIII, 329 pages. 1997.

Vol. 1656: B. Biais, T. Björk, J. Cvitanic, N. El Karoui, E. Jouini, J. C. Rochet, Financial Mathematics. Bressanone, 1996. Editor: W. J. Runggaldier. VII, 316 pages. 1997.

Vol. 1657: H. Reimann, The semi-simple zeta function of quaternionic Shimura varieties. IX, 143 pages. 1997.

Vol. 1658: A. Pumarino, J. A. Rodrıguez, Coexistence and Persistence of Strange Attractors. VIII, 195 pages. 1997.

Vol. 1659: V, Kozlov, V. Maz'ya, Theory of a Higher-Order Sturm-Liouville Equation. XI, 140 pages. 1997.

Vol. 1660: M. Bardi, M. G. Crandall, L. C. Evans, H. M. Soner, P. E. Souganidis, Viscosity Solutions and Applications. Montecatini Terme, 1995. Editors: I. Capuzzo Dolcetta, P. L. Lions. IX, 259 pages. 1997.

Vol. 1661: A. Tralle, J. Oprea, Symplectic Manifolds with no Kähler Structure. VIII, 207 pages. 1997.

Vol. 1662: J. W. Rutter, Spaces of Homotopy Self-Equivalences – A Survey. IX, 170 pages. 1997.

Vol. 1663: Y. E. Karpeshina; Perturbation Theory for the Schrödinger Operator with a Periodic Potential. VII, 352 pages. 1997.

Vol. 1664: M. Väth, Ideal Spaces. V, 146 pages. 1997.

Vol. 1665: E. Giné, G. R. Grimmett, L. Saloff-Coste, Lectures on Probability Theory and Statistics 1996. Editor: P. Bernard, X, 424 pages, 1997.

Vol. 1666: M. van der Put, M. F. Singer, Galois Theory of Difference Equations. VII, 179 pages. 1997.

Vol. 1667: J. M. F. Castillo, M. González, Three-space Problems in Banach Space Theory. XII, 267 pages. 1997.

Vol. 1668: D. B. Dix, Large-Time Behavior of Solutions of Linear Dispersive Equations. XIV, 203 pages. 1997.

Vol. 1669: U. Kaiser, Link Theory in Manifolds. XIV, 167 pages. 1997.

Vol. 1670: J. W. Neuberger, Sobolev Gradients and Differential Equations. VIII, 150 pages. 1997.

Vol. 1671: S. Bouc, Green Functors and G-sets. VII, 342 pages. 1997.

Vol. 1672: S. Mandal, Projective Modules and Complete Intersections. VIII, 114 pages. 1997.

Vol. 1673: F. D. Grosshans, Algebraic Homogeneous Spaces and Invariant Theory. VI, 148 pages. 1997.

Vol. 1674: G. Klaas, C. R. Leedham-Green, W. Plesken, Linear Pro-p-Groups of Finite Width. VIII, 115 pages. 1997.

Vol. 1675: J. E. Yukich, Probability Theory of Classical Euclidean Optimization Problems. X, 152 pages. 1998.

Vol. 1676: P. Cembranos, J. Mendoza, Banach Spaces of Vector-Valued Functions. VIII, 118 pages. 1997.

Vol. 1677: N. Proskurin, Cubic Metaplectic Forms and Theta Functions. VIII, 196 pages. 1998.

Vol. 1678: O. Krupková, The Geometry of Ordinary Variational Equations. X, 251 pages. 1997.

Vol. 1679: K.-G. Grosse-Erdmann, The Blocking Technique. Weighted Mean Operators and Hardy's Inequality. IX, 114 pages. 1998.

Vol. 1680: K.-Z. Li, F. Oort, Moduli of Supersingular Abelian Varieties. V, 116 pages. 1998.

Vol. 1681: G. J. Wirsching, The Dynamical System Generated by the 3n+1 Function. VII, 158 pages. 1998.

Vol. 1682: H.-D. Alber, Materials with Memory. X, 166 pages. 1998.

Vol. 1683: A. Pomp, The Boundary-Domain Integral Method for Elliptic Systems. XVI, 163 pages. 1998.

Vol. 1684: C. A. Berenstein, P. F. Ebenfelt, S. G. Gindikin, S. Helgason, A. E. Tumanov, Integral Geometry, Radon Transforms and Complex Analysis. Firenze, 1996. Editors: E. Casadio Tarabusi, M. A. Picardello, G. Zampieri. VII, 160 pages. 1998.

Vol. 1685: S. König, A. Zimmermann, Derived Equivalences for Group Rings. X, 146 pages. 1998.

Vol. 1686: J. Azéma, M. Émery, M. Ledoux, M. Yor (Eds.), Séminaire de Probabilités XXXII. VI, 440 pages. 1998.

Vol. 1687: F. Bornemann, Homogenization in Time of Singularly Perturbed Mechanical Systems. XII, 156 pages. 1998.

Vol. 1688: S. Assing, W. Schmidt, Continuous Strong Markov Processes in Dimension One. XII, 137 page. 1998.

Vol. 1689: W. Fulton, P. Pragacz, Schubert Varieties and Degeneracy Loci. XI, 148 pages. 1998.

Vol. 1690: M. T. Barlow, D. Nualart, Lectures on Probability Theory and Statistics. Editor: P. Bernard. VIII, 237 pages. 1998.

Vol. 1691: R. Bezrukavnikov, M. Finkelberg, V. Schechtman, Factorizable Sheaves and Quantum Groups. X, 282 pages. 1998.

Vol. 1692: T. M. W. Eyre, Quantum Stochastic Calculus and Representations of Lie Superalgebras. IX, 138 pages. 1998.

Vol. 1694: A. Braides, Approximation of Free-Discontinuity Problems. XI, 149 pages. 1998.

Vol. 1695: D. J. Hartfiel, Markov Set-Chains. VIII, 131 pages. 1998.

Vol. 1696: E. Bouscaren (Ed.): Model Theory and Algebraic Geometry. XV, 211 pages. 1998.

Vol. 1697: B. Cockburn, C. Johnson, C.-W. Shu, E. Tadmor, Advanced Numerical Approximation of Nonlinear Hyperbolic Equations. Cetraro, Italy, 1997. Editor: A. Quarteroni. VII, 390 pages. 1998.

Vol. 1698: M. Bhattacharjee, D. Macpherson, R. G. Möller, P. Neumann, Notes on Infinite Permutation Groups. XI, 202 pages. 1998.

Vol. 1700: W. A. Woyczyński, Burgers-KPZ Turbulence, XI, 318 pages. 1998.

Vol. 1701: Ti-Jun Xiao, J. Liang, The Cauchy Problem of Higher Order Abstract Differential Equations, XII, 302 pages. 1998.

General Remarks

Lecture Notes are printed by photo-offset from the master-copy delivered in camera-ready form by the authors. For this purpose Springer-Verlag provides technical instructions for the preparation of manuscripts.

Careful preparation of manuscripts will help keep production time short and ensure a satisfactory appearance of the finished book. The actual production of a Lecture Notes volume normally takes approximately 8 weeks.

Authors receive 50 free copies of their book. No royalty is paid on Lecture Notes volumes.

Authors are entitled to purchase further copies of their book and other Springer mathematics books for their personal use, at a discount of 33,3 % directly from Springer-Verlag.

Commitment to publish is made by letter of intent rather than by signing a formal contract. Springer-Verlag secures the copyright for each volume.

Addresses:

Professor A. Dold
Mathematisches Institut
Universität Heidelberg
Im Neuenheimer Feld 288
D-69120 Heidelberg, Germany

Professor F. Takens
Mathematisch Instituut
Rijksuniversiteit Groningen
Postbus 800
NL-9700 AV Groningen
The Netherlands

Professor Bernard Teissier
École Normale Supérieure
45, rue d'Ulm
F-7500 Paris, France

Springer-Verlag, Mathematics Editorial
Tiergartenstr. 17
D-69121 Heidelberg, Germany
Tel.: *49 (6221) 487-410